16th International Conference on Liquid and Amorphous Metals (LAM-16)

EPJ Web of Conferences Volume 151 (2017)

Bonn, Germany
4-9 September 2016

Editors:

D. Holland-Moritz **F. Kargl**

ISBN: 978-1-5108-4703-3

Printed from e-media with permission by:

Curran Associates, Inc.
57 Morehouse Lane
Red Hook, NY 12571

Some format issues inherent in the e-media version may also appear in this print version.

Printed by Curran Associates, Inc. (2017)

For additional information, please contact EDP Sciences – Web of Conferences
at the address below.

EDP Sciences – Web of Conferences
17, Avenue du Hoggar
Parc d'Activité de Courtabœuf
BP 112
F-91944 Les Ulis Cedex A
France

Phone: +33 (0) 1 69 18 75 75
Fax: +33 (0) 1 69 28 84 91

contact-edps@webofconferences.org

Additional copies of this publication are available from:

Curran Associates, Inc.
57 Morehouse Lane
Red Hook, NY 12571 USA
Phone: 845-758-0400
Fax: 845-758-2633
Email: curran@proceedings.com
Web: www.proceedings.com

TABLE OF CONTENTS

Proceedings of the 16th International Conference on Liquid and Amorphous Metals (LAM-16)

Editors: *Dirk* Holland-Moritz[1,*] and *Florian* Kargl[1]

[1] Institut für Materialphysik im Weltraum, Deutsches Zentrum für Luft- und Raumfahrt (DLR), 51170 Köln, Germany

Preface

The 16th International Conference on Liquid and Amorphous Metals (LAM-16) was held at the town hall of Bonn - Bad Godesberg, Germany, from September 04-09, 2016. It was part of a series of conferences that has a long standing history dating back to the liquids metal conference LM-1 in Brookhaven (1966). The LM-1 was followed by the liquid metals conferences in Tokyo (1972) and Bristol (1976). The scope of the conference series was complemented by topics of amorphous metals in the following LAM conferences held in Grenoble (1980), Los Angeles (1983), Garmisch-Partenkirchen (1986), Kyoto (1989), Wien (1992), Chicago (1995), Dortmund (1998), Yokohama (2001), Metz (2004), Ekaterinburg (2007), Rome (2010) and Beijing (2013). The LAM conferences aim to provide a forum for scientists interested in liquid and amorphous metals in order to discuss recent advances and future directions in this intriguing field of condensed matter physics, chemistry and materials sciences. Apart from papers on metallic systems, the LAM conferences

* Corresponding author: dirk.holland-moritz@dlr.de

traditionally also welcome contributions on liquid and amorphous semi-conductors as well as on molten salts.

Fig. 1. Participants of the LAM-16 conference gathered at the band-stand of the town-hall of Bonn - Bad Godesberg.

135 scientists from 19 different countries, 9 accompanying persons and 3 non-scientific members of the organization team participated in the LAM-16 conference. The conference program comprised 5 plenary talks, 25 invited talks and 72 contributed talks. Different from preceding LAM-conferences the invited and contributed talks were held in parallel sessions in two rooms, allowing for a comparatively large number of talks. Most of the thematically focused sessions were introduced by an invited talk on the topic. Altogether we had 30 oral sessions on the following topics:

- Glass Transition
- Extreme Conditions
- Demixing / Liquid-Liquid Phase Transformations
- Electronic and Magnetic Properties
- Applications / Materials Development
- Structure and Dynamics
- Expanded Liquids / Electronic Properties
- Heat Treatment / Devitrification
- Properties of Metallic Glasses
- Dynamics
- Structure
- Solidification
- Excitations in Liquids and Glasses
- Containerless Processing
- Thermophysical Properties
- Modelling / Simulations
- Structure and Properties of Glasses

In addition to the oral sessions 48 poster contributions were presented in a vivid poster session on Tuesday evening. The posters were displayed in the foyer of the town-hall during 3 conference days allowing for a view of the posters also during conference breaks. Two young scientists were awarded with a best poster award. The second prize won Ms. M. Sc. Nadège Meyer for her poster entitled 'Shear Viscosity of Liquid Potassium and Cesium: A Simulation Study'. The first prize was awarded to Ms. M. Sc. Beatrix Gonzalez del Rio for her poster entitled 'Orbital-free *ab initio* molecular dynamics study of the free liquid surface of Sn. From pseudopotential generation to structural and dynamic properties'.

In these conference proceedings 30 papers are collected that provide an overview over the research field of liquid and amorphous metals. The papers have been accepted for publication after a peer-reviewing process. We would like to thank the group of renowned experts that served as referees of the manuscripts for their support and their critical suggestions that helped improving the quality of the papers.

Finally we would like to express our gratitude to the sponsors of the LAM-16 conference: German Aerospace Center (DLR), Deutsche Forschungsgemeinschaft (DFG) and Leybold GmbH. We also would like to thank the LAM committees and the members of the organizing team for their effort that helped making LAM-16 conference a success.

Conference Organization

Local Organizing Committee

D. Holland-Moritz (Chairman)
F. Kargl (Co-Chairman)

R. Bartkowski	P. Fopp	P. Knölke-Pohl
A. Meyer	B. Nowak	M. Reinartz
S. Zimmermann		

International Program Committee

T. Bryk (Ukraine)	B. Lin (USA)	M. L. Saboungi (France)
S. De Panfilis (Italy)	A. Mikula (Austria)	W. Schirmacher (Germany)
D. Holland-Moritz (Germany)	J. Morris (USA)	V. Sidorov (Russia)
K. Khishchenko (Russia)	P. Pershan (USA)	A. P. Tsai (Japan)

International Advisory Board

A. Di Cicco (Italy)	S. Hosokawa (Japan)	P. Salmon (UK)
C. Dong (China)	K. F. Kelton (USA)	T. Scopigno (Italy)
B. Gelchinski (Russia)	H. Liu (China)	K. Tamura (Japan)
D. J. Gonzalez (Spain)	W. C. Pilgrim (Germany)	J. F. Wax (France)
P. Häussler (Germany)	P. Popel (Russia)	M. Yao† (Japan)
B. K. Hoshino (Japan)	C. D. L. Price (France)	

National Committee

R. Busch	P. Häussler	W. Schirmacher
F. Faupel	D. M. Herlach	T. Voigtmann
G. Gerbeth	W. C. Pilgrim	G. Wilde

edp sciences | EPJ Web of Conferences

Statement of Peer review

In submitting conference proceedings to EPJ Web of Conferences, I certify to the Publisher that I adhere to the **Policy on Publishing Integrity** of the journal in order to safeguard good scientific practice in publishing.

1. All articles have been subjected to peer review administered by the proceedings editors.
2. Reviews have been conducted by expert referees, who have been requested to provide unbiased and constructive comments aimed, whenever possible, at improving the work.
3. Proceedings editors have taken all reasonable steps to ensure the quality of the materials they publish and their decision to accept or reject a paper for publication has been based only on the merits of the work and the relevance to the journal.

Title, date and place of the conference

16th International Conference on Liquid and Amorphous Metals (LAM-16), September 04-09, 2016, Bonn - Bad Godesberg, Germany

Proceedings editor(s):

Dirk Holland-Moritz and Florian Kargl

Date and editor's signature

August 02, 2017

Deutsches Zentrum für
Luft- und Raumfahrt e.V.
Institut für Materialphysik
im Weltraum
51170 Köln

Institut für Materialphysik im Weltraum
DLR
Prof. Dr. F. Kargl

The structure of liquid metals probed by XAS

Adriano Filipponi[1,*], *Andrea* Di Cicco[2], *Fabio* Iesari[2], and *Angela* Trapananti[3]

[1]Dipartimento di Scienze Fisiche e Chimiche, Università dell'Aquila, I-67100 L'Aquila, Italy
[2]Scuola di Scienze e Tecnologie, Sezione di Fisica, Camerino I-62032, Macerata, Italy
[3]CNR-Istituto Officina dei Materiali-Perugia, I-06123 Perugia, Italy

Abstract. X-ray absorption spectroscopy (XAS) is a powerful technique to investigate the short-range order around selected atomic species in condensed matter. The theoretical framework and previous applications to undercooled elemental liquid metals are briefly reviewed. Specific results on undercooled liquid Ni obtained using a peak fitting approach validated on the spectra of solid Ni are presented. This method provides a clear evidence that a signature from close packed triangular configurations of nearest neighbors survives in the liquid state and is clearly detectable below $k \approx 5$ Å^{-1}, stimulating the improvement of data-analysis methods that account properly for the ensemble average, such as Reverse Monte Carlo.

1 Introduction

The development of methods that provide access to the undercooled liquid state, combined with the availability of improved x-ray and neutron radiation sources, has stimulated a novel interest in experimental investigations of liquid metals. Containerless materials processing techniques such as electromagnetic levitation [1] or aerodynamic levitation [2] were exploited to perform x-ray or neutron diffraction [3–5] experiments in a clean environment reaching a deep undercooling in metallic melts.

X-ray absorption spectroscopy (XAS) [6, 7] is a powerful technique to probe the short-range structure around selected atomic species in condensed matter including the liquid state [8, 9]. The typical bulk sample thickness required to optimize transmission experiments for transition metal K-edges is of the order of 10-20 μm therefore suitable samples can be prepared as a dispersion of micrometric sized grains embedded in a supporting matrix [10]. Even if this environment is not as clean as in the case of levitation techniques, a deep undercooling can be usually achieved because of the small droplet size, in particular when the supporting framework does not act as a strong heterogeneous nucleant. Our contribution to the field includes a number of structural investigations on undercooled liquid elemental metals with high melting points such as Pd [11], Rh [12], Cu [13], Ni [14] at high temperature and low melting point metals such as Sn [15] and Ga [16] at high pressure.

The undercooling capabilities can be assessed using x-ray absorption temperature scans [17] and these latter measurements exploited to determine the specific crystal nucleation rate [18]. Recent Ni data [14] were further investigated to assess the possible occurrence of homogeneous nucleation [19] with the support of kinetic Monte Carlo simulations [20].

*e-mail: adriano.filipponi@aquila.infn.it

The focus of this paper is on the nature of the short-range structural information in x-ray absorption spectra measured in the liquid phase and in particular on the assessment of the presence and intensity of signals associated with triplet correlations. The presence of a signature from close packed equilateral triangular configurations has been previously emphasized in the Cu case [13]. In this paper Ni K-edge XAS spectra at $T = 1493$ K in the solid and undercooled liquid phases [14] are analyzed with a peak fitting approach.

2 Theoretical framework

XAS has been extensively reviewed [6, 7]. The photo-absorption cross section is dominated by one-electron dipole allowed transitions in which a core state electron ($1s$ for K-edges) is excited to unoccupied continuum states. The photon energy $\hbar\omega$ determines the photo-electron wavevector modulus $k = \frac{1}{\hbar}\sqrt{2m(\hbar\omega - E_0)}$ where E_0 is the continuum threshold energy. The process occurs in the fs timescale, therefore the measured cross section probes an ensemble average over the instantaneous atomic positions and cannot access dynamical information. The final state photo-electron wavefunction is a continuum state at $10 - 1000$ eV above E_0 in a space containing the photoabsorber and the surrounding atoms. Current approximations for the cross-section calculation adopt an effective potential for the photoelectron which includes Coulomb and exchange-and-correlation terms from the surrounding charge density. The effect of the surrounding atoms is taken into account within a multiple scattering formalism. In the most effective approaches the usage of a muffin-tin approximation for the potential allows to separate the effects of the atomic species, represented by spherically symmetric potentials and related phase shifts, that are independent from the actual atomic positions, from the geometrical information. The origin of the oscillations in the cross-section, usually referred to as x-ray absorption fine structure (XAFS), is associated with interference effect from scattered waves from the surrounding atoms and therefore contains a valuable information on the short-range structure. Moreover, the strong non-linear relation between geometry and cross-section makes the dimensionless oscillating term $\chi(k) = \sigma/\sigma_0 - 1$ sensitive to contributions beyond the pair correlations (σ_0 represents the atomic cross-section). It is possible to write a useful expansion of the signal in terms of irreducible pair, triplet, and higher order terms:

$$\chi(k) = \sum_i \gamma^{(2)}(0,i) + \sum_{(i,j)} \gamma^{(3)}(0,i,j) + \sum_{(i,j,l)} \gamma^{(4)}(0,i,j,l) + \cdots \quad (1)$$

Each of the $\gamma^{(n)}(0,i,\dots)$ signal is an oscillating function with a leading phase proportional to kR_p R_p being the length of the shortest (or dominant) path involving all atoms in the configuration. This includes the dominant pair contributions with $R_p = 2R$, where R is the corresponding shell distance. The actual expressions for the $\gamma^{(n)}(0,i,\dots)$ signals are not a simple functions and involve series expansions over angular momenta and related phase shifts for the atoms involved [21, 22]. In the case of disordered systems the sums in Eq. (1) can be evaluated using appropriate integrals over the pair, triplet, ..., atomic distributions [8].

3 Validity assessment of the underlying approximations

The simplest strategy to verify the accuracy of the present computational scheme is to attempt to model spectra of known structures in which the effect of the configurational average does not represent an unknown. Useful comparisons can be performed with the spectra of molecules in the gas phase [23] or of simple crystalline structures [22]. For this specific investigation is it clearly important to verify how the theory can account for the spectrum of crystalline f.c.c. Ni. The comparison is usually performed on the signal, conveniently plotted as $k^n\chi(k)$ where n is a suitable integer power chosen to magnify the higher k range of the spectra (for graphical reasons). The corresponding Fourier transform magnitude is evaluated with an $\exp(i2kR)$ exponential in such a way that the conjugated variable R corresponds roughly to the physical distance (apart from a phase shift contribution which shifts the peaks to lower R distances by about 0.4 Å).

The XAFS signals in an f.c.c. structure are dominated by the first shell single scattering contribution with a low frequency in k space and peaking around the first shell distance in R space. The intermediate range order instead contributes with a number of pair and triplet signals associated with triangular arrangements of first neighbor bonds at 60°, 90°, 120°, and 180°. The longer distances in the latter three triangular configurations correspond to second, third and fourth neighbors shells. All the above signals can be calculated for a reference configuration and the effect of atomic vibrations accounted for by a suitable complex damping term. The fit of the solid Ni $k\chi(k)$ spectrum, measured at $T = 1493$ K [14], including all pair

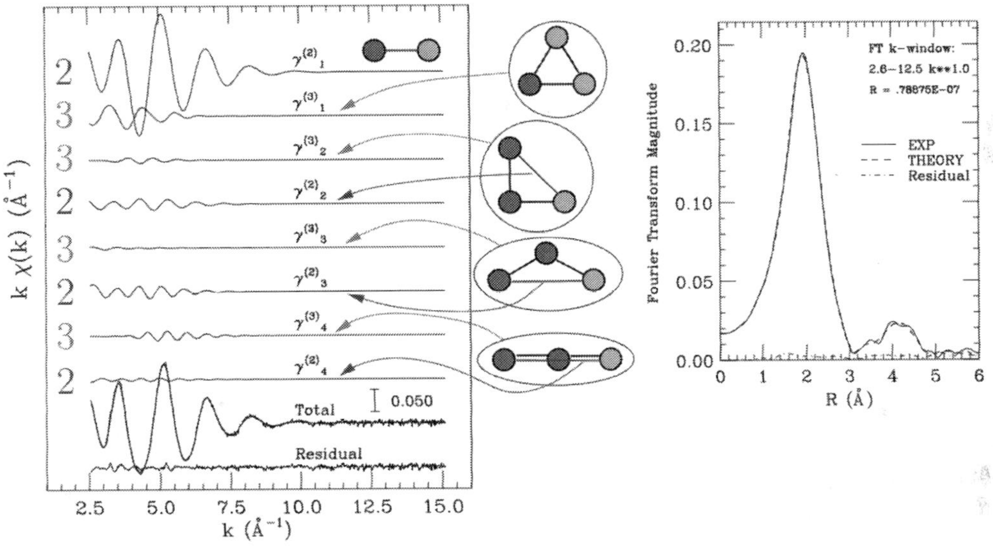

Figure 1. XAFS data analysis of the f.c.c. solid Ni at $T = 1493$ K, the various pair $\gamma^{(2)}$ or triplet $\gamma^{(3)}$ oscillating contributions to the $k\chi(k)$ spectrum associated with the first shell or triangular atomic configurations involving first neighbors are reported in the left panel. The Fourier transform magnitude of the experimental, theoretical, and residual signal are shown in the right hand side panel.

and triplet contributions required to model the spectrum up to $R \approx 5$ Å and the related Fourier transformations are reported in figure 1. The relevant triangular structures are indicated in the graphical insets: the inequivalent possible positions of the photoabsorber atoms are reported as red circles. The inter-atomic distances are schematically indicated as black or blue segments, the latter representing the successive coordination shells in the f.c.c. structure with associated $\gamma^{(2)}_{1,2,3,4}$ signals. The dominant first shell signal $\gamma^{(2)}_1$ is due to the 12 first neighbors. The distance distribution was modeled with a slightly asymmetric Gamma distribution with skewness $\beta = 0.33(5)$, average distance $R = 2.485(5)$ Å and variance $\sigma^2 = 0.033(4)$ Å^2 compatible with the phonon contribution at this temperature. The other signals associated with triangular configuration are displayed separately for the pair $\gamma^{(2)}_{2,3,4}$ or irreducible triplet $\gamma^{(3)}_{1,2,3,4}$ contributions. The relatively strong triplet signal from equilateral triangles can be appreciated. In the case of 90° and especially 120° triangles, the triplet signal is much weaker than the corresponding pair signal from the (second or third) shell, the opposite occurs for aligned atomic configurations (fourth shell) due to the well known focusing effect. The vibrational damping of all these signals can be calculated assuming Gaussian three-dimensional distributions of the geometrical parameters defining the triangles. Their inclusion can be done avoiding a proliferation of fitting parameters since they involve first neighbor bonds whose R and σ^2 parameters are already constrained by the $\gamma_1^{(2)}$ signal and only the angular fluctuations (of the order of 1°) and vibrational correlations have to be included. The fitted correlations are negligible apart form the one between the vibrations of adjacent aligned bonds that results $\rho_{R,R'} = -0.19(2)$.

The fit is excellent and the residual difference signal (experiment−model) is substantially flat and dominated by random noise (or residual weak higher frequency neglected contributions). The possibility to model accurately the spectra of a known Ni structure provides a strong evidence for the validity of the underlying approximations and the possibility of applications to cases in which the structure is unknown.

4 A peak fitting analysis of the spectrum of undercooled liquid Ni

The analysis of XAS spectra of liquid structures requires a proper account for the configurational average and limitations in the application of a peak fitting approach were clearly stated [8]. The peak fitting approach, however, can be adopted as a preliminary and effective tool to investigate the information content in a signal, and this analysis is presented here for the case of undercooled liquid Ni at $T_u = 1493$ K [14]. The experimental spectrum is slightly weaker and shifted with respect to the one of the f.c.c. crystal at the same T.

A central issue in this kind of data analysis is that the signal should be fitted with structural contributions compatible with a disordered three-dimensional close-packed structure in which the atomic environments may differ considerably from the one in the f.c.c. crystal and include distorted or defected icosahedral configurations, resulting in an angular distribution between nearest neighbor bonds similar to the one shown in Fig. 3 of Ref. [14]. Multiple scattering calculations however show that for all intermediate angles (similarly to the triplet signals for the 90° and 120° triangles in the f.c.c. structure) the triplet contributions are negligible, so a detectable signal may only arise from the limit 60° (and possibly 180°) configurations that are present in any close-packed disordered configuration. The question addressed here is if the experimental signal can be modeled by accounting only for the pair signals associated with the radial distribution function, or if any signal from triplet configuration survives and can be detected. The answer is given by the fit reported in figure 2.

The first signal (on the left hand side panel from the top) accounts for the dominant short-range part of the radial distribution, compatible with the previous analysis [14]. This signal however in not able to account completely for the experimental data since it leaves out a higher frequency oscillation that is out of phase with respect to an hypothetical pair signal with the same frequency. By attempting to include the dominant triplet configurations associated with equilateral triangles or 180° alignments, similarly to the fit of the f.c.c. Ni spectrum of figure 1, an excellent fit can be obtained. The 180° alignments (the reported $\eta_2^{(3)}$ signal includes both pair and triplet contributions) give only a weak signal due to the increased disorder in the liquid phase, and the present analysis cannot be regarded as conclusive. On the other hand a clear signal survives from the triplet equilateral configurations involving first neighbors. This signal has a reduced intensity with respect to the f.c.c. one, as a consequence of the increased bond distance variance $\sigma^2 = 0.044(5)$ Å^2 of the first $g(r)$ peak with respect to the f.c.c. case. In addition its intensity was further reduced by about 80% indicating a reduction of the number or equilateral triangles/atom in the disordered structure. This analysis

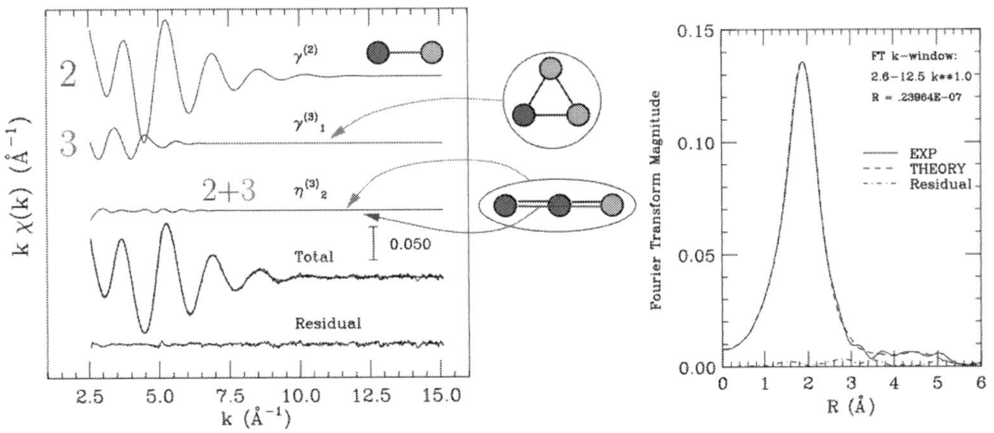

Figure 2. XAFS data analysis of the spectrum of undercooled liquid Ni at $T_u = 1493$ K accounting for all short-range pair contributions $\gamma^{(2)}$, equilateral triangular configurations $\gamma_1^{(3)}$, and configurations at 180° $\eta_2^{(3)}$ (including both pair and triplet). The Fourier transform magnitude of the experimental, theoretical, and residual signals are shown in the right hand side panel.

was performed taking account of the dominant peak of the triplet distribution modeled with an uncorrelated Gaussian shape. While this model is clearly inadequate to account for the exact entire triangle distribution in a disordered liquid structure, the presence of this signal provides a clear signature of the close packed nature of the liquid structure.

5 Conclusions

X-ray absorption spectroscopy provides a unique probe for the structure of elemental or multi-component liquids with accessible x-ray absorption edges. The structural information content is complementary to the one provided by x-ray or neutron diffraction experiments because of the greater sensitivity to the short-range characteristics of the radial distribution function that can be reliably constrained by a XAFS spectrum. Moreover, and especially in the case of close packed liquid metal structures, a clear signature from triplet correlations associated with close packed equilateral triangular structures can be detected, as this work demonstrates for the Ni case. These structures are at the basis of the polytetrahedral atomic arrangements [24] expected in the case of close packed elemental liquids and the access to experimental information on triplet correlations is clearly a unique and valuable characteristic of XAS. Present evidence shows that, for the Ni case, the signal accounts for about the 10-20% of the total signal and is confined to the low-k range $k \lesssim 5$ Å^{-1}.

Our current data analysis approach based on the Reverse Monte Carlo simulation of a three dimensional model structure compatible with the experimental spectrum was so far mostly confined to $k \geq 4$ Å^{-1} [13, 14]. The contribution from three body configurations detectable in the experimental data was accounted for by subtracting from the experimental data the three body signal for a Gaussian three-dimensional distribution of equilateral triangles formed by first neighbours, whose compatibility with the simulated average structure was verified as a self-consistency check. The structure models were anyway analyzed to extract information of the bond-angle distributions between first-neighbor bonds and using structural indicators of the local symmetry (W_6 invariants or common neighbor analysis) that revealed the presence of an increased icosahedral short-range order upon undercooling. Ongoing efforts are devoted to implement the direct inclusion of the triplet signal in the RMC optimization procedure and to the extension of the fitting k range down to $k \simeq 2.5$ Å^{-1} where the theoretical modeling was shown to be still accurate. It is believed that these computational efforts will better constrain the resulting three-dimensional structures and provide a deeper insight into their microscopic details.

References

[1] C. Notthoff, H. Franz, M. Hanfland, D.M. Herlach, D. Holland-Moritz, W. Petry, Rev. Sci. Instrum. **71**, 3791 (2000)
[2] L. Hennet, I. Pozdnyakova, A. Bytchkov, V. Cristiglio, P. Palleau, H.E. Fischer, G.J. Cuello, M. Johnson, P. Melin, D. Zanghi et al., Rev. Sci. Instrum. **77**, 053903 (2006)
[3] T. Schenk, D. Holland-Moritz, V. Simonet, R. Bellissent, D.M. Herlach, Phys. Rev. Lett. **89**, 075507 (2002)
[4] K.F. Kelton, G.W. Lee, A.K. Gangopadhyay, R.W. Hyers, T.J. Rathz, J.R. Rogers, M.B. Robinson, D.S. Robinson, Phys. Rev. Lett. **90**, 195504 (2003)
[5] G.W. Lee, A.K. Gangopadhyay, K.F. Kelton, R.W. Hyers, T.J. Rathz, J.R. Rogers, D.S. Robinson, Phys. Rev. Lett. **93**, 037802 (2004)
[6] J.J. Rehr, R.C. Albers, Rev. Mod. Phys. **72**, 621 (2000)
[7] J.A. Van Bokhoven, C. Lamberti, eds., *X-Ray Absorption and X-Ray Emission Spectroscopy: Theory and Applications* (Wiley, New York, 2016), ISBN 978-1-118-84423-6
[8] A. Filipponi, J. Phys.: Condens. Matter **13**, R23 (2001)
[9] A. Filipponi, P. D'Angelo, in *X-Ray Absorption and X-Ray Emission Spectroscopy: Theory and Applications*, edited by J.A. Van Bokhoven, C. Lamberti (Wiley, New York, 2016), chap. 25, pp. 745–771
[10] L. Ottaviano, A. Filipponi, A. Di Cicco, Phys. Rev. B **49**, 11749 (1994)
[11] A. Filipponi, A. Di Cicco, S. De Panfilis, Phys. Rev. Lett. **83**, 560 (1999)
[12] A. Di Cicco, G. Aquilanti, M. Minicucci, A. Filipponi, J. Rybicki, J. Phys.: Condens. Matter **11**, L43 (1999)
[13] A. Di Cicco, A. Trapananti, S. Faggioni, A. Filipponi, Phys. Rev. Lett. **91**, 135505 (2003)
[14] A. Di Cicco, F. Iesari, S. De Panfilis, M. Celino, S. Giusepponi, A. Filipponi, Phys. Rev. B **89**, 060102(R) (2014)
[15] A. Di Cicco, A. Trapananti, E. Principi, S. De Panfilis, A. Filipponi, Appl. Phys. Lett. **89**, 221912 (2006)
[16] R. Poloni, S. De Panfilis, A. Di Cicco, G. Pratesi, E. Principi, A. Trapananti, A. Filipponi, Phys. Rev. B **71**, 184111 (2005)
[17] A. Filipponi, M. Borowski, P.W. Loeffen, S. De Panfilis, A. Di Cicco, F. Sperandini, M. Minicucci, M. Giorgetti, J. Phys.: Condens. Matter **10**, 235 (1998)
[18] S. De Panfilis, A. Filipponi, J. Appl. Phys. **88**, 562 (2000)
[19] A. Filipponi, A. Di Cicco, S. De Panfilis, P. Giammatteo, F. Iesari, Acta Mater. **124**, 261 (2017)
[20] A. Filipponi, P. Giammatteo, J. Chem. Phys. **145**, 211913 (2016)
[21] A. Filipponi, A. Di Cicco, C.R. Natoli, Phys. Rev. B **52**, 15122 (1995)
[22] A. Filipponi, A. Di Cicco, Phys. Rev. B **52**, 15135 (1995)
[23] A. Filipponi, P. D'Angelo, J. Chem. Phys. **109**, 5356 (1998)
[24] F. Spaepen, Nature **408**, 781 (2000)

Pressure dependence of the static structure of liquid GeTe based on *ab initio* molecular dynamics simulations

Akihide Koura[1,*], and *Fuyuki* Shimojo[1]

[1]Depertment of Physics, Kumamoto University, Kumamoto 860-8555, Japan

Abstract. We have investigated the pressure dependence of the static structure of liquid GeTe based on *ab initio* molecular dynamics simulations. The pressure range is between ambient pressure and 250 GPa, and their temperatures between 1000 K and 4000 K, which keep the liquid state. In this study, we found two transition stages caused by the compression. At the first stage, below 12 GPa, atomic distances elongate and Peierls-type distortion is dissolved with increasing pressure. At the second stage, above 12 GPa, atomic distances shorten and the electronic states shows metallic.

1 Introduction

Germanium tellurides (GeTe) are interesting materials because of their semi-conducting properties and the fast phase transition, e.g., Ge-Sb-Te. Crystalline GeTe at room temperature has A7 structure, which is distorted from fcc rocksalt (B1) to expand (111) direction. This distortion is named as Peierls distortion. In this state, Ge atoms and Te atoms are threefold coordinated each other. When the temperature increases up to about 700 K, GeTe changes from A7 to B1 structure, where Ge and Te atoms are sixfold coordinated each other. When the temperature increases up to about 1000 K, crystalline GeTe melts and its coordination number becomes approximately 3-4, i.e., the Peierls-type A7 structure appears again [1].

On the other hand, compression also induces the structural change of crystalline GeTe. When pressure increases up to around 3 GPa, GeTe changes from A7 to rocksalt structure. It has been suggested that, under further compression, GeTe eventually becomes bcc (CsCl-type) structure at 40-50 GPa, through several structural changes [2]. The pressure dependence of the structure of the liquid state was also reported by X-ray diffraction measurements [3], which suggest that the structural change in the liquid state is different from that in the crystalline state. However, the details are not clarified yet.

* Corresponding author: koura@kumamoto-u.ac.jp

In this study, we investigate the pressure dependence of the static structure of liquid GeTe based on *ab initio* molecular dynamics (AIMD) simulations, in which the electronic structure is calculated in the framework of the density functional theory with the generalized gradient approximation. The temperature and pressure ranges are from 1000 to 4000 K and from 0 to 250 GPa, respectively.

2 Numerical details

We have carried out the calculations by using AIMD simulations, in which the electronic states were calculated by the projector-augmented-wave (PAW) method [4,5] within the framework of the density functional theory (DFT) in which the generalized gradient approximation (GGA) [6] was used for the exchange-correlation energy. The plane wave cutoff energies are 9 and 90 Ry for the electronic pseudo-wave functions and the pseudo-charge density, respectively. The energy functional was minimized using an iterative scheme [7,8]. The Γ point was used for Brillouin-zone sampling. Projector functions of the *s*, *p*, and *d* types were generated for the 4*s*, 4*p*, and 4*d* states of Ge, and the 5*s*, 5*p*, and 5*d* states of Te. By investigating the energy dependence of the logarithmic derivatives of the pseudo-wave functions, we verified that our data sets have good transferability, and do not possess any ghost states in the energy range considered.

Molecular-dynamics simulations were carried out at pressures from approximately 0 to 250 GPa and the temperature T = 1000-4000 K. The number densities ρ, the pressures P and the temperature T used in our simulations are listed in Table I. They were determined from the averaging result of the isothermal-isobaric MD simulations at least 1500 MD steps (4.35 ps). We used a 128-atom systems in a cubic supercell with periodic boundary conditions. Using Nosé-Hoover thermostat technique [9,10], the equations of motion were solved via an explicit reversible integrator [111] with a time step of Δt = 2.9 fs. The quantities of interest were obtained by averaging over 29 ps after an initial equilibration taking at least 1.5 ps.

Table 1. Calculated conditions. Pressure P, temperature T, number density ρ, and the ratio of volume V/V_0, where V_0 is the volume at ambient condition.

P (GPa)	T (K)	ρ (10^{-2} Å^{-3})	V/V_0
0.0	1000	2.99	1.000
1.4	1000	3.35	0.892
4.1	1000	3.72	0.803
5.9	1000	3.89	0.769
11.9	1000	4.32	0.692
23.9	1500	4.80	0.623
47.8	2000	5.50	0.542
96.0	2500	6.49	0.460
129.4	2500	7.04	0.424
211.9	3500	7.99	0.374
249.7	4000	8.35	0.358

3 Results and discussion

3.1 Static structure factors

Figure 1 shows the pressure dependence of the static structure factors (a) $S_X(k)$ using X-ray form factor and (b) $S_n(k)$ neutron scattering length. Our results shown by solid lines are in good agreement with experimental results [3] illustrated by circles at wide pressure range.

The height of the first sharp diffraction peak (FSDP) of $S_n(k)$ at about $k = 1.0$ Å^{-1} decreases with increasing pressure, and the FSDP disappears at 4.1 GPa. The disappearance of the FSDP suggests the change of the medium range order. The first peak position shifts from $k = 2.2$ to 3.0 Å^{-1} with increasing pressure.

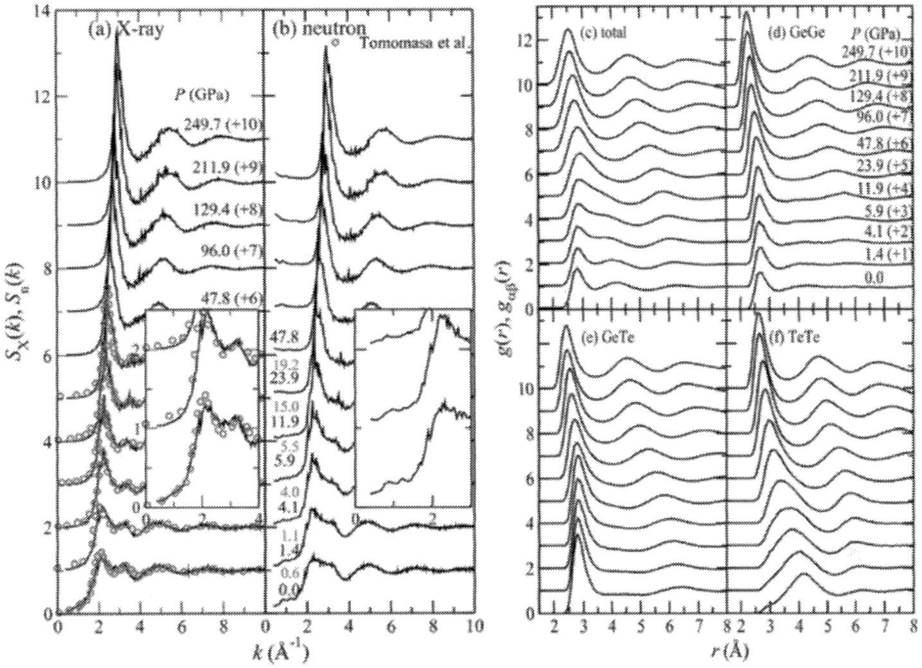

Figure 1. The pressure dependence of the static structure factors (a) $S_X(k)$ and (b) $S_n(k)$ using X-ray form factor and neutron scattering length, respectively. Circles show the result of experiment [3]. The pressure dependence of the (c) total and (d)-(f) partial pair distribution functions $g(r)$ and $g_{\alpha\beta}(r)$, respectively. Insets in pannels (a) and (b) indicate zooming up each FSDP.

3.2 Pair distribution functions

The pressure dependence of the total and partial pair distribution functions $g(r)$ and $g_{\alpha\beta}(r)$ are shown in Fig. 1(c) and 1(d)-(f), respectively.

The first and second peaks of $g(r)$ exist at approximately $r = 2.8$ and 4.2 Å. The second peak shifts toward smaller r with increasing pressure, and it becomes a shoulder of the first peak between 4.1-5.9 GPa. Under further compression, the first peak position shifts toward smaller r. The partial correlation functions $g_{GeGe}(r)$ and $g_{GeTe}(r)$ have their first peaks at about $r = 2.6$ and 2.8 Å.

The pressure dependence of the first peak positions of $g(r)$ and $g_{\alpha\beta}(r)$ is shown in Fig. 2. Figure 2 suggests that there are two stages compression process. The first stage appears below 12 GPa. In this stage, the peak position of $g(r)$ elongates from $r = 2.8$ to 2.9 Å. While peak positions of $g_{GeGe}(r)$ and $g_{GeTe}(r)$ do not change so much. On the other hand, that of $g_{TeTe}(r)$ shortens drastically, however, this change is caused by the difficulty to determine precisely the peak position of $g_{TeTe}(r)$ and growth of the first peak at around $r = 3$ Å. In the second stage, above 12 GPa, all positions of the first peaks shift monotonically to smaller r with increasing pressure. The temperature effect of the change between 12-24 GPa is discussed in the section 3.8.

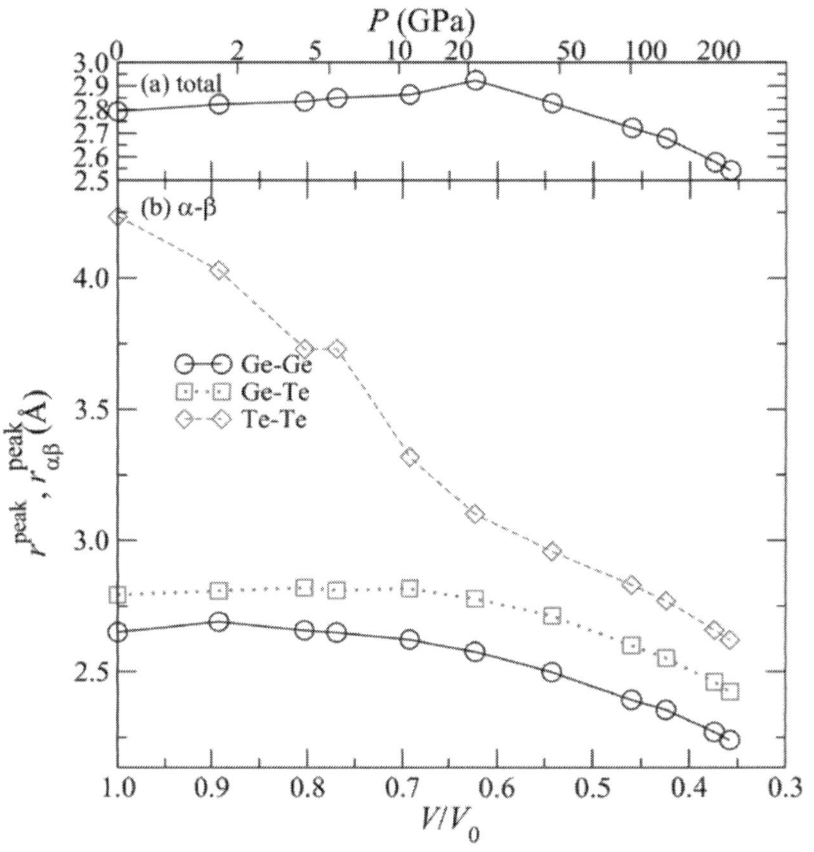

Figure 2. The pressure dependence of the position of the first peak of $g(r)$ and $g_{\alpha\beta}(r)$. The position of $g_{TeTe}(r)$ below 11 GPa is expedient because their positions are unclear (see Fig.1(f)).

3.3 Coordination numbers

Figure 3 shows the pressure dependence of the coordination numbers (a)-(b) N_α and (c)-(e) $N_{\alpha\beta}$. Solid lines with circles and dotted lines with squares correspond to cutoff distances r_c = 3.0 and 3.2 Å, respectively. The former, r_c = 3.0 Å, and the latter, r_c = 3.2 Å, correspond to the short bond length caused by Peierls distortion in the crystalline state, where the coordination number should be 3, and the first minimum position of $g(r)$, respectively. Dashed lines with diamonds show values obtained by integration of $g_{\alpha\beta}(r)$ until the first peak position r_{peak}, and then the values are doubled, i.e., if the shape of first peak of $g_{\alpha\beta}(r)$ would be symmetric, the pressure dependence of dashed lines would show a similar tendency as that of the other lines.

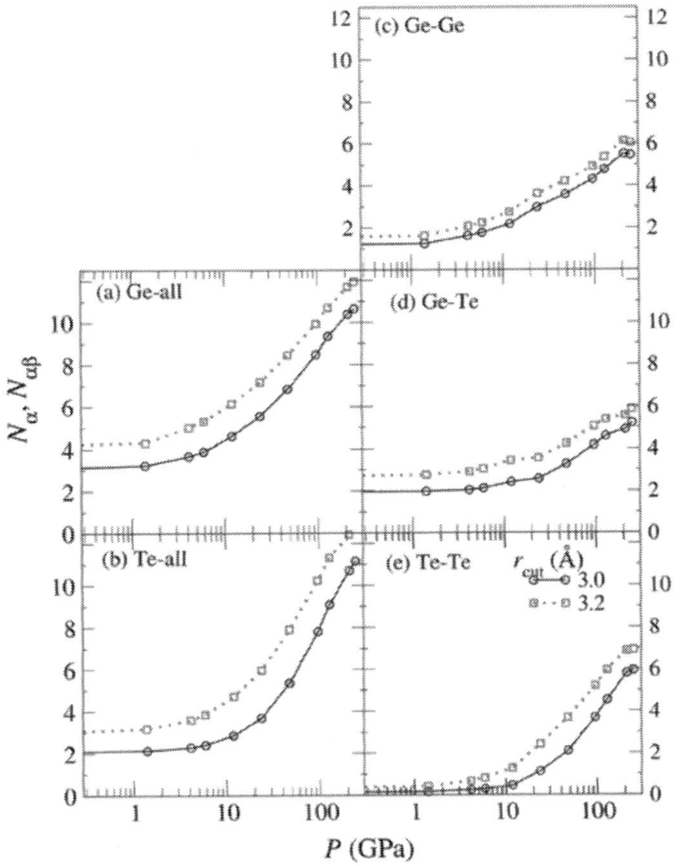

Figure 3. The pressure dependence of coordination numbers (a)-(b) N_α and (c)-(e) $N_{\alpha\beta}$. Solid lines with circles, dotted lines with squares, and dashed lines with diamonds depend on the cutoff distances r_c = 3.0, 3.2, and r_{peak}, respectively. The value of the last one are obtained by integration of $g_{\alpha\beta}(r)$ until r_{peak}, and then the values are doubled.

In Fig. 3(a), the solid line shows that N_{Ge} at ambient pressure is almost 3, however, N_{GeTe} at ambient pressure is 2. If GeTe is in the crystalline state, N_{GeTe} should be 3. While, in the liquid state, although N_{GeTe} is only 2, Ge-Ge homopolar bond appears (N_{GeGe} = 1) and this supplements the lack of Ge-Te bonds. With increasing pressure, N_α and $N_{\alpha\beta}$ does not change much below 12 GPa. However, above 12 GPa, both N_{α} and $N_{\alpha\beta}$ increase rapidly with compression. The appearance of a difference between $N_{\alpha\beta}$ using r_{peak} and other r_c suggests the existence of a new local order at high pressure.

3.4 Three-body bond angle distributions

Figure 4 shows the pressure dependence of the three-body bond angle distributions $B_{\alpha\beta\alpha}(\theta)$. A cutoff distance of $r_c = 3.0$ Å is used. The top panels (a1) and (b1) correspond to the pressure range from 0.0 to 11.9 GPa, and the bottom panels (a2) and (b2) show $B_{\alpha\beta\alpha}(\theta)$ at the pressures above 11.9 GPa, respectively.

At ambient pressure, which shown by black solid lines in the top panels, $B_{GeTeGe}(\theta)$ and $B_{TeGeTe}(\theta)$ have each peak at about $\theta = 90°$. The profile of $B_{GeTeGe}(\theta)$ is slightly wider than that of $B_{TeGeTe}(\theta)$. This is a result of high mobility of Ge atoms. With increasing pressure, the

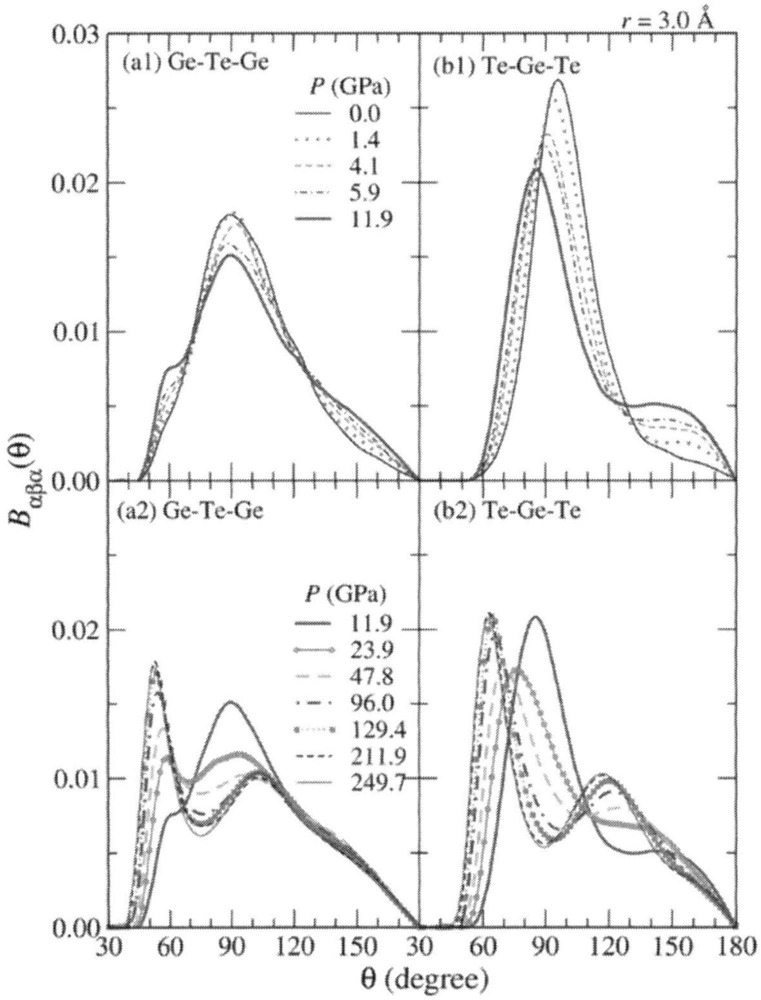

Figure 4. The pressure dependence of three-body bond angle distributions (a1)-(a2) $B_{GeTeGe}(\theta)$ and (b1)-(b2) $B_{TeGeTe}(\theta)$. The upper pannels (a1) and (b1) show that below 12 GPa. The bottom pannels (a2) and (b2) correspond to pressures above 12 GPa.

height of the main peak of $B_{\alpha\beta\alpha}(\theta\sim 90°)$ decreases, and the peak position of $B_{TeGeTe}(\theta)$ shifts toward smaller θ. Under further compression, the height of the main peak at approximate $\theta = 90°$ decreases and a new peak at $\theta = 60°$ appears above 24 GPa. The bond angle $\theta = 60°$ suggests that the liquid GeTe at 250 GPa is similar to bcc crystalline state.

3.5 Mulliken charges

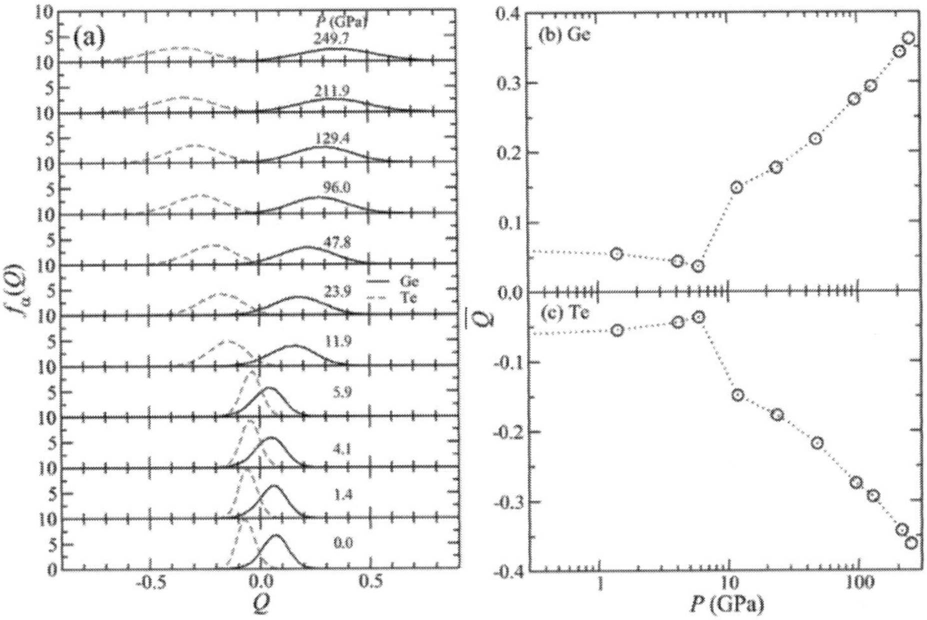

Figure 5. Pressure dependence of (a) grossnet charge densities $f_\alpha(Q)$ and (b)-(c) their average values $\bar{Q}$.

Figure 5 shows the pressure dependence of (a) the charge distribution $f_\alpha(Q)$ and (b)-(c) their average values $\bar{Q}$. At ambient and low pressure range, $f_{Ge}(Q)$ and $f_{Te}(Q)$ exist at almost $Q = 0$. With increasing pressure, the distribution $f_\alpha(Q)$ and their averages $\bar{Q}$ do not change so much below 12 GPa. However, for furthermore compression, $f_\alpha(Q)$ become wider and $\bar{Q}$ leave the range of $Q \approx 0$. In other words, covalent type liquid GeTe changes to more ionic type at approximately 12 GPa.

3.6 Electronic density of states

At ambient pressure, the total electronic density of states $D(E)$ shows that liquid GeTe has semiconductor-type properties because $D(E)$ at $E = 0$ (Fermi energy) has a dip in Fig.6. With increasing pressure, the dip at $E = 0$ becomes shallow, and it completely vanishes at 12GPa, i.e., a semiconductor-metal transition occurs at about 12 GPa.

3.7 Bond overlap populations

Figure 7(a)-(c) shows the time averaged bond overlap populations $p_{\alpha\beta}(\bar{O})$. This figure focuses on the conditions which the structural and/or electronical properties changed

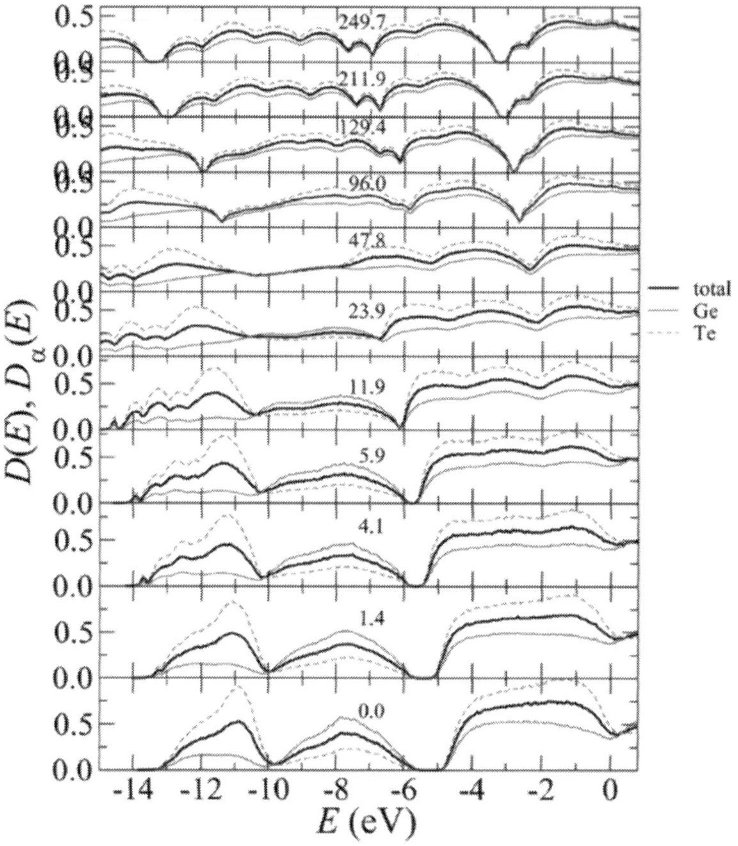

Figure 6. Pressure dependence of the total and partial electornic densities of states $D(E)$ and $D_\alpha(E)$, respectively. The bold line shows $D(E)$. The thin solid and dashed lines correspond to $D_{Ge}(E)$ and $D_{Te}(E)$, respectively.

drastically. Here, the positive and negative regions of the horizontal axis show the covalent and anti bondings, respectively.

At ambient pressure, Ge-Te covalent bonds exist and Ge-Ge homopolar covalent bonds also appear in the liquid state. The average of the bond strength of Ge-Ge atoms is in the same order as that of Ge-Te atoms, while the number of Te-Te homopolar bonds is small. With increasing pressure, Ge-Ge and Ge-Te anti-bondings clearly appear at approximately 12 GPa. The number of Te-Te covalent bonds increases during the compression process.

In crystalline GeTe, Ge atoms and Te atoms have three short bonds to each other. The strength of the short bonds is much stronger than that of the other three long bonds. In the liquid state, the numbers of strong bond $n^{\mathrm{SB}}_{\alpha\beta}$ are estimated by integration of $p_{\alpha\beta}(\overline{O})$ for $\overline{O} >$ 0.2, where $p_{\mathrm{GeTe}}(\overline{O})$ has a minimum at ambient pressure. The result of the estimation $n^{\mathrm{SB}}_{\alpha\beta}$ is shown by triangles with dashed-dotted lines in Figs.7(d)-(f). Focusing on numbers $n^{\mathrm{SB}}_{\alpha\beta}$ at ambient conditions, $n^{\mathrm{SB}}_{\mathrm{GeTe}}$ is 2.49 which is less than 3. However, there are Ge-Ge and Te-Te homopolar bonds, whose numbers $n^{\mathrm{SB}}_{\mathrm{GeGe}}$ and $n^{\mathrm{SB}}_{\mathrm{TeTe}}$ are 1.08 and 0.29, respectively. As a result, n^{SB}_{Ge} and n^{SB}_{Te} are 3.57 and 2.78, i.e., environments around Ge and Te atoms are mostly kept as Peierls-type in the liquid state at ambient pressure. With increasing pressure, n^{SB}_{GeTe} does not change much except for a small hump at about 12 GPa. The numbers of homopolar bonds n^{SB}_{GeGe} and n^{SB}_{TeTe} also do not change below about 12 GPa. However, both of them increase with increasing pressure above approximately 12 GPa. On the other hand, the numbers of covalent and anti bonding are calculated by integration of $p_{\alpha\beta}(\overline{O})$ for $\overline{O} > 0$ and $\overline{O} < 0$, respectively. The former, $n^{CB}_{\alpha\beta}$, and the latter, $n^{AB}_{\alpha\beta}$, are shown by closed circles with dashed lines and open circles with solid lines in Figs.7(d)-(f), respectively. With increasing pressure, n^{CB}_{GeTe} do not change below about 12 GPa. Under further compression, n^{CB}_{GeTe} decreases with increasing pressure. Hence, the difference between n^{CB}_{GeTe} and n^{SB}_{GeTe} decreases in the high pressure region. This difference corresponds to the weak bonding in the Peierls-type distortion, i.e., the decrease of the difference means the disappearance of the Peierls-type distortion in the high pressure region, where strong bonding is not distinguishable from

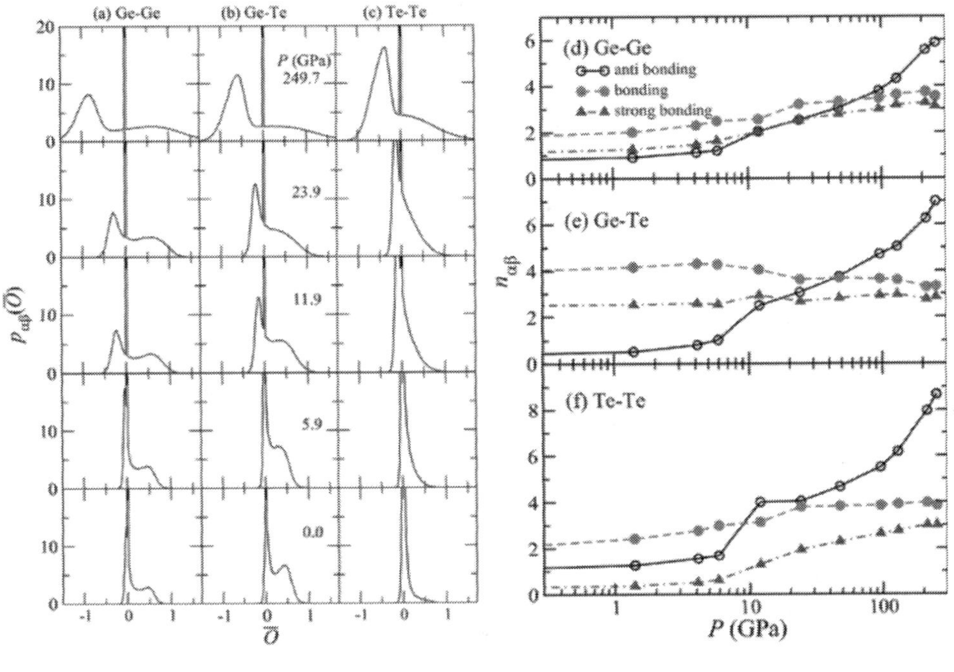

Figure 7. (a)-(c) Pressure dependence of the time averaged bond overlap populations $p_{\alpha\beta}(\overline{O})$ and (d)-(f) numbers of neighboring atoms $n_{\alpha\beta}$ obtained by integration of $p_{\alpha\beta}(\overline{O})$. Solid lines with open circles, dashed lines with closed circles, and dashed-dotted lines with closed triangles show anti-, covalent-, and strong covalent-bondings, respectively.

weak one (see Figs.7(a)-(c)). While, n^{CB}_{GeGe} increases gently in this pressure range, the difference between n^{CB}_{GeGe} and n^{SB}_{GeGe} slightly decreases. For Te-Te bonding, n^{CB}_{TeTe} increases monotonically below about 12 GPa. However, it becomes constant above 12 GPa. Above 12 GPa, the numbers of covalent bonding $n^{CB}_{\alpha\beta}$ for all α-β pairs are almost 4, i.e., Ge and Te atoms have 8 covalent bonds, which is similar to the bcc crystalline structure. Numbers of anti bonding $n_{\alpha\beta}^{\mathrm{AB}}$ slightly increase with increasing pressure below 12 GPa. At about 12 GPa, $n_{\mathrm{GeTe}}^{\mathrm{AB}}$ and $n_{\mathrm{TeTe}}^{\mathrm{AB}}$ jump up to 2.49 and 4.00, respectively. Under further compression, $n_{\alpha\beta}^{\mathrm{AB}}$ increase with increasing pressure.

3.8 Temperature dependence

In order to discuss the pressure dependence of the structure carefully, additional calculations were performed using the same volume as 12 for GPa. The temperature increases from 1000 to 1250 K, and further to 1500 K. Static structure factors $S_X(k)$, partial correlation functions $g_{\alpha\beta}(r)$, and coordination numbers N_α and $N_{\alpha\beta}$ were obtained by averaging 5000 MD steps (14.2 ps) after an initial equilibration taking at least 1.5 ps.

In Figs. 8(a)-(d), solid and dotted lines show the results at T = 1250 K and 1500 K, respectively. It is found that $S_X(k)$ at 1250 and 1500 K are quite similar with that at 1000 K in Fig.8(a). On the other hand, for $g_{\alpha\beta}(r)$, the heights of the first peaks seem to be slightly lower and the width of $g_{\mathrm{GeGe}}(r)$ becomes slightly wider than that at T = 1000 K. Positions of $g_{\mathrm{GeTe}}(r)$ and $g_{\mathrm{TeTe}}(r)$ at high temperatures slightly shift toward smaller r. However, the temperature effect for $S_X(k)$ and $g_{\alpha\beta}(r)$ is considerably smaller than the pressure induced structural change.

The temperature dependence of the coordination numbers N_α and $N_{\alpha\beta}$ at about 12 GPa is shown in Fig.8(e)-(i). The solid line with open circles and dashed line with open squares

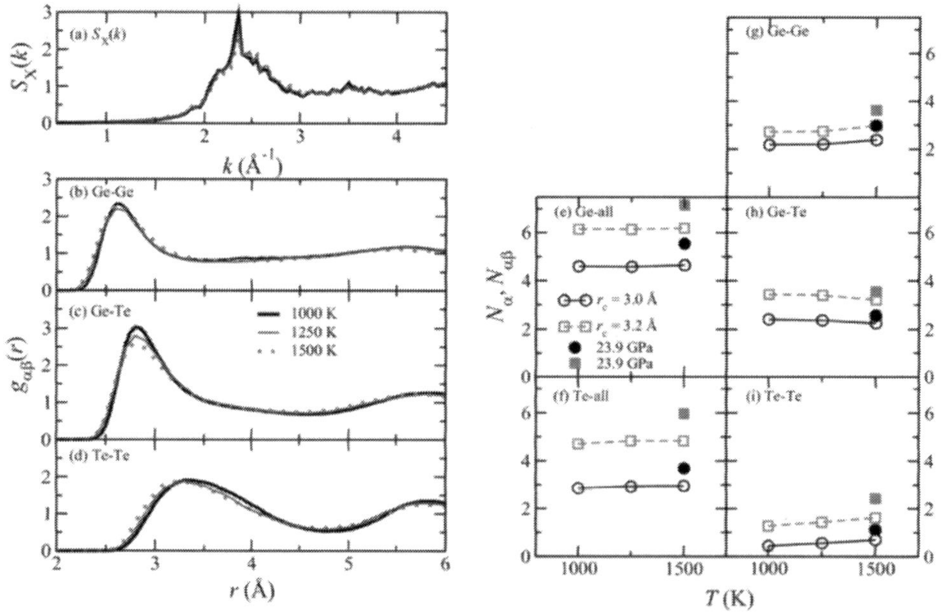

Figure 8. (a) Temperature dependence of the static structure factors $S_X(k)$ and (b)-(d) that of partial distribution functions $g_{\alpha\beta}(r)$. Bold solid lines show the same as shown in Fig.1. Thin solid and dotted lines indicate the result at T = 1250 and 1500 K, respectivaly. (e)-(i) Temperature dependence of coordination numbers N_α and $N_{\alpha\beta}$ at about 12 GPa. Circles and squares correspond to cutoff distances r_c = 3.0 and 3.2Å, respectively. Closed symbols show them at about 24 GPa.

 DOI: 10.1051/epjconf/201715101002

show coordination numbers at about 12 GPa using cutoff distances r_c = 3.0 and 3.2 Å, respectively. Closed symbols indicate results at about 24 GPa. It is clearly seen that the temperature effect is much smaller than the pressure effect.

4 Conclusion

We have investigated the pressure dependence of the static structure of liquid GeTe under pressure based on *ab initio* molecular dynamics simulations. Our simulation shows that the structure of the liquid changes in two stages under pressure. In the first stage below 12 GPa, the Peierls-type distortion in the liquid state was completely dissolved. In this pressure range, the nearest neighbour distance becomes longer, and the semiconducting properties are maintained. It seems that semiconductor-metallic transition occurs at approximately 12 GPa. In the second stage above 12 GPa, the nearest neighbour distance shortens. The coordination number reaches about 13.7 at 250 GPa, which is quite similar to that of CsCl-type crystalline state with taking the second neighbour into account.

Acknowledgements

One of the authors (A.K.) thanks the Murata Science Foundation for the financial support to attend the LAM-16 conference. This work was supported by JSPS KAKENHI Grant Number JP16K05478. The authors acknowledge the Supercomputer Center, Institute for Solid State Physics, University of Tokyo and the Research Institute for Information Technology, Kyushu University for the use of their facilities.

References

1. J. Y. Raty, V. Godlevsky, P. Ghosez, C. Bichara, J. P. Gaspard, and J. R. Chelikowsky, Phys. Rev. Lett. **85**, 1950 (2000).
2. Gap-Soek Do, et al., Phys. Rev. B **82**, 054121 (2010).
3. M. Tomomasa, T. Higaki, T. Hayakawa, A. Chiba, and K. Tsuji, J. Phys.: Conf. Ser. **121**, 022007 (2008).
4. P. E. Blöchl, Phys. Rev. B **50**, 17953 (1994).
5. G. Kresse and D. Joubert, Phys. Rev. B **59**, 1758 (1999).
6. J. P. Perdew, K. Bruke, and M. Ernzerhof, Phys. Rev. Lett. **77**, 3865 (1996).
7. F. Shimojo, R. K. Kalia, A. Nakano, and P. Vashishta, Comput. Phys. Commun. **140**, 303 (2001).
8. G. Kresse and J. Hafner, Phys. Rev. B **49**, 14251 (1994).
9. S. Nosé, Molecular Physics **52**, 255 (1984).
10. W. G. Hoover, Phys. Rev. A **31**, 1695 (1985).
11. M. Tuckerman, B. J. Berne, and G. J. Martyna, J. Chem. Phys. **97**, 1990 (1992).

 DOI: 10.1051/epjconf/201715101003

The Packing of Helical and Zigzag Chains and Distribution of Interstitial Voids in Expanded Liquid Se near the Semiconductor to Metal Transition

Kenji Maruyama[1,*], *Satoshi* Hiroi(Sato)[2], *Hirohisa* Endo[3], *Hideoki* Hoshino[4], *Takashi* Odagaki[5], and *Friedrich* Hensel[6]

[1]Department of Chemistry, Faculty of Science, Niigata University, Niigata 50-2181, Japan
[2]Toyota Technological Institute, Nagoya 468-8511, Japan
[3]Department of Physics, Faculty of Science, Kyoto University, Kyoto 606-8224, Japan
[4]Faculty of Education, Hirosaki University, Hirosaki 036-8560, Japan
[5]Research Institute for Science Education, Kyoto 603-8346, Japan
[6]Physikalische Chemie, Philipps-Universität Marburg, 35032 Marburg, Germany

Abstract. The reverse Monte Carlo (RMC) and Voronoi-Delaunay (VD) void analyses were applied to study the modification of chain geometries near the semiconductor (SC) to metal (M) transition in expanded liquid Se along the isochore of d = 3.4 g/cm^3.
Fluctuations of dihedral angles with increasing temperature and pressure cause modification of the helical (H) chain to the planar zigzag (Z) chain conformations. The distribution of voids size (r_V) supported by chain segments and distances to the 4th ~ 6th neighbor atoms on the chain segments provide information on the stacking of planar zigzag chains compensated by empty space (L-voids, r_V ~ 3.6 Å) which leads to the formation of metallic domains. Near SC-M transition region the number fraction N_Z/N_H for Z and H chain segments increases.

1 Introduction

Expanded liquid (l-) Se transforms from semiconducting to metallic state near the critical point accompanying the volume contraction. The reverse Monte Carlo (RMC) simulation[1] and Voronoi-Delaunay (VD) void analyses[2,3] on the basis of structural data at various temperatures and pressures are useful tools to clarify the relationship between the packing of chains and voids in the intermediate scale near the SC-M transition.

2 Results and Discussion

X-ray scattering measurements have been performed with the white X-ray beamline BL28B2 at SPring-8. The monocrystalline saphire cell which has been developed by Tamura and Inui [4] at Hiroshima University was used to maintain the liquid Se as thin film. The apparatus containing the sample cell and electrinc furnace is described in Ref. [5]. The scattering intensity was obtained with the energy dispersion method. After the corrections for absorption,

*e-mail: maruken@sc.niigata-u.ac.jp

incoherent scattering, background from He gas and sample cell, and thereafter the normalization with the incident X-ray spectrum, the structure factors $S(Q)$ were obtained. The X-ray scattering measurements of expanded l-Se were performed along T and P indicated by red circles (see Fig. 1). The solid lines show the contours of constant DC conductivity (σ) in expanded l-Se. The green line denotes the isochore of d = 3.4 g/cm^3. The analyses along the isochore give a distinctive response to a change of chain geometry in expanded l-Se near the SC-M transition.

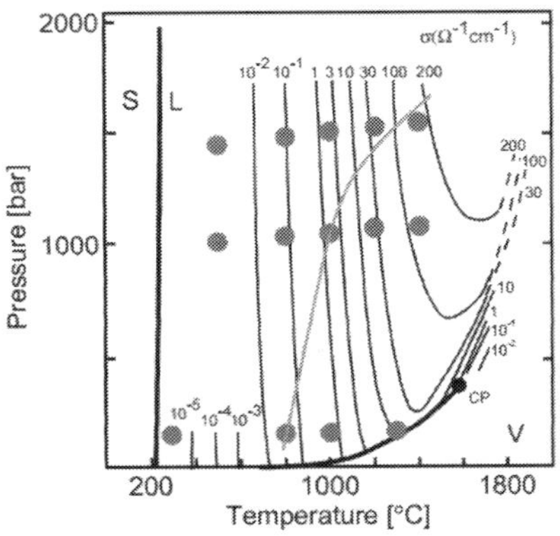

Figure 1. Pressure-temperature phase diagram[6,7] of selenium showing contours of constant DC electrical conductivity in the liquid and supercritical fluid. Green line shows the isochore of d = 3.4 g/cm^3.

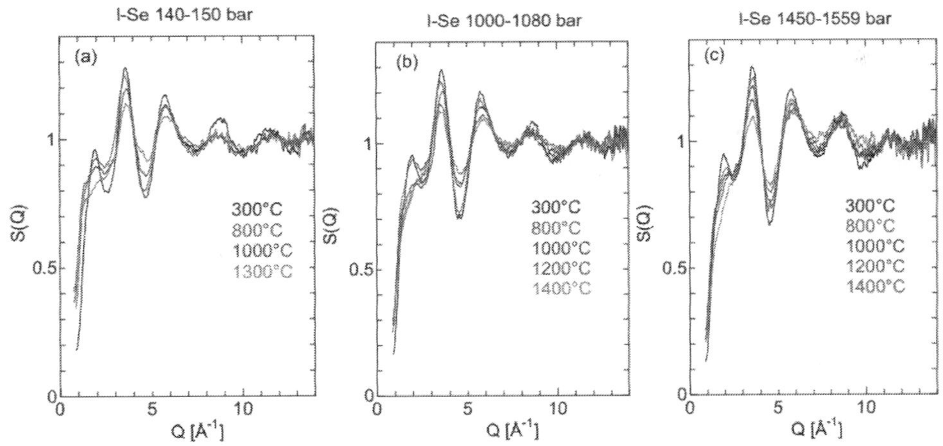

Figure 2. The temperature variations of the structure factor S(Q) for l-Se under 140-150 bar (a), 1000-1080 bar (b), and 1450-1559 bar (c).

Fig. 2 shows temperature variations of the structure factor $S(Q)$ for l-Se under 140 ~ 150 bar (a), 1000 ~ 1080 bar (b) and 1450 ~ 1559 bar (c).

The radial distribution functions $g(r)$ for l-Se along d ~ 3.4 g/cm^3 at 800 °C (147 bar), 1000 °C (1052 bar), 1200 °C (1536 bar) and 1400 °C (1559 bar) are shown in Fig. 3. The peaks of $g(r)$ at 2.4 Å give the bond lengths along a chain. The first peak exhibits a decrease in intensity at high temperature. The dip around r ~ 3 Å becomes less deep above 1100°C

 DOI: 10.1051/epjconf/201715101003

near the SC-M transition, which may be associated with a change of chain geometry (helical → zigzag) as discussed later.

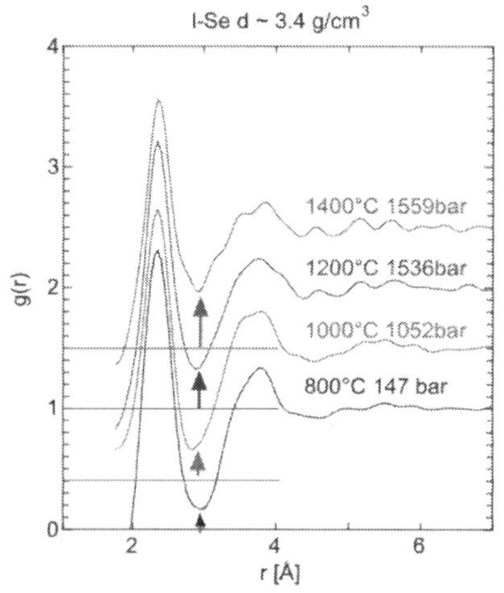

Figure 3. The pair distribution functions $g(r)$ of l-Se along the isochore of d ~ 3.4 g/cm^3. The arrows indicate the depth of the dip of $g(r)$ around r ~ 3 Å.

The RMC simulation and VD void analyses have been applied to clarify the relationship between the packing of chains and interstitial voids in the intermediate scale. The void radius r_v is given by that of the sphere circumscribing the DV tetrahedron. Fig. 4 (a) shows the size distribution of voids $P(r_V)$ in l-Se along d ~ 3.4 g/cm^3. $P(r_V)$ exhibits peaks or shoulders at r_V ~ 2.8 Å and r_V ~ 2.3 Å and additional peak around r_V ~ 3.6 Å as illustrated in Fig. 4(b). The peak at r_V ~ 2.8 Å gives a measure of the interstices between neighboring chains which is defined by the tetrahedron formed between three neighboring Se atoms along a chain and an atom belonging to the neighboring chains as illustrated in Fig. 5(a). As illustrated in Fig. 5(b) the peak at r_V ~ 2.3 Å is assigned to the radius estimated for the stacked zigzag chains driven by inter-chain coupling, which leads to the formation of metallic domain (M-domain)

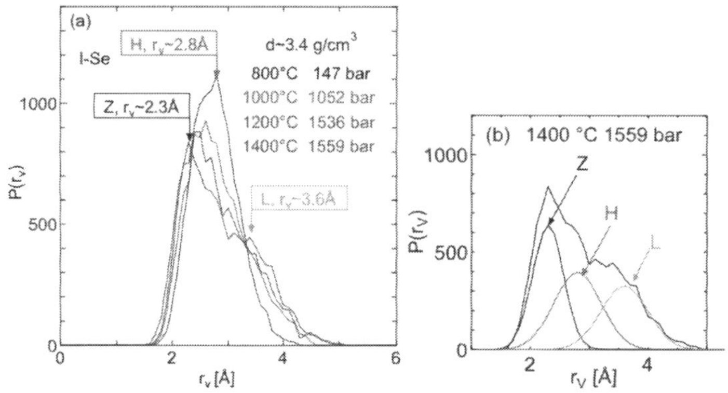

Figure 4. (a) The size distribution of void, $P(r_V)$ in the l-Se along the isochore of d ~ 3.4 g/cm^3. as a function of the void radius, r_V, of VD circusmsphere. (b) Gaussian curve fitting is shown in 1400 °C and 1559 bar.

 DOI: 10.1051/epjconf/201715101003

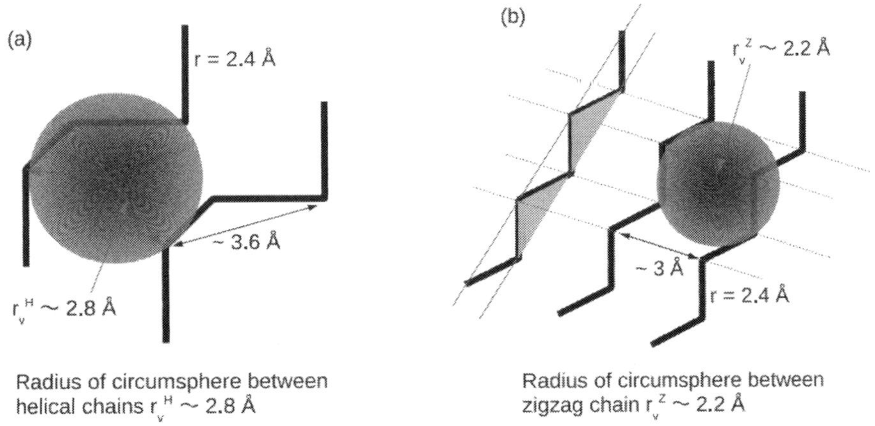

Figure 5. Schematic helical (a) and zigzag (b) chain configurations of expanded l-Se around the void.

The geometrical requirement for stacked planar zigzag chain is compensated by empty space (L-voids, $r_V \sim 3.6$ Å). The temperature variations of number fraction of voids N with $r_V \sim 2.8, 2.3$ and 3.6 Å along $d \sim 3.4$ g/cm^3 is shown in Fig. 6. The proportions for zigzag chains increase near the SC-M transition region (~ 1100 °C). The shortening of chains, fluctuation of dihedral angles and branched chains with increasing temperature and pressure causes modification of helical chain to planar zigzag chain configurations. The distributions of helical (H) and zigzag (Z) chains are characterized by the distance to nth neighbor atom (n: 4 ~ 6) along a chain. It is interesting to note that the distribution of distance assigned to H and Z give distinct peak as shown in Fig. 7, which suggests that chain geometry in expanded l-Se near the SC-M transition contains a stable planar zigzag conformation[8,9].

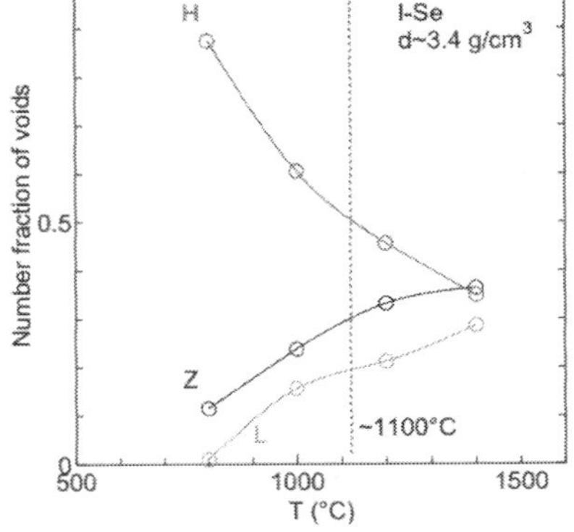

Figure 6. The tenperature variations of the number fraction of voids in expanded l-Se at $d \sim 3.4$ g/cm^3 derived from the Gaussian fitting of the size distribution of voids, as shown in Fig.4(b).

Fig. 8 shows the temperature variations of number fractions, N_Z/N_H, for zigzag to helical chain configurations characterized by the distances to neighbor atoms ($n = 4$) along H and Z chains. The proportion of Z chains with metallic character increases near the SC-M transition, which is consistent with the results in Fig. 6.

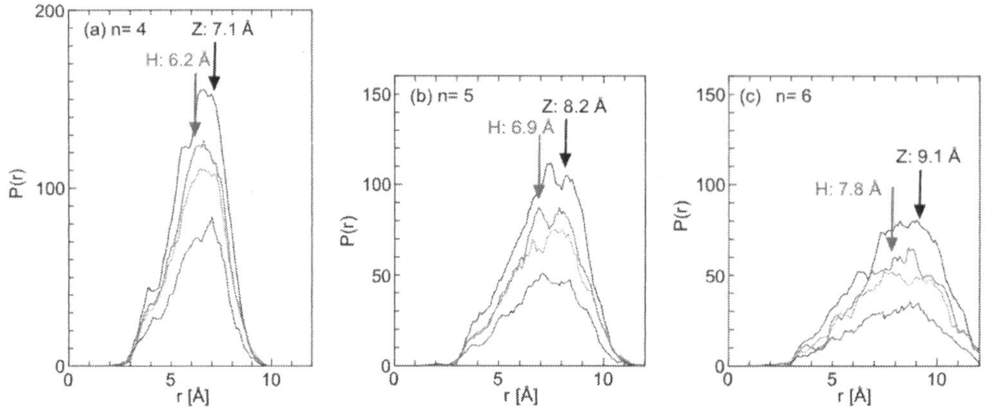

Figure 7. Distribution of atomic distances *P*(*r*) along a chain: black curve indicate *P*(*r*) for 800°C (147bar), red for 1000°C (1052bar), green for 1200°C (1536bar) and blue for 1400°C (1559bar). (a)distribution of $n = 4$, (b) $n = 5$, (c) $n = 6$. Red arrows indicate the position assigned to helical chain and black arrow position assigned to zigzag chain.

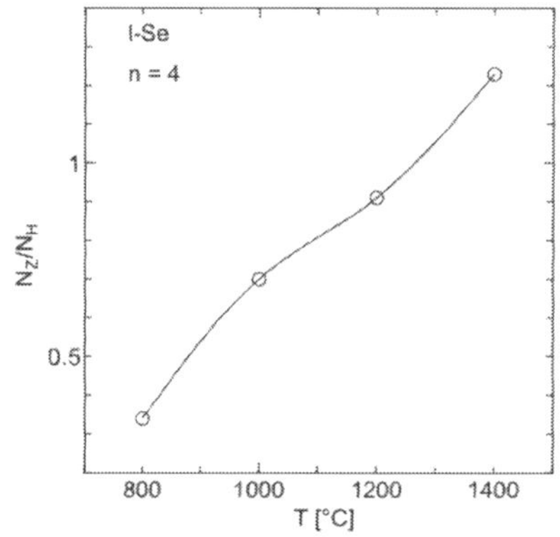

Figure 8. The temperature variatons of number fraction N_Z/N_H for Z and H chain configuratins characterized by distance to 4th neighbor atom along a chain as shown in Fig. 7(a).

3 Conclusion

X-ray scattering measurements were performed on the expanded liquid Se near the SC-M transition along the isochore of $d = 3.4$ g/cm^3. RMC and VD void analyses were applied to clarify the modification of chain geometries in intermediate scale in expanded liquid Se near the SC-M transition region.

The helical chains are modified to the stacked planar zigzag chain which leads to the formation of metallic domains. The distributions of H and Z chains are characterized by the distribution of void sizes supported by chain segments and the distances to neighbor atoms along a chain. It is found that near the SC-M transition region number fraction of N_Z/N_H for Z to H chain segments increases and is suggestive of micro-segregation [10] of helical and zigzag chain segments.

References

[1] R. L. McGreevy, J. Phys.: Condens. Matter **13**, R877 (2001)
[2] S. L. Chan and S. R. Elliot, Phys. Rev. B**43**, 4423 (1991)
[3] P.A. Madden and M. Wilson, Phys. Rev. Lett. **80**, 532 (1998)
[4] K. Tamura and M. Inui, J. Phys.: Condens. Matter **13**, R337 (2001)
[5] K. Tamura and S. Hosokawa, Ber. Bunsenges. Phys. Chem. **96**, 681 (1992)
[6] H. Hoshino, R. W. Schmutzler and F. Hensel, Ber. Bunsenges. Phys. Chem. **80**, 27 (1976)
[7] F. Hensel and W. W. Warren, Jr., *Fluid Metals* (Princeton Univ. Press, Princeton, 1999) chap. 5
[8] A. Ikawa and H. Fukutome, J. Phys. Soc. Jpn. **59**, 1002 (1990)
[9] K. Maruyama, H. Endo, H. Hoshino, Y. Kajihara, M. Nakada and S. Sato, J. Phys.: Condens. Matter **22**, 455103 (2010)
[10] H. Kajikawa, S. Takahashi, M. Iwakoshi, T. Hoshino and M.Yao, J. Phys. Soc. Jpn. **76**, 014604 (2007)

EPJ Web of Conferences **151**, 01004 (2017) DOI: 10.1051/epjconf/201715101004
LAM-16

Structural Analysis of Molten $NaNO_3$ by Molecular Dynamics Simulation

Shuta Tahara[1,*], *Hiroshi* Toyama[1], *Hironori* Shimakura[2], and *Takanori* Fukami[1]

[1]Department of Physics and Earth Sciences, Faculty of Science, University of the Ryukyus, Okinawa 903-0213, Japan

[2]Niigata University of Pharmacy and Applied Life Sciences, Niigata 956-8603, Japan

Abstract. MD simulation for molten $NaNO_3$ has been performed by using the Born-Mayer-Huggins-type potentials. The new structural features of molten $NaNO_3$ are investigated by several analytical methods. The coordination-number and bond-angle distributions are similar to those of simple molten salts such as NaCl except for the variation caused by the different size of the anion and cation. Na^+ ions are attracted toward O^- ions, and get separated from N^+ ions by Coulomb interactions. The distribution of the dihedral angle between NO_3^- plannar ionic molecules has also been investigated.

1 Introduction

$NaNO_3$ consists of Na^+ ions and NO_3^- ionic molecules, where three O^- ions in a NO_3^- molecule form an equilateral triangle around the central N^+ ion with a covalent bond between N^+ and O^- ions. $NaNO_3$ is a well-known component of explosives. Recently, the molten state of $NaNO_3$ has also been used for fabricating heat-storage materials for application in solar power generation [1-5]. Adya et al. for the first time investigated the microscopic atomic configuration and diffusion in molten $NaNO_3$ by means of X-ray diffraction measurement and molecular dynamics (MD) simulation, and compared its partial structure factors and partial pair distribution functions with those of molten $NaNO_2$ and the eutectic mixture of the $NaNO_3$-$NaNO_2$ system showed good additivity for the intermolecular structure because the ionic radius of NO_3 and NO_2 ions are similar owing to the presence of a lone pair on the nitrite ion [6]. After Adya et al. [6], several researchers have reported the MD simulation results of molten alkali nitrates [7-12]. For example, Kato et al. have reported cation dependence of the detailed structural features and ionic dynamics in molten alkali nitrates [11]. They have pointed out that the diffusion constant of small cation is larger than that of large anion. In this study, we have performed MD simulations for molten $NaNO_3$, and obtained new structural information by using several analytical methods.

* Corresponding author: tahara@sci.u-ryukyu.ac.jp

2 Molecular dynamics simulation

The simulation cell contained 1080 ions (Na^+: 216, N^+: 216, and O^-: 648), and NO_3^- was assumed as a rigid ionic molecule in the simulation process. The side length of the simulation cell was 25.3358 Å, which corresponds to the experimental density at the respective temperature of 623 K. The system was annealed at 923 K, and then quenched to 623 K. A time step was set as 5 fs. MD simulation was carried out by using LAMMPS software [13].

The pair potentials used for the simulation were Born-Mayer-Huggins type

$$\phi_{ij} = z_i z_j e^2/r + A_{ij} b \exp[-\varrho(\sigma_i + \sigma_j - r)] \quad (1).$$

The first term on the right-hand side of eq. (1) is Coulomb interaction between ionic charges, where z and e denote valence and unit charges, respectively. The second term denotes repulsion between the ions arising from the overlap of the outer shell of electrons, where A, b, ϱ, and σ denote the Pauling parameter, a constant, a softness parameter, and an ionic radius, respectively. These potential parameters were taken from Adya et al. [6].

3 Results and discussion

3.1 Partial pair distribution functions

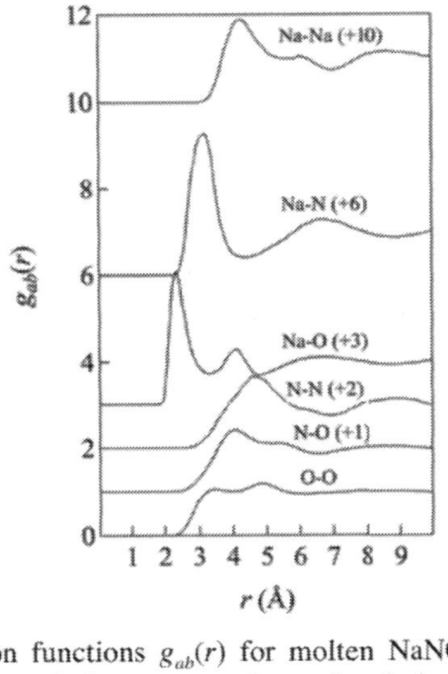

Fig. 1. Partial pair distribution functions $g_{ab}(r)$ for molten $NaNO_3$ at 623 K. The inner molecule interactions for N-O and O-O correlations are not shown for clarity.

Figure 1 shows partial pair distribution functions $g_{ab}(r)$ for molten $NaNO_3$. The $g_{ab}(r)$ values well reproduce earlier simulation results obtained by Adya et al. [6]. We defined the first-coordination length r_c for Na-Na, Na-N, and N-N correlations by using the first-minimum positions in $g_{ab}(r)$, as summarized in Table 1, and these values are used in the analyses described in Sec. 3.3.

 DOI: 10.1051/epjconf/201715101004

Table 1. First-coordination length r_c for Na-Na, Na-N, and N-N correlations.

	Na-Na	Na-N	N-N
r_c [Å]	5.4	4.5	6.7

3.2 Visualization of ionic distribution

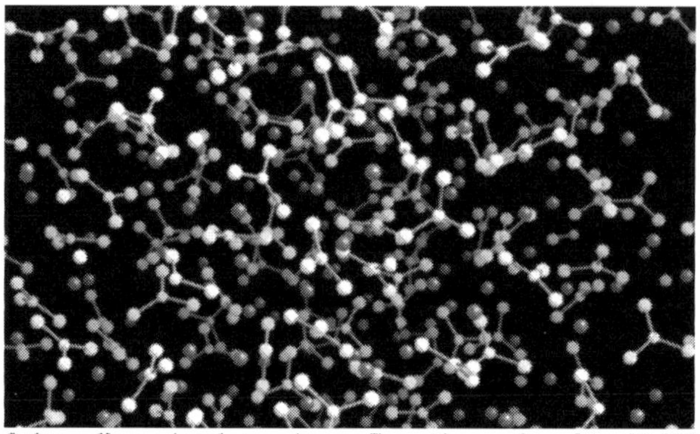

Fig. 2. Snapshot of three-dimensional atomic configurations of molten $NaNO_3$ obtained by MD simulation. Gray spheres denote Na^+ ions. Small and large white spheres denote N and O ions, respectively. N and O ions in a molecule are connected with sticks.

The snapshot of a three-dimensional atomic configuration of molten $NaNO_3$ obtained by MD simulation is visualized in Fig. 2. In this figure, it can be seen that NO_3 molecules are formed by equilateral triangles made of one N^+ and three O^- ions. Na^+ ions are distributed between NO_3^- molecules owing to Coulomb attractions between Na^+ and O^- ions, which is consistent with the relative shorter first-correlation length of Na-N correlation compared to that of Na-Na and N-N correlations, as shown in Fig. 1. Such a structural feature is similar to that of simple molten salts such as NaCl. However, the distributions of Na^+ and NO_3^- ions are slightly asymmetric, because the spectrum patterns of $g_{ab}(r)$ for Na-Na and N-N correlations do not fully correspond with each other. For example, the first peak in g_{NaNa} is narrower than that in $g_{NN}(r)$ and has a shoulder at approximately $r = 6$ Å, which would be due to the size difference between the Na^+ ion and the NO_3^- ionic molecule.

3.3 Coordination-number and bond-angle distributions

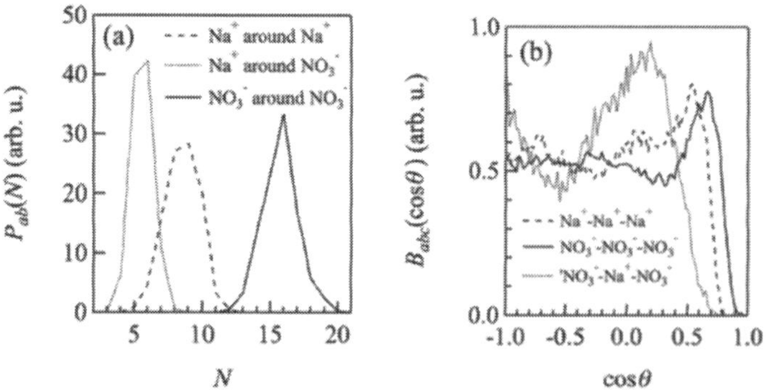

Fig. 2. (a) Coordination-number and (b) bond-angle distributions of Na^+ and NO_3^- ions in molten $NaNO_3$ at 623 K.

The coordination-number distributions $P_{ab}(N)$ (a, b = N^+, NO_3^-) of molten $NaNO_3$ are shown in Fig. 2(a). Since the N^+ ion is located at the center of the NO_3^- molecule, the N^+ ion is regarded as a NO_3^- ionic molecule in this statistics, for example, the coordination number of NO_3^- molecules around a NO_3^- molecule is the same as that of N^+ ions around a N^+ ion. The mean coordination number N_{ab} is summarized in Table 2. The N_{+-} value (= 5.6) for molten $NaNO_3$ is approximately intermediate between that of molten NaCl (N_{+-} = 4.85) and CsCl (N_{+-} = 6.1) [14]. This N_{+-} value is consistent with that reported by Kato et al. (N_{+-} = 5.7), but larger than that for molten $LiNO_3$ (5.0) [11]. The N_{++} value for molten $NaNO_3$ is smaller than the N_{--} value because the first correlation length of Na-Na correlation defined in Sec. 3.1 is shorter than that of N-N correlation. When we consider the shoulder at the high-r side of the first peak in $g_{NaNa}(r)$, the N_{--} and N_{++} are similar to each other. Therefore, the relation between N_{--} and N_{++} largely depends on the definition of the first correlation length of N-N and Na-Na correlations. The N_{--} values of molten NaCl and CsCl are approximately 14 and 17 [14], respectively; therefore, the N_{--} value of molten $NaNO_3$ is also similar to that of simple molten salts.

Fig. 2(b) shows the bond-angle distribution B_{abc} of molten $NaNO_3$. The peaks of B_{+++}, B_{-+-}, and B_{---} are found at approximately cosθ = 0.51 (θ = 59°), 0.17 (θ = 80°), and 0.65 (θ = 49°), respectively, which is also similar to those of simple alkali halides [14]. For molten NaCl and CsCl, the spectral patterns of B_{+++} and B_{---} are almost the same. The difference in the peak positions of B_{+++} and B_{---} for molten $NaNO_3$ reflects the difference in the size of the cation and anion. According to Kato et al. [11], B_{+-+} for molten $NaNO_3$ can be estimated to be 60-90°, which is consistent with our analysis results. Smaller B_{+-+} has been observed for molten $LiNO_3$ [11] due to the smaller size of the Li ion than the Na ion. The B_{+++} and B_{---} values of molten $NaNO_3$ have not been reported as yet.

Table 2. Mean coordination number N_{ab} (ab = ++, +-, --), where + and - denote the Na^+ ion and the NO_3^- ionic molecule, respectively.

	++	+-	--
N_{ab}	8.5	5.6	15.7

3.4 Distribution of Na^+ ions around a NO_3^- ionic molecule

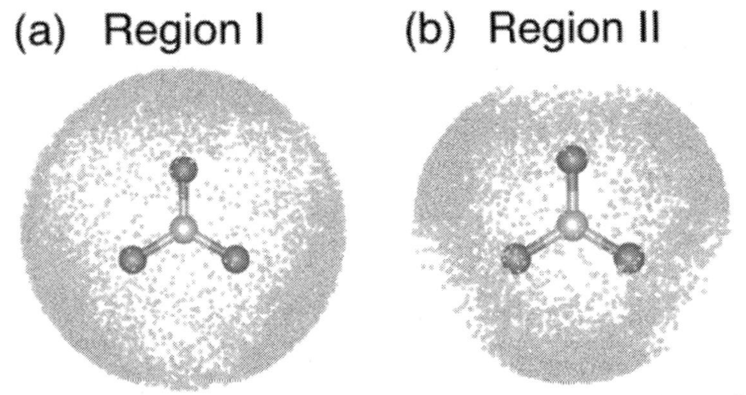

Fig. 3. Statistical distribution of Na^+ ions around a NO_3^- ionic molecule in regions (a) I (3.1 < r_{NaN} < 3.4 Å) and (b) II (2.8 < r_{NaN} < 3.1 Å) for molten $NaNO_3$ at 623K. The ionic sizes depicted are smaller than the real size for clarity.

The statistical coordinates of Na^+ ions around a NO_3^- molecule within the spherical regions I (3.1 < r_{NaN} < 3.4 Å) and II (2.8 < r_{NaN} < 3.1 Å) were obtained (Fig. 3); regions I and II

DOI: 10.1051/epjconf/201715101004

correspond to the top and bottom of the first peak in $g_{NaN}(r)$, respectively. In region I, Na^+ ions are gathered around O^- ions because of the Coulomb attraction. In contrast, the distribution of Na^+ ions in the region II is reverse of that in the region I, because the repulsion of Na-O pairs due to the second term of the right-hand side of eq. (1) becomes larger than the Coulomb attraction. In both of the regions I and II, Na^+ ions are not observed near the N^+ ion due to the repulsion between Na^+ and N^+ ions; the entire right-hand side of eq. (1) shows this repulsion. Such strucutural features are consistent with the analysis results of the cation distribution function around a NO_3^- molecule reported by Kato et al. [11]. The cationic distributions shown in Fig. 3 would help readers in imaging the cation distribution.

3.5 Molecular-orientation correlation

Since NO_3^- is a planar ionic-molecule, the orientation of the molecule can be determined by a normal vector of the plane. The information on the molecular-orientation correlations between NO_3^- molecules is obtained by considering the statistics of dihedral angles θ calculated by the inner product of two normal vectors, where θ is an angle between the normal vectors of NO_3^- molecules. When $\theta = 0$, two molecules are parallel to each other. However, when $\theta = 90°$, the molecules are perpendicular to each other. In this study, two spherical regions III ($4.4 < r_{NN} < 4.6$ Å) and IV ($3.0 < r_{NN} < 3.2$ Å) around the N^+ ion in a central NO_3^- molecule are defined as regions that correspond to the top and bottom of the first peak in $g_{NN}(r)$, respectively. Distance between N^+ ions is regarded as the distance between NO_3^- molecules. The dihedral angle formed by a central molecule and another molecule in the region III (or IV) is calculated as shown in Fig. 4.

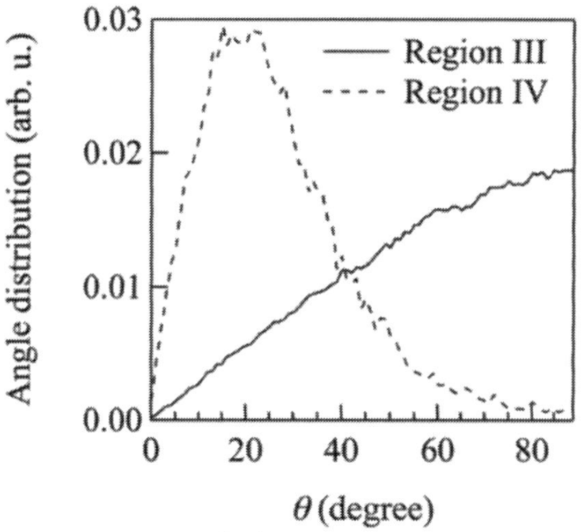

Fig. 4. Distribution of the dihedral angle θ (degree) between NO_3^- ionic molecules.

For the region III, the distribution monotonically increases with increasing θ, and no characteristic peak is observed. In contrast, for the region IV, a characteristic peak is observed at approximately $\theta = 20°$. It is difficult for two NO_3^- ionic molecules to come close to each other within the region III because of their size and shape; however, two NO_3^- molecules can close if the dihedral angle of the molecules becomes small.

4 Summary

The MD simulation for molten $NaNO_3$ was carried out by using the Born-Mayer-Huggins-type potentials. After confirming that the $g_{ab}(r)$ value obtained by the MD simulation well reproduced the results reported in literature, we attempted to obtain the new structural information. We visualized three-dimensional ionic configurations, and showed that Na^+ ions are distributed between NO_3 ions, which is similar to simple molten salts such as NaCl. The coordination-number and bond-angle distributions are similar to those of simple molten salts, except for the variation caused by the different size of the cation and anion. Na^+ ions are basically attracted by O^- ions, and are separated from N^+ ions. When two NO_3^- ions are close to each other, it is need that the dihedral angle of the molecules becomes small.

References

1. C. Y. Zhao, Z. G. Wu, Sol. Energy Mater. Sol. Cells, **95**, 3341 (2011)
2. T. Wang, S. Viswanathan, D. Mantha, R. G. Reddy, Sol. Energy Mater. Sol. Cells, **102**, 201 (2012)
3. D. Mantha, T. Wang, R. G. Reddy, Sol. Energy Mater. Sol. Cells, **118**, 18 (2013)
4. A.G. Fernández, S. Ushak, H. Galleguillos, F. J. Pérez, Sol. Energy Mater. Sol. Cells, **132**, 172 (2015)
5. P. Zhang, X. Xiao, Z. N. Meng, M. Li, Appl. Energy **137**, 758 (2015)
6. A. K. Adya, R. Takagi, K. Kawamura, M. Mikami, J. Mol. Phys. **62**, 227 (1987)
7. T. Kato, K. Machida, M. Oobatake, S. Hayashi, J. Chem. Phys. **89**, 3211 (1988)
8. T. Kato, K. Machida, M. Oobatake, S. Hayashi, J. Chem. Phys. **89**, 7471 (1988)
9. T. Kato, K. Machida, M. Oobatake, S. Hayashi, J. Chem. Phys. **92**, 5506 (1990)
10. T. Kato, K. Machida, M. Oobatake, S. Hayashi, J. Chem. Phys. **93**, 3970 (1990)
11. T. Kato, S. Hayashi, M. Oobatake, K. Machida, J. Chem. Phys. **99**, 3966 (1993)
12. G. Vohringer, J. Richter, Z. Naturforsch. **56 a**, 337 (2001)
13. S. Plimpton, J. Comput. Phys. **117**, 1 (1995)
14. R. L. McGreevy, L. Pusztai, Proc. Royal Soc. **430**, 241 (1990)

Ab initio molecular dynamics study of collective dynamics in liquid Tl: Thermo-viscoelastic analysis

Taras Bryk[1,2,*] and *Taras* Demchuk[1,**]

[1]Institute for Condensed Matter Physics, NAS of Ukraine, Svientsitskii Street 1, 79011 Lviv, Ukraine
[2]Institute for Applied Mathematics and Fundamental Sciences, Lviv Polytechnic National University, S. Bandera Str. 12, 79013 Lviv, Ukraine

Abstract. We studied collective dynamics of pure liquid metal Tl using a combination of *ab initio* molecular dynamics (AIMD) simulations and a thermo-viscoelastic model applied to calculations of dynamic eigenmodes and dispersion of collective excitations in particular. We found that for liquid Tl at ambient pressure the transverse current spectral functions obtained directly in *ab initio* simulations for wave numbers larger than first pseudo-Brillouin-zone boundary contain two low- and high-frequency peaks that is an evidence of emergence of the unusually high-frequency transverse modes as it was observed before in liquid Li at very high pressures. The thermo-viscoelastic dynamic model shows perfect reproduction of the simulation-derived longitudinal current autocorrelation functions, and the acoustic eigenmodes are in nice agreement with the peaks of the longitudinal current spectral functions up to the first pseudo-Brillouin-zone boundary. The deviation of the dynamic eigenmodes from peak positions at higher wave numbers gives evidence of L-T coupling effects.

1 Introduction

Collective dynamics in disordered systems is one of the most fascinating topics of modern condensed matter physics. In particular the increasing amount of information from inelastic X-ray (IXS) scattering experiments and computer simulations on liquids opens new insight in understanding and new problems in description of collective dynamics in liquids on atomistic scale. The theory of collective model in liquids is working well in frames of linearized hydrodynamics, which is able to describe correctly dynamic processes in liquids, like sound propagation, thermal diffusivity, etc only on macroscopic scales where atomistic structure of liquid is not taken into account[1, 2]. The simulations and scattering experiments, instead, allow probing the correlations on atomistic scale, and a challenge for generalized hydrodynamic theory is to describe emerging effects in liquid dynamics due to non-hydrodynamic effects on molecular scale like structural relaxation[3], cage effect, shear waves, or optic-like excitations in binary liquids[4] and molten salts[5] etc.

The standard generalized hydrodynamic approach is based on the memory function formalism[1, 2] and for simple liquids it provides quite good description of time-dependent

*e-mail: bryk@icmp.lviv.ua
**e-mail: demchuk@icmp.lviv.ua

processes. Recently there appeared reports on IXS experiments on liquid metals Ga[6, 7], Sn [8], Na [9], Zn, Cu and Fe[10] in which a contribution from transverse excitations to the experimental dynamic structure factors were discussed as a consequence of some longitudinal-transverse (L-T) coupling. From the simulation side there was nothing similar observed from classical simulations based on effective pair potentials between particle, however when the *ab initio* simulations were applied it was found that for liquid Sn the spectral functions can have two-peak structure [11]. Also, it was observed in AIMD that at very high pressure in liquid Li there emerged unusual high-frequency peaks on transverse current correlation functions[12], which give evidence of two low- and high-frequency contribution to the transverse dynamics at large wave numbers, which can be also a consequence of the suggested before L-T coupling in liquid metals. Later on similar findings were reported from AIMD study of collective excitations in liquid Fe[13]. At the date there is no clear explanation by theory of the suggested L-T coupling effects.

On the theoretical side there is practically absence of theoretical analysis of AIMD-derived time correlation functions, while the dispersions of collective excitations are routinely estimated from the peak (or shoulder) positions of the L and T current spectral functions despite low statistics which is inherent in AIMD simulations. Therefore other schemes which would be based in particular on the analysis of time correlation functions in order to avoid numerical time Fourier-transformations are very timing. In particular those theoretical schemes of analysis should include temperature fluctuations in order to be consistent in the long-wavelength limit with the hydrodynamic theory as well as offer explanations based on emergence of non-hydrodynamic collective processes[14] outside the hydrodynamic region. Recently an analysis of collective dynamics in liquid Na[15] implied contributions to the density-density time correlation functions coming from non-hydrodynamic heat waves[16], which exist on nanoscales and take part in heat transport in liquids and solids.

The aim of this study is to study features in longitudinal and transverse collective excitation spectra of liquid Tl (as a metallic system having similarly to Ga three valence electrons per ion) from *ab initio* simulations and rationalize them using thermo-viscoelastic approach of generalized hydrodynamics. The remaining paper has the following structure: in the next section we supply information on the *ab initio* simulations and analysis used in this study. Section III contains the spectra calculations and discussion on the obtained results. The conclusions of this study are given in the last section.

2 Simulations and methodology of thermo-viscoelastic analysis

We performed *ab initio* simulations of liquid Tl at T=577 K using a collection of 300 particles interacting with the electron subsystem via PAW-potentials [17, 18]. The exchange-correlation functional was taken in Perdew-Burke-Ernzerhof (PBE) form [19], and because of quite large size of simulated system only Γ-point of the Brillouin zone was taken for calculations of electron density. The initial configuration was taken from classical simulations of liquid Tl[20]. The time step in *ab initio* simulations was 2 fs, and production runs were over 39000 configurations. Each time step the positions, velocities and forces acting on particles were stored, as well as every 10 steps we stored the instantaneous configuration of the electron density. The ionic pair distribution functions $g_{ii}(r)$ and electron-ion distribution $g_{ie}(r)$ are shown in Fig.1. The obtained in *ab initio* simulations pair distribution function is very similar as the one from classical simulations [20] with the main peak located at 3.29Å. Although the temperature in the AIMD simulations was right at the melting point the pair distribution functions are typical as for the liquid systems, without any signs of supercooling on the second maximum or possible solid-like clusters. The advantage of AIMD simulations is in a possibility to study simultaneously with atomistic structure the distributions of electron density. The electron-ion correlations for valence electrons in liquid Tl are represented by the ion-electron distribution function $g_{ie}(r)$ (Fig.1), which in contrast to $g_{ii}(r)$ does not start from zero. The main maximum of the ion-electron distribution function is located at 0.93Å, that is also a consequence of the preserved nodal structure of electron wave functions in PAW formalism [17] in contrast to the nodeless pseudo-wave functions when the simple norm-conserving pseudopotentials are used in AIMD. The calculated static structure factor $S(k)$ has the same location of the first sharp diffraction peak, however with larger amplitude, than the experimental static structure factor for liquid Tl[21]. This fact can be a consequence of overbinding in local structure due to the applied PBE exchange-correlation, similar as this is usually observed in AIMD simulations of water [22].

We calculated the density-density $F_{nn}(k,t)$ and current-current $F_{JJ}^{L/T}(k,t)$ time correlation functions by averaging them over all possible directions of **k**-vectors having the same absolute value. The smallest wave number accessible from our AIMD simulations was $k_{min} = 0.301\text{Å}^{-1}$. The simplest methodology to roughly estimate the frequencies of collective excitations is from the peak positions of the longitudinal (L) and transverse (T) current spectral functions $C^{L/T}(k,\omega)$, which are obtained from numerical time-Fourier transformation of the corresponding simulation-derived current-current time correlation functions $F_{JJ}^{L/T}(k,t)$.

More precise analysis of collective dynamics and calculations of dispersion of collective excitations consists in application of generalized hydrodynamic models in order to estimate contributions from different relaxing and propagating modes to the time correlation functions (or spectral functions) of interest. In particular, in [23] a thermo-viscoelastic analysis based on generalized collective modes (GCM) approach in connection with the *ab initio* simulations was suggested. In the thermo-viscoelastic approach a five-variable dynamic model is considered:

$$\mathbf{A}^{(5)}(k,t) = \left\{n(k,t), J^L(k,t), e(k,t), \dot{J}^L(k,t), \dot{e}(k,t)\right\}, \tag{1}$$

where $n(k,t)$, $J^L(k,t)$ and $e(k,t)$ are the spatial Fourier-components of particle density, density of longitudinal current and energy density, respectively. The overdots in 1 denote the time derivative of corresponding dynamic variable, which can be usually represented in analytical form [24]. The set of dynamic variables $\mathbf{A}^{(5)}(k,t)$ is applied for solving the generalized Langevin equation[1] in matrix form in terms of dynamic eigenmodes for so-called generalized hydrodynamic matrix $\mathbf{T}(k)$[24]. The corresponding eigenvectors allow to estimate contribution of particular dynamic mode to the time correlation function of interest, and the quality of the theory is usually verified by comparison of theoretical and simulation-derived time correlation functions [23, 25]. An important issue is the number of exact sum rules fulfilled for theoretical time correlation functions (short-time behavior), and the thermo-viscoelastic dynamic model provides the same first time derivatives at the origin up to the fourth order for density-density time correlations functions from theory and simulations, fulfilling in general six sum rules within the GCM scheme. For the longitudinal current-current time correlation functions this GCM model provides six sum rules too [23].

The 5×5 generalized hydrodynamic matrix $\mathbf{T}(k)$ is expressed in frames of the GCM approach[24] in the following way

$$\mathbf{T}(k) = \mathbf{F}(k,t=0)\tilde{\mathbf{F}}^{-1}(k,z=0) \tag{2}$$

via the 5×5 matrices of static correlation functions $\mathbf{F}(k,t=0)$ and of Laplace-transformed time correlation functions in Markovian approximation $\tilde{\mathbf{F}}(k,z=0)$. For each wave number

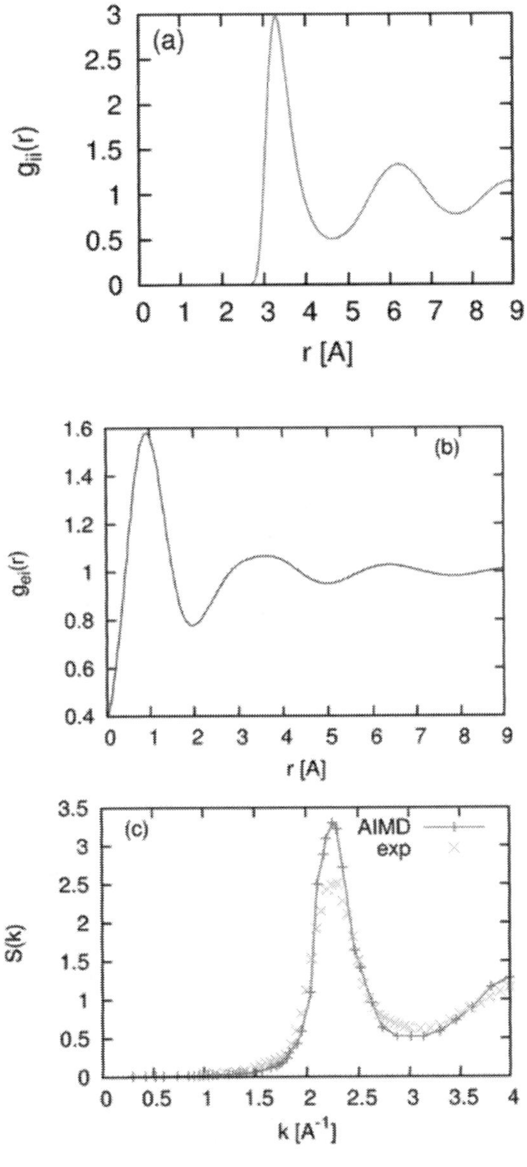

Figure 1. Obtained in AIMD ion-ion pair distribution function $g_{ii}(r)$ (a), ion-electron distribution function $g_{ie}(r)$ (b) and static structure factor S(k) (c) for liquid Tl at 577 K. Experimental static structure factor [21] in (c) is shown by cross symbols.

one has to calculate from *ab initio* simulations the following matrix elements of matrices [23]

$$\mathbf{F}(k, t=0) = \begin{pmatrix} f_{nn} & 0 & f_{ne} & -ikf_{JJ} & 0 \\ 0 & f_{JJ} & 0 & 0 & -if_{Je} \\ f_{ne} & 0 & f_{ee} & -if_{Je} & 0 \\ ikf_{JJ} & 0 & if_{Je} & f_{JJ} & 0 \\ 0 & if_{Je} & 0 & 0 & f_{\grave{e}\grave{e}} \end{pmatrix}, \tag{3}$$

and

$$\tilde{\mathbf{F}}(k, z = 0) = \begin{pmatrix} \tau_{nn} f_{nn} & \frac{i}{k} f_{nn} & \tau_{ne} f_{ne} & 0 & f_{ne} \\ \frac{i}{k} f_{nn} & 0 & \frac{i}{k} f_{ne} & f_{JJ} & 0 \\ \tau_{nn} f_{ne} & \frac{i}{k} f_{ne} & \tau_{ee} f_{ee} & 0 & f_{ee} \\ 0 & -f_{JJ} & 0 & 0 & i f_{\dot{J}e} \\ -f_{ne} & 0 & -f_{ee} & i f_{\dot{J}e} & 0 \end{pmatrix}, \quad (4)$$

where the indices in matrix elements correspond to dynamic variables from the set (1). For simplicity the purely imaginary matrix elements in (3) and (4) are shown explicitly. In *ab initio* simulations in contrast to classical molecular dynamics it is extremely difficult to sample dynamic variables of energy density and its first time derivative. Therefore the corresponding matrix elements in (3) and (4) which involve these two dynamic variables for calculation were treated as fitting parameters. All the matrix elements with $n(k, t)$, $J^L(k, t)$ and $\dot{J}^L(k, t)$ were directly calculated from *ab initio* simulations. The fitting procedure was applied simultaneously to the AIMD-derived density-density and longitudinal current-current time correlation functions having just six fitting parameters in matrices (3) and (4). Having some trial values for the fitting parameters (matrix elements) at each iteration the eigenvalues and associated eigenvectors of the 5×5 generalized hydrodynamic matrix $\mathbf{T}(k)$ were obtained, the GCM-theoretical time correlation functions were constructed from the eigenvalues and associated eigenvectors [24] and compared to the AIMD-derived functions.

3 Results and discussion

The time correlation functions for liquid Tl obtained directly in AIMD simulations were analyzed by the thermo-viscoelastic dynamic model (1). The quality of the fit procedure is demonstrated in Fig. 2 for longitudinal current-current time correlation functions with two wave numbers. One can see perfect recovering by the thermo-viscoelastic theory the damping oscillations in the time dependence of current correlations. As it was stated in [23] the nice quality of the fit is provided by the exact sum rules for short-time behavior and time moments of the corresponding AIMD-derived function. The proposed fitting scheme in that sense has advantages over the fitting of spectral functions with spectral shapes of a single or two damped harmonic oscillators (DHO)[26].

We also performed the numerical time-Fourier transformation for AIMD-derived longitudinal and transverse time correlation functions in order to check their peak structure. The longitudinal current-current spectra $C^L(k, \omega)$ showed a single-peak shape for all studied wave numbers. The transverse current spectral function had a well-defined single-peak shape for wave numbers less than the first pseudo-Brillouin zone boundary at $\sim 1.15\text{Å}^{-1}$, however for larger wave numbers we observed a noisy two-peak structure (see Fig.3). Note that similar two-peak structure of the transverse current spectral function was observed before for liquid Li at very high pressure, while for ambient pressure in liquid Li the two-peak structure was absent[12]. Here we stress that we obtained for liquid Tl at the melting point and ambient pressure an evidence of the similar extra high-frequency excitations contributing to transverse dynamics.

The behavior of peak positions of L- and T-current spectral functions is shown in Fig.4. The dispersion of T-excitations shows a tendency very similar as was observed before for the case of liquid Li at high-pressure [12] when the quite standard dispersion of the shear waves splits into two low- and high-frequency branches upon wave number reaching the pseudo-Brillouin zone boundary. The low-frequency transverse branch practically overlaps with the peaks of the longitudinal current spectral function for wave numbers $k > 1.7\text{Å}^{-1}$ that supports an idea of the existing L-T coupling on atomic scale[6]. The thermo-viscoelastic

 DOI: 10.1051/epjconf/201715102001

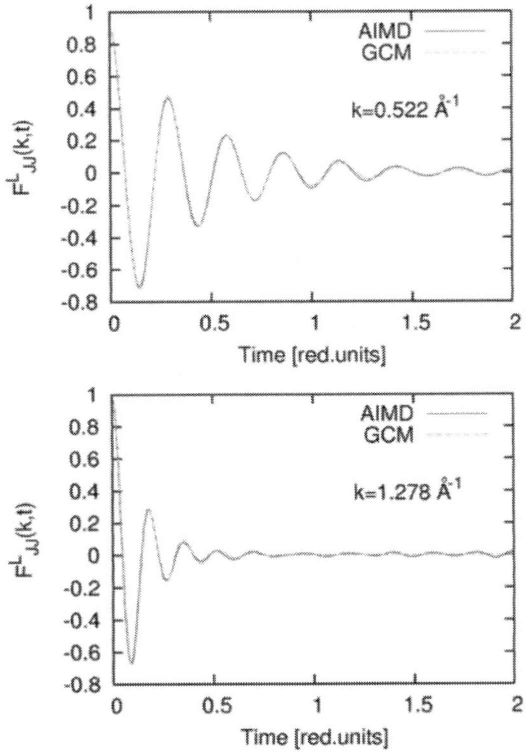

Figure 2. Longitudinal current-current time correlation functions obtained from *ab initio* molecular dynamics simulations (solid red line) and from the thermo-viscoelastic model $\mathbf{A}^{(5)}(k,t)$ (1) (dashed green line). Time scale for reduction of units τ=2.1665 ps

dynamic model results in dispersion for acoustic modes, which perfectly recovers the peak positions of the L-current spectral function but only up to the maximum of the dispersion curve. Further increase of wave numbers results in a high-frequency deviation of the acoustic eigenvalues from the peak positions. Namely in this region of wave numbers emerges the second, high-frequency, branch of transverse excitations. It is obvious that the thermo-viscoelastic dynamic model does not contain any mechanism of possible L-T coupling, and perhaps therefore we observed the discrepancy between numerical and theoretical dispersion of longitudinal propagating modes. A further development of the thermo-viscoelastic dynamic model is needed in order to describe the L-T coupled spectra on atomic scale.

4 Conclusion

We performed *ab initio* simulations of liquid Tl at the melting point with the purpose to study dispersion of longitudinal and transverse collective excitations. We calculated the dispersion by two ways: from peak positions of the AIMD-derived longitudinal and transverse current spectral functions, and for the longitudinal case - using analysis of AIMD-derived time correlation functions by the five-variable thermo-viscoelastic dynamic model. We found a nice agreement between the two methodologies for the acoustic dispersion in the region of wave numbers up to the first pseudo-Brillouin zone boundary.

 DOI: 10.1051/epjconf/201715102001

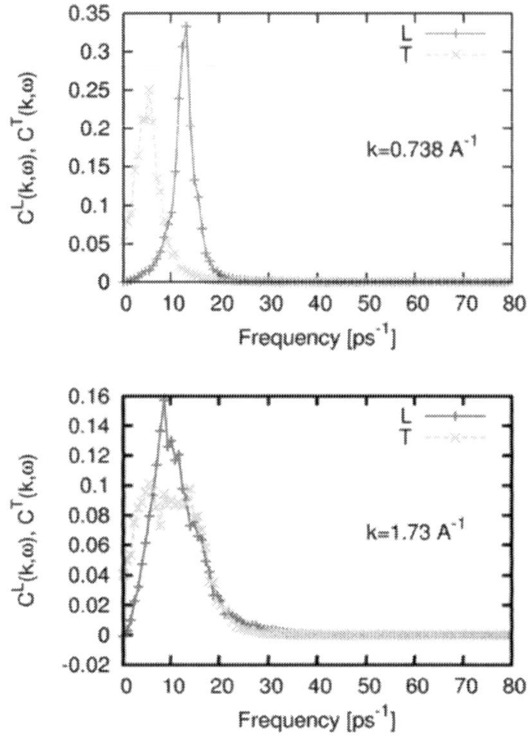

Figure 3. Longitudinal (L) and transverse (T) current spectral functions obtained from numerical time-Fourier transformation of corresponding time correlation functions with two wave numbers

For larger wave numbers the numerically obtained transverse current spectral functions gave evidence for existing two branches of collective excitations contributing to transverse dynamics. Namely in this range of wave numbers we observed a high-frequency deviation of the eigenvalues corresponding to acoustic modes from relevant peak positions of the L-current spectral functions. This deviation is attributed to the existing L-T coupling effects, which were not taken into account in the thermo-viscoelastic model.

These results imply a need in elaborating dynamic models which would explicitly treat the coupling between longitudinal and transverse processes on atomic scales. However by date there is no clear understanding of what kind of microscopic processes are responsible for the L-T coupling, because the local L-T coupling is prohibited by symmetry reasons. These effects can emerge as a consequence of rotational motion of bonded units like dimers with short lifetime. However even in that case it is a challenge how to introduce the short-time bondings into generalized hydrodynamic theory. Further *ab initio* simulations should answer the question on the role of local bonding in liquid (not really simple) metals in collective dynamics on atomic scale.

The computing time allocation for the ab initio simulations in frames of the UA-GRID-project is gratefully acknowledged. The calculations have been performed using the ab-initio total-energy and molecular dynamics program VASP (Vienna ab-initio simulation program) developed at the Institute für Materialphysik of the Universität Wien [27–29].

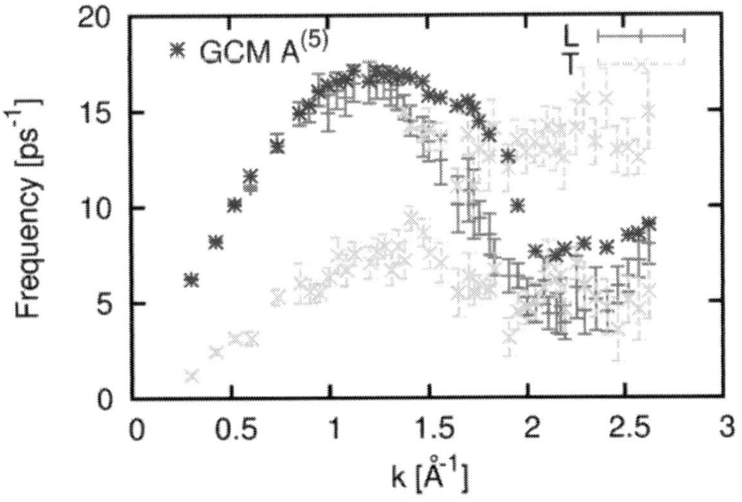

Figure 4. Wavenumber dependence of the peak positions of the longitudinal (L) and transverse (T) current spectral functions for liquid Tl at T=577 K. The acoustic eigenvalues obtained from the five-variable analysis within the thermo-viscoelastic model are shown by blue asterisks

References

[1] J.-P.Boon and S.Yip, *Molecular Hydrodynamics* (New-York: McGraw-Hill, 1980)

[2] J.-P. Hansen and I.R. McDonald, *Theory of Simple Liquids* (Academic Press - London, 3rd ed., 2006)

[3] T.Bryk and I.Mryglod, Condens. Matter Phys., **11**, 139 (2008)

[4] T.Bryk and I.Mryglod, Phys.Lett. A, **261**, 349 (1999).

[5] T.Bryk and I.Mryglod, J.Phys.:Cond.Matt., **16**, L463 (2004).

[6] S. Hosokawa, M. Inui, Y. Kajihara, K. Matsuda, T. Ichitsubo, W.-C. Pilgrim, H. Sinn, L. E. González, D.J. González, S. Tsutsui, and A.Q.R. Baron, Phys. Rev. Letters, **102**, 105502 (2009)

[7] V.M. Giordano and G. Monaco, Phys. Rev. B, **84**, 052201 (2011).

[8] S. Hosokawa, S. Munejiri, M. Inui, Y. Kajihara, W.-C. Pilgrim, Y. Ohmasa, S. Tsutsui, A. Q. R. Baron, F. Shimojo, K. Hoshino, J.Phys.: Condens. Matter, **25**, 112101, (2013)

[9] V.M. Giordano, G. Monaco, PNAS, **107**, 21985 (2010).

[10] S. Hosokawa, M. Inui, Y. Kajihara, S. Tsutsui, A. Q. R. Baron, J.Phys.: Condens. Matter **27**, 194104, (2015)

[11] S. Munejiri, F. Shimojo, K. Hoshino, Phys. Rev. B, **86**, 104202 (2012)

[12] T. Bryk, G. Ruocco, T. Scopigno, A.P. Seitsonen, J. Chem. Phys., **143**, 110204 (2015)

[13] M. Marqués, L.E. González, D.J. González, Phys. Rev. B, **92**, 134203 (2015)

[14] T. Bryk, Eur. Phys. J. Special Topics, **196**, 65 (2011).

[15] T. Bryk, J.-F. Wax, J. Chem. Phys., **144**, 194501 (2016); J. Chem. Phys., **144**, 219903 (2016)

[16] D.D. Joseph, L. Preziosi, Rev. Mod. Phys., **61**, 41 (1989)

[17] P.E. Blöchl, Phys. Rev. B, **50**, 17953 (1994)

[18] G. Kresse and D. Joubert, Phys. Rev. B, **59**, 1758 (1999)

[19] J.P. Perdew, K. Burke, M. Ernzerhof, Phys. Rev. Lett. **77**, 3865 (1996)
[20] T. Bryk, J.-F. Wax, J. Chem. Phys., **132**, 074504 (2010)
[21] Y.Waseda, http://res.tagen.tohoku.ac.jp/$waseda/scm/LIQ/all/tl.html
[22] T. Bryk, A.P. Seitsonen, Condens. Matter Phys., **19**, 23604 (2016)
[23] T. Bryk and G. Ruocco, Mol. Phys., **111**, 3457 (2013)
[24] I.M. Mryglod, I.P. Omelyan, and M.V. Tokarchuk, Mol. Phys. **84**, 235 (1995)
[25] T. Bryk and I. Mryglod, Phys. Rev. E, **64**, 032202 (2001)
[26] P. Verkerk, J.Phys.:Condens.Matter, **13**, 7775 (2001)
[27] G. Kresse and J. Hafner, Phys. Rev. B **47**, 558 (1993); ibid. **49**, 14251 (1994).
[28] G. Kresse and J. Furthmüller, Comput. Mat. Sci. **6**, 15 (1996).
[29] G. Kresse and J. Furthmüller, Phys. Rev. B **54**, 11169 (1996).

Experimental evidence for two different dynamical regimes in liquid rubidium

Franz Demmel[1,*] and *Christoph* Morkel[2,**]

[1]ISIS Facility, Rutherford Appleton Laboratory, Didcot, OX11 0QX, UK
[2]Physikdepartment E21, TU München, 85748 Garching, Germany

Abstract. We present evidence for changes in the dynamics of liquid rubidium with rising temperature. The thermal expansion of this liquid alkali metal shows a changing derivative with temperature in a temperature range of about 400-500 K. With neutron scattering the amplitude at the structure factor maximum demonstrates a changing slope with increasing temperature. A derived averaged structural relaxation time can be understood that an additional relaxation process sets in upon cooling. The deduced generalized viscosity and high frequency shear modulus indicate a change in dynamics in the same temperature range. All these findings point to a change in dynamics of the equilibrium liquid metal state with a dynamical crossover from a viscous to a fluid-like liquid metal well above the melting point.

1 Introduction

During the past decades the collective dynamics of liquid metals has been intensively investigated near the melting point, e.g. for an earlier and a more recent review on the dynamics of liquid metals see [1, 2]. Liquid alkali metals have been studied in particular, because they show distinct inelastic excitations [3–7]. One common finding was the upward bending of the frequency dispersion above the values expected from hydrodynamics, which is called *positive dispersion*. This enhancement in frequency can be described as a viscoelastic reaction of the fluid. At low frequencies the relaxations in the liquid can follow the density fluctuations and the sound waves propagate with the adiabatic sound velocity. At high frequencies the liquid response becomes solid-like with a transition to a high frequency sound velocity. A formal description can be obtained within the so-called viscoelastic model, which incorporates the two different dynamic regimes separated by a relaxation time τ, referred to as Maxwell's relaxation time.

However, the temperature dependence of the collective dynamics was rarely studied. One of the few studies explored the changes of the collective excitations with temperature in liquid sodium up to the boiling point in four coarse temperature steps [6]. Further investigations on changes in the dynamics with rising temperature up to the boiling point and beyond focused on the self-diffusion of liquid alkali metals and the underlying changes in the dynamics with decreasing density [8].

*e-mail: franz.demmel@stfc.ac.uk
**e-mail: christoph.morkel@frm2.tum.de

Less into the focus of research have been the changes in dynamics between the melting point and a few hundred degree above it, when the density changes only about 10 %. However, recent studies demonstrated distinct changes in the rubidium dynamics on atomic length scales with increasing temperature from the melting point $T_{melt} = 312.65\ K$ up to about twice the melting temperature [9–12]. The alkali metal rubidium has been studied for several decades and a lot of data, macroscopic and microscopic, are available over a wide temperature range from experiment and simulation (e.g.[13–17]). Therefore, rubidium represents an ideal test case to investigate changes in liquid metal dynamics with rising temperature in detail. With neutrons there is a long tradition of studying liquid rubidium due to its favorable scattering properties, namely the negligible incoherent cross section, low absorption and a reasonable large coherent cross section. Most of the experiments were dedicated to investigate inelastic properties of liquid rubidium [3, 18]. Towards the end point of the liquid gas phase boundary, the critical point, measurements of the structure and collective dynamics have been performed [19–21]. More specifically, a change in the current spectra has been related to the onset of dimerization in the proximity of the metal-nonmetal transition. However, most of the previous microscopic experimental or simulation work on liquid rubidium were concerned with quite coarse steps in temperature away from the melting point.

Here we assess the obtained data on liquid rubidium in a novel way, extract average structural relaxation times from the amplitude at zero energy transfer and derive the elastic shear modulus from the analysis of the acoustic-type excitations with rising temperature. We will relate the observed changes to the temperature dependence of a macroscopic parameter, the thermal expansion. The comparison of the collected data represents a coherent picture about the changes in dynamics between melting point and twice T_{melt}.

2 Experimental details

The coherent dynamic structure factor $S(Q, \omega)$ of liquid rubidium has been measured by neutron scattering. Rubidium is a nearly coherent scatterer with an incoherent contribution of less than 0.04 barn compared to the coherent cross section of 6.24 barn [22]. Rubidium (99.9 % purity) was filled in a cylindrical aluminium can with 16 mm diameter (9 % scatterer) inside a glove box. The cell has been electron beam welded to keep the sample vacuum tight. The structure factor maximum of liquid rubidium occurs at $Q_0 = 1.52$ Å^{-1}. The melting temperature of rubidium is $T = 312.65\ K$. Three experiments with differing energy resolutions have been performed to follow the changes in dynamics at the structure factor maximum. A first series of experiments has been performed at the three axis spectrometer with multi-analyzer-detector unit of the Forschungsreaktor München FRM [23]. Constant-Q scans at thirteen temperatures have been measured with an energy resolution of $FWHM = 1.3\ meV$. A second series of experiments has been performed at the thermal three axis spectrometer IN3 at the Institute Laue Langevin, Grenoble. At a final energy of 14.66 meV an energy resolution of 0.8 meV was obtained. A third series of experiments was undertaken at the IRIS-spectrometer of the ISIS Facility. A configuration with an end energy of 7.38 meV was chosen, which provided an excellent energy resolution of 0.055 meV (FWHM). More experimental details can be found in [9, 12, 23]. Furthermore, inelastic experiments to follow the collective excitations at 320 K, 420 K, 500 K and 600 K have been performed and analysed [11]. More details about the extensive data analysis and comparison with MD-simulations can be found there.

3 Results and Discussion

At first we consider a macroscopic quantity, the volumetric thermal expansion coefficient: $\alpha = \frac{1}{V}(\frac{dV}{dT})_P$. In Fig 1 we plot the thermal expansion and the mass density of liquid rubidium obtained through macroscopic measurements [24]. Apparently the slope of the thermal expansion changes with rising temperature in a temperature range from 400 K to 500 K. The inset shows the measured mass density, which already indicates this change with a changing slope. Within fluctuation theory the thermal expansion is related to correlated fluctuations

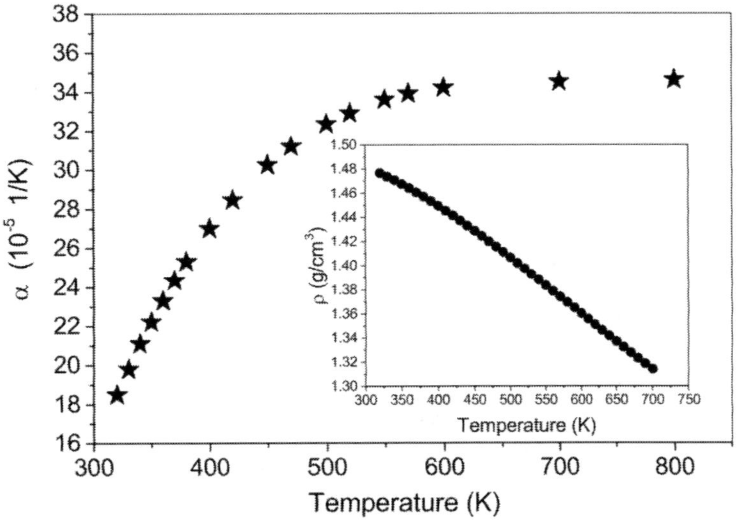

Figure 1. The thermal expansion coefficient of liquid rubidium is presented over a large range of temperatures [24]. The inset shows the change of mass density with increasing temperature.

of volume and entropy [25]. It appears that the correlations between volume and entropy fluctuations are changing much more in a temperature range up to about 400-500 K compared to the higher temperatures. That change in fluctuation correlations hints to a changing microscopic dynamics within that temperature range.

Now we want to relate these fluctuations on a macroscopic scale with density fluctuations on an atomic length scale through coherent neutron scattering. Coherent neutron scattering provides insight into the collective movements of the particles. In the hydrodynamic limit the scattering function is given by a combination of three Lorentzians, neglecting a small asymmetry term [26]. For larger momentum transfers or when length scales of atomic diameters are probed the quasielastic line shows a narrowing around the structure factor maximum, known as deGennes narrowing [27]. In the time domain the intermediate scattering function for density fluctuations $F(Q,t)$ demonstrates a slowing down at the structure factor maximum. In a simple picture it costs time for a density fluctuation to relax on a next-neighbor length scale due to a necessary rearrangement of the surrounding particles. That necessary time cost appears in the frequency domain as a narrowing of the quasielastic line. On this length scale the most direct access to the evolution of structural relaxation is provided.

We analysed the amplitude at the structure factor maximum $S(Q_0, \omega = 0)$ from three different measurements on liquid rubidium, all performed with different energy resolutions. Fig. 2 depicts the derived amplitudes $S(Q_0, \omega = 0)$. The inset illustrates the change of the spectral amplitude with rising temperature. From 320 K to 400 K the amplitude changes

 DOI: 10.1051/epjconf/201715102002

about twice as much as between 400 K and 550 K. With increasing temperature the amplitude decreases in a continuous but nonlinear manner. This change was emphasized through a linear fit through the temperature points above 400 K. Such a linear fit has no theoretical basis and is merely a guide to the eye. However, there is obviously a change in the slope around $T \approx 400$ *K*. Because we are probing structural relaxations at nearest neighbor distances that change in dynamics is evidence for a change in dynamics. Yet, the amplitude $S(Q_0, \omega = 0)$

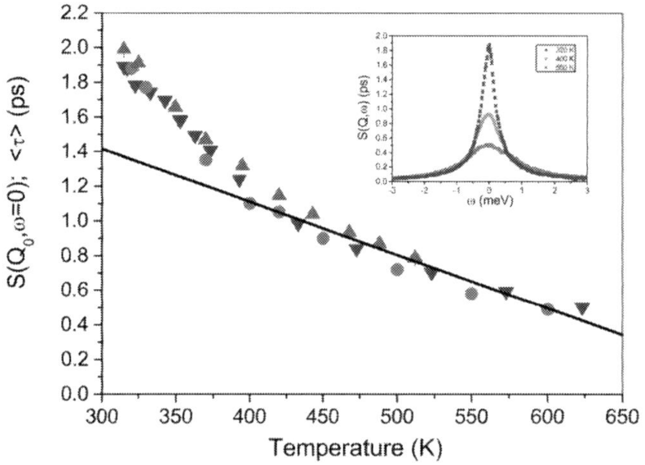

Figure 2. The amplitude $S(Q_0, \omega = 0)$ and the average relaxation time $\langle\tau\rangle$ is presented. The different symbols relate to the different experiments: triangle down (FRM), triangle up (IN3, ILL) and circles (IRIS, ISIS Facility). The line is a fit to the high temperature values and represents a mere guide to the eye.

can be considered in a different way. $S(Q_0, \omega = 0)$ is related to the intermediate scattering function $F(Q, t)$ through a Fourier transformation in the following way:

$$S(Q_0, \omega = 0) = \int F(Q, t) exp(i\omega t) dt \mid_{\omega=0} = \int F(Q, t) dt \tag{1}$$

and the last integral can be regarded as a definition for an average relaxation time:

$$\langle\tau\rangle = \int F(Q, t) dt \tag{2}$$

This average relaxation time $\langle\tau\rangle$ is identical to our measured amplitude $S(Q_0, \omega = 0)$ and can be interpreted as a decay time of density fluctuations on next neighbor distances. That approach delivers only an average relaxation time and no details about the actual decay of the intermediate scattering function. It might be interpreted as the average time a particle is associated with its nearest neighbors. The changing slope of $\langle\tau\rangle$ with decreasing temperature could then be understood that the stronger increase of the relaxation time is caused by a further relaxation process which sets in at around 400-450 K on cooling. Hence, the change in the average relaxation time at next nearest neighbor distances indicates a change in underlying dynamics in the temperature range 400-450 K.

Near the melting point an additional slow relaxation process has been demonstrated by MD-simulation and explained through mode-coupling theory [28]. That slow process was brought into connection with structural freezing and was predicted to increase strongly in the supercooled state. Sophisticated neutron scattering experiments proved the existence of this slow relaxation process near the melting point [10] and a later investigation showed that this slow process disappears at about 400 K [12], in agreement with the changing slope of the average relaxation time. To directly demonstrate a further relaxation process within an average relaxation time $\langle\tau\rangle$ would need a single precise measurement over a large range of temperatures, which is planned for the near future.

Furthermore, the amplitude $S(Q, \omega = 0)$ is connected to a generalized longitudinal viscosity coefficient $\eta_l(Q)$. The longitudinal viscosity is a measure for the diffusion of momentum parallel to the velocity of the particles and can be regarded as being sensitive to long lasting atomic-level stress correlations, in our case, on next neighbor distances. The evaluation of $\eta_l(Q)$ showed a step-like increase of the generalised viscosity with decreasing temperature at around 400 K [9]. In addition, MD-simulations of generalized viscosities at the structure factor maximum demonstrated a similar increase with decreasing temperature [29]. The increase in generalized viscosity and the increase of the amplitude $S(Q_0, \omega = 0)$, related to the set-in of an additional slow relaxation process, are evidence for a change in the dynamics to a more viscous, solid-like state upon cooling at around 400-450 K .

If the dynamics changes to a more solid-like behavior this should be mirrored in changes of elastic moduli, like the high frequency bulk or shear modulus. Near the melting point the dynamics has been related to high frequency elastic moduli of liquid rubidium [30]. Previously we analysed the inelastic collective dynamics of liquid rubidium in detail and derived from spectra of the current correlation functions dispersion curves over a wide range of momentum vectors [11]. From these dispersion curves phase velocities $c(Q) = \omega/Q$ can be calculated for the acoustic-type excitations. At high frequencies the instantaneous response of the system is probed and a high frequency phase velocity can be defined: $c_\infty(Q) = \frac{\omega_l}{Q}$, where ω_l is related to the fourth frequency moment and is directly linked to the interaction potential.

This high frequency sound velocity has been related to high frequency elastic moduli [26]:

$$c_\infty(Q) = \sqrt{\frac{1}{nM}(\frac{4}{3}G_\infty(Q) + K_\infty(Q))} \tag{3}$$

with n the particle density, M the mass of the particle, G_∞ the high frequency shear modulus and K_∞ the high frequency bulk modulus. Due to the isotropy of the system and in the low Q limit this equation can be written: $c_\infty(Q) = \sqrt{\frac{3G_\infty(Q)}{nM}}$ and hence a high frequency shear modulus G_∞ can be derived from the measured data. The $c_\infty(Q)$ values from the measurement [11] have been averaged between $Q = 0.45 - 0.55$ Å^{-1} and then G_∞ has been calculated accordingly. Fig 3 shows the obtained shear modulus G_∞ with rising temperature. The value for solid rubidium was derived from ultrasound velocity measurements using the longitudinal sound velocity [31]. A small change occurs from the solid state to the liquid state near the melting point. There are only four data points in the liquid state, however, a change in the temperature dependence of the shear modulus is evident for the higher two temperature points. Apparently, above 420 K the rigidness of liquid rubidium decreases with a higher rate than at lower temperature. That softening of the generalised elastic modulus agrees with a reduction in generalised viscosity in this temperature range, although both experiments are monitoring different spatial correlations [9].

DOI: 10.1051/epjconf/201715102002

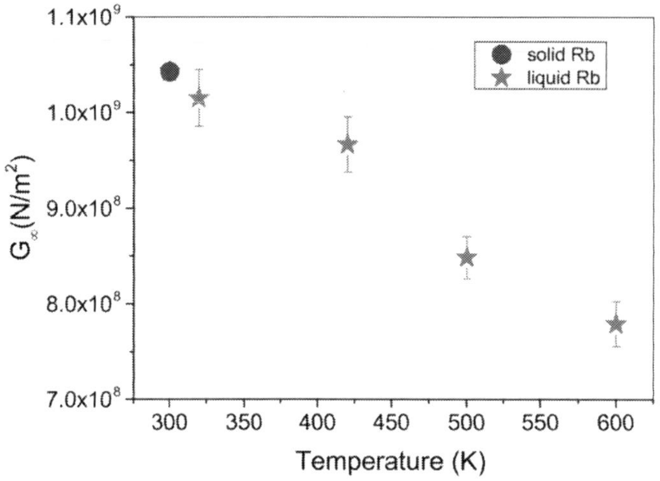

Figure 3. The derived high frequency shear modulus G_∞ for the liquid and the solid state is plotted.

4 Conclusion

For liquid rubidium the thermal expansion coefficient changes its temperature dependence between 400 K and 500 K significantly. On a microscopic scale the amplitude $S(Q_0, \omega = 0)$, related to dynamics on next-neighbor distances, changes its slope in the same temperature range. The average relaxation time of density fluctuations at this momentum transfer follows the same trend. An explanation for this change in dynamics might be the set-in of an additional slow relaxation process upon cooling, which has been related to structural freezing. These changes signal a transition between different dynamical regimes in liquid rubidium. Furthermore, in the same temperature range a decrease of generalised viscosity and a softening of the high-frequency shear modulus was observed upon heating. These evidences can be summarized that liquid rubidium shows two different dynamical regimes, a low temperature viscous one and a high temperature fluid one. A similar behavior in the amplitude and generalised viscosity was observed in liquid lead and aluminium [32, 33]. Whether such a transition from a more solid-like to a fluid-like response is a universal behavior in liquid metals needs further investigations.

References

[1] J.R.D. Copley and S. W. Lovesey Rep. Prog. Phys. **38** 461 (1975)
[2] T. Scopigno, G. Ruocco and F. Sette Rev. Mod. Phys. **77** 881 (2005)
[3] J.R.D. Copley and J.M. Rowe Phys. Rev. Lett. **32** 49 (1974)
[4] T. Bodensteiner, C. Morkel, W. Gläser and B. Dorner Phys. Rev. A **45** 5709 (1992)
[5] H. Sinn, F. Sette, U. Bergmann, C. Halcoussis, M. Krisch, R. Verbeni and E. Burkel Phys. Rev. Lett. **78** 1715 (1997)
[6] W.C. Pilgrim, S. Hosokawa, H. Saggau, H. Sinn and E. Burkel J. Non-Cryst. Solids **250–252** 96 (1999)
[7] L.E. Bove, B. Dorner, C. Petrillo, F. Sacchetti and J.B. Suck Phys. Rev. B **68** 024208 (2003)

[8] W.-C. Pilgrim and C. Morkel, J. Phys.: Condens Matter **18** R585 (2006)
[9] F. Demmel, A. Diepold, H. Aschauer and C. Morkel, Phys. Rev. B **73** 104207 (2006)
[10] F. Demmel, P. Fouquet, W. Häussler and C. Morkel, Phys. Rev. E **73** 032202 (2006)
[11] F. Demmel, D. Pasqualini and C. Morkel, Phys. Rev. B **74** 184207 (2006); F. Demmel, A. Diepold, H. Aschauer and C. Morkel, J. Non-Cryst. Solids **353** 3164 (2007)
[12] F. Demmel and C. Morkel, Phys. Rev. E **85** 051204 (2012)
[13] A. Rahman, Phys Rev. Lett. **32** 52 (1974)
[14] G. Kahl and S. Kambayashi, J.Phys.: Condens. Matter **6** 10897 (1994)
[15] D. Pasqualini, R. Vallauri, F. Demmel, Chr. Morkel and U. Balucani, J Non-Cryst. Solids **250–252** 76 (1999)
[16] F. Shimojo and Y. Zempo, Phys. Rev. B **52** 9320 (1994); S. Munejiri, F. Shimojo and K. Hoshino, J. Phys.: Condens Matter **12** 4313 (2000)
[17] J.F. Wax, R. Albaki and J.L. Bretonnet, Phys. Rev. B **65** 014301 (2001)
[18] P. Chieux, J. Dupuy-Philon, J. Jal and J.B. Suck, J. Non-Cryst. Solids **205–207**, 370 (1996); G. Pratesi, J.B. Suck and P.A. Egelstaff, J. Non-Cryst. Solids **250–252** 91 (1999)
[19] K. Matsuda, K. Tamura and M. Inui, Phys. Rev. Lett. **98**, 096401 (2007)
[20] R. Winter, C. Pilgrim, F. Hensel, C. Morkel and W. Gläser, J Non-Cryst. Solids **156-158** 9–14 (1993)
[21] W.C. Pilgrim, M. Ross, L.H. Yang and F. Hensel, Phys. Rev. Lett. **78** 3685 (1997)
[22] L. Köster, K. Knopf and W. Waschkowski, Z. Phys A **301** 215–226 (1981)
[23] F. Demmel, A. Fleischmann and W. Gläser, Nucl. Instr. Meth. A **416** 115–122 (1998)
[24] W. Ohse *Handbook of thermodynamic and transport properties of alkali metals* (Blackwell scientific Publications Academic Press, Oxford 1985)
[25] L.D. Landau and E.M. Lifshitz *Statistical Physics* p353 (Pergamon Press, Oxford 1970)
[26] J.P. Boon and S. Yip *Molecular Hydrodynamics* McGraw Hill (1980)
[27] P.G. deGennes, Physica **25**, 825 (1959)
[28] U. Balucani and R. Vallauri, Phys. Rev. A **40** 2796 (1989)
[29] D. Bertolini, F. Demmel and A. Tani, Phys. Rev. B **76** 094204 (2007)
[30] Chr. Morkel, T. Bodensteiner and H. Gemperlein, Phys. Rev. E **47** 2575 (1993)
[31] W. Schaffs *Landolt Börnstein*, Neue Serie Band 5, p236, Springer, Berlin-Heidelberg, (1967)
[32] F. Demmel, W. S. Howells and C. Morkel, J.Phys.: Condens. Matter **20** 205106 (2008); F. Demmel, W. S. Howells, C. Morkel and W.C. Pilgrim, Z Phys Chem **224** 83 (2010)
[33] F. Demmel, A. Fraile, D. Szubrin, W.C. Pilgrim and C. Morkel, J.Phys.: Condens. Matter **27** 455102 (2015)

EPJ Web of Conferences **151**, 02003 (2017) DOI: 10.1051/epjconf/201715102003
LAM-16

Dynamic properties of liquid Ni revisited

B. G. del Rio[1,*], *L. E.* González[1], and *D. J.* González[1]

[1]Departamento de Fisica Teórica, Facultad de Ciencias, Universidad de Valladolid, Valladolid, SPAIN

Abstract. Liquid Ni has previously been studied by different approaches such as molecular dynamics simulations and experimental techniques including inelastic neutron and X-ray scattering. Although some puzzling results, such as the shape of the sound dispersion curve for $q \leq 1.0$ Å^{-1}, have already been sorted out, there still persist some discrepancies, among different studies, for greater q-values. We have performed *ab initio* simulation calculations which show how those differences can be reconciled. Moreover, we have found that the transverse current spectral functions have some features which, so far, had previously been shown by high pressure liquid metals.

1 Introduction.

Transition metals such as iron and nickel play a relevant role in geophysical science because of their presence in many planetary cores, including the Earth's core. Although iron is the dominant constituent of Earth's solid inner core and liquid outer core, however the presence of Ni (along with other light elements, i.e. sulfur, silicon) must be accounted for in order to achieve an accurate description of the different phenomena accurring in the core.

The static structure factor of liquid Ni (l-Ni) near its triple point has been measured by X-ray diffraction (XD) [1] and neutron diffraction (ND) [2, 3] techniques. As for its microscopic dynamics, the first experimental study was carried out by Johnson *et al* [4] who performed inelastic neutron scattering (INS) measurements of l-Ni at T = 1870 K and determined the dynamic structure factor, $S(q,\omega)$, within a range 2.2 $\leq q \leq$ 4.4 Å^{-1}. Years later, Bermejo *et al* [5] performed INS measurements of l-Ni at T= 1763 K and within the range $0.8 \leq q \leq 3.5$ Å^{-1}; therefrom, several dynamical quantities were determined. Most interesting was the dispersion relation (DR), which suggested that isothermal behaviour was attained for $q \leq 0.8$ Å^{-1}. More recently, Cazzato *et al* [6] have performed inelastic X-ray scattering (IXS) measurements on l-Ni at T= 1767 K and the $S(q,\omega)$ was determined within a range $0.2 \leq q \leq 0.8$ Å^{-1}. The associated DR was found to be rather different from the previous one, with an adiabatic dynamical regime undoubtedly holding down to at least $q \approx 0.2$ Å^{-1}.

Molecular dynamics simulations, both classical (CMD) and *ab initio* (AIMD), have also been applied to l-Ni, with most studies focused on thermodynamic states around the triple point. Among the CMD simulation studies, we mention those by Alemany

*e-mail: beatriz@metodos.fam.cie.uva.es

et al [7] and Ruiz-Martin *et al* [8] which used, in both cases, interatomic potentials constructed within the framework of the embedded atom model [9]. More recently AIMD simulation studies have been performed by Jakse *et al* [10]. These simulation studies have evaluated a range of static and dynamic properties of l-Ni and, oddly enough, they have yielded DRs very similar to each other and to the IXS data, but qualitatively different from the INS one.

Prompted by these disparities we have performed an AIMD study on several static and dynamic properties of l-Ni near its triple point. Section 2 briefly describes the theory underlying the present AIMD simulations, and section 3 reports and discusses some structural and dynamical results. Finally some conclusions are drawn.

2 Computational method.

The total potential energy of N ions with valence Z, enclosed in a volume Ω, and interacting with $N_e = NZ$ valence electrons is written, within the Born-Oppenheimer approximation, as the sum of the direct ion-ion coulombic interaction energy, $E_{i-i}[\{\vec{R}_l\}]$, plus the ground state energy of the electronic system subjected to the ionic external potential $V_{\text{ext}}(\{\vec{R}_l\})$,

$$E(\{\vec{R}_l\}) = E_{i-i}[\vec{R}_l] + E_g[n_g(\vec{r}), V_{\text{ext}}(\{\vec{R}_l\})] , \qquad (1)$$

where $n_g(\vec{r})$ is the ground state electronic density and $\vec{R}_l$ are the ionic positions. According to Density Functional Theory, $n_g(\vec{r})$ minimizes the energy functional

$$E[n(\vec{r})] = T_s[n] + E_{\text{ext}}[n] + E_H[n] + E_{\text{xc}}[n] , \qquad (2)$$

where $T_s[n]$ is the non-interacting electronic kinetic energy, $E_{\text{ext}}[n]$ is the electron-ion interaction energy, $E_H[n]$ is the electrostatic Hartree energy and $E_{\text{xc}}[n]$ is the electronic exchange-correlation energy for which we have used the local density approximation, as parametrized by Perdew and Zunger [11]. The ion-electron interaction has been described by means of an ultrasoft pseudopotential [12].

AIMD calculations have been performed for l-Ni at a thermodynamic state near its triple point, characterized by a number density, $\rho = 0.0792$ Å^{-3} and temperature $T =$ 1773 K [1]. We have used 120 atoms (i.e. $120\times(9d+1sp) = 1200$ valence electrons) in a cubic supercell. After thermalization, the microcanonical AIMD simulations lasted for 21000 time steps, which amounted to 115.5 ps of simulation time. We have used a plane-wave representation with an energy cutoff of 25.0 Ryd and the single Γ point was used in sampling the Brillouin zone.

3 Results and discussion.

3.1 Static properties

The AIMD simulations provide a direct evaluation of both the pair distribution function, $g(r)$, and the static structure factor, $S(q)$. The calculated $g(r)$ (not shown) has the typical shape for a liquid, and shows a good agreement with experiment [1]. By integrating the $g(r)$ up to the position of its first minimum, $R_{\text{min}} = 3.40$ Å, we have evaluated the number of nearest neighbors obtaining a value ≈ 12.6 atoms, which is typical of the simple liquid metals near their triple point [13].

Figure 1 shows the calculated $S(q)$ along with the XD [1] and ND [2, 3] data. The experimental $S(q)$'s have a symmetric main peak at $q_p \approx 3.10$ Å^{-1} and its height

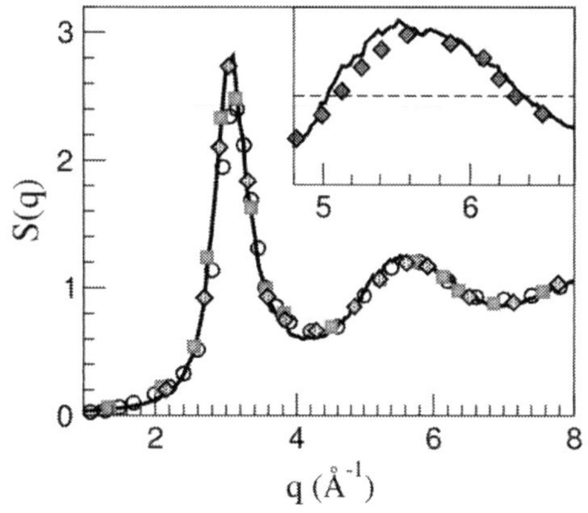

Figure 1. Static structure factor, $S(q)$, of l-Ni at T=1773 K. Full line: present AIMD calculations. Open circles: XD data from Waseda [1]. Open squares and full diamonds: ND data from Johnson *et al* [2] at T= 1875 K and Schenk *et al* [3] at T= 1765 K, respectively. The inset shows an enlarged second maximum.

varies between $S(q_p) \approx 2.4$ and ≈ 2.75 according to the XD and ND data respectively. Moreover, the ND data show a weak shoulder on the second maximum of the $S(q)$ at $q \approx 6.0$ Å^{-1}.

As evinced in Figure 1, the AIMD $S(q)$ shows good agreement with experiment; moreover, its second maximum also displays a small shoulder at $q \approx 6.0$ Å^{-1}. This type of asymmetric shape of the second peak of $S(q)$ has been experimentally observed in several liquid transition metals and it has been related to an important presence of icosahedral local order [10].

We have also estimated the isothermal compressibility, κ_T, of l-Ni by the relation $S(q \to 0) = \rho k_B T \kappa_T$ where k_B is Boltzmann's constant. The calculated $S(q)$ has been extrapolated to $q \to 0$ yielding $S(q \to 0) = 0.018 \pm 0.002$, which leads to $\kappa_T = 0.95 \pm 0.10$ (in units of 10^{11} m^2 N^{-1}) which is close to the experimental data $\kappa_T \approx 1.04 \pm 0.02$ [14].

3.2 Dynamic properties

The collective dynamics of density fluctuations in a liquid is described by the intermediate scattering function, $F(q,t)$, defined as

$$F(q,t) = \frac{1}{N}\left\langle \left(\sum_{j=1}^{N} e^{-i\vec{q}\cdot\vec{R}_j(t+t_0)} \right) \left(\sum_{l=1}^{N} e^{i\vec{q}\cdot\vec{R}_l(t_0)} \right) \right\rangle , \qquad (3)$$

The Fourier Transform (FT) of the $F(q,t)$ into the frequency domain leads to the dynamic structure factor, $S(q,\omega)$, which has experimental relevance because it is directly related to the scattered intensity measured in the INS or IXS experiments. Another interesting magnitude associated with the density fluctuations is the current due to the overall motion of the particles, i.e.

$$\vec{j}(q,t) = \sum_{j=1}^{N} \vec{v}_j(t) \; \exp[i\vec{q}\cdot\vec{R}_j(t)] , \qquad (4)$$

where $\vec{v}_j(t)$ is the velocity of particle j at time t. The current is usually split into longitudinal ($\vec{j}_L(q,t)$) and transverse ($\vec{j}_T(q,t)$) components with respect to $\vec{q}$. Therefrom, the longitudinal, $C_L(q,t)$, and transverse, $C_T(q,t)$, current correlation functions are obtained as

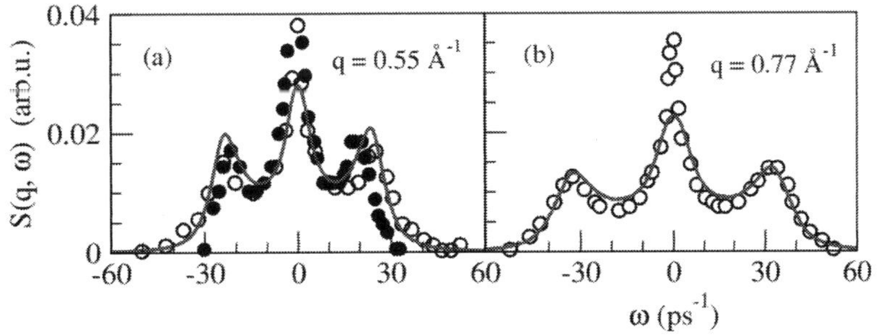

Figure 2. Dynamic structure factors, $S(q,\omega)$, of liquid Ni at T= 1773 K. (a) Filled and open circles: IXS data [6] for $q = 0.50$ and 0.60 Å^{-1} respectively. Line: AIMD results for $q = 0.55$ Å^{-1} after convolution with the experimental resolution function and inclusion of the detailed balance condition. (b) Same as before but experimental data for $q = 0.80$ and the AIMD results for 0.77 Å^{-1}.

$$C_L(q,t) = \frac{1}{N}\,\langle \vec{j}_L(q,t)\cdot \vec{j}^{\,*}_L(q,0)\rangle \qquad\qquad C_T(q,t) = \frac{1}{2N}\,\langle \vec{j}_T(q,t)\cdot \vec{j}^{\,*}_T(q,0)\rangle \quad (5)$$

The corresponding time FT give the associated spectra, $C_L(q,\omega)$ and $C_T(q,\omega)$ respectively, with $C_L(q,\omega) = (\omega^2/q^2)S(q,\omega)$.

The AIMD results for $F(q,t)$ show an oscillatory behaviour up to $q \approx (4/5)\ q_p$, with the amplitude of the oscillations becoming weaker for increasing q-values; moreover, this oscillatory shape is superposed on a rather weak diffusive component. At $q \approx q_p$, the $F(q,t)$ exhibit a slow decay, known as "de Gennes narrowing", which is induced by the strong spatial correlations at around those q-values. The AIMD $S(q,\omega)$ show side-peaks, indicative of collective density excitations, up to $q \approx (3/5)q_p$; therefrom the side- peaks evolve into shoulders lasting until $q \approx (4/5)q_p$ and for greater q's the corresponding $S(q,\omega)$ show a monotonic decreasing behavior. Figure 2 provides a comparison with the IXS data of Cazzato *et al* [6] and although it is performed for slightly different q-values, we observe a reasonable good agreement, specially concerning the position and amplitude of the side-peaks.

We have used Eqs. 4-5 to evaluate the current correlation functions $C_{L,T}(q,t)$ and their spectra $C_{L,T}(q,\omega)$. Our AIMD calculated $C_L(q,\omega)$ show one peak for each q-value and from the frequencies of those peaks, the DR for the longitudinal modes, $\omega_l(q)$, has been obtained and is plotted in Figure 3. In the figure, we have also included the experimental DRs derived from the IXS data [6], $\omega_l^{\mathrm{IXS}}(q)$, and from the INS data [5], $\omega_l^{\mathrm{INS}}(q)$.

In Ref. [6], the DR $\omega_l^{\mathrm{IXS}}(q)$ was extracted by fitting the experimental scattered intensity to an expression composed of a lorentzian (accounting for the quasielastic part) plus the damped harmonic oscillator (DHO) model (for the inelastic part). A similar procedure was applied in Ref. [5] to obtain the DR $\omega_l^{\mathrm{INS}}(q)$, although an additional lorentzian function was included to account for the incoherent part. In both cases, the frequencies assigned to the DR were the DHO frequencies obtained from the fits.

A first noticeable feature is the discrepancy between $\omega_l^{\mathrm{IXS}}(q)$ and $\omega_l^{\mathrm{INS}}(q)$ in the narrow q-region where they overlap. This was attributed by Cazzato *et al* [6] to shortcomings of the experimental INS setup at these wavevectors. Indeed, the calculated

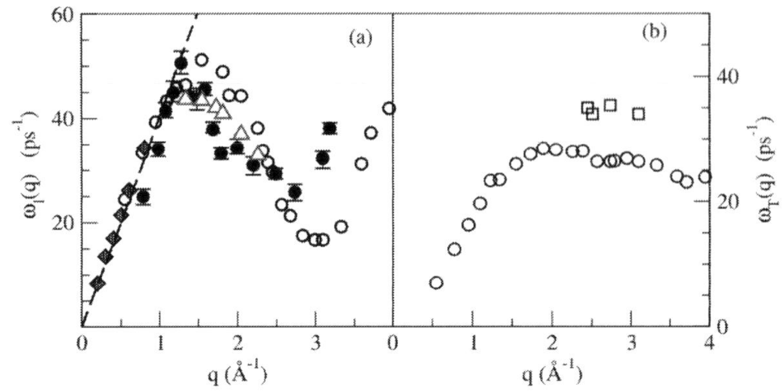

Figure 3. (a) Longitudinal DR for l-Ni at $T = 1773$ K. Circles: AIMD longitudinal DR from the positions of the peaks in the spectra $C_L(q,\omega)$. Filled circles with error bars: INS data from Bermejo *et al* [5]. Diamonds: IXS data from Cazzato *et al* [6]. Broken line: linear dispersion with the hydrodynamic sound velocity, $c_s = 4030$ m/s. Triangles: DR obtained by fitting the AIMD total (neutron weighted) dynamic structure factor to a model of two Lorentzians plus one DHO. (b) Circles and squares: AIMD transverse dispersion from the positions of the peaks in the spectra $C_T(q,\omega)$.

AIMD DR, $\omega_l(q)$, shows a good agreement with $\omega_l^{\rm IXS}(q)$. In the low-q region, the slope of the $\omega_l(q)$ curve gives a q-dependent adiabatic sound velocity which in the limit $q \to 0$ reduces to the bulk adiabatic sound velocity, c_s. We have used the AIMD results for $\omega_l(q)$, within the range $q \leq 1.0$ Å^{-1}, to evaluate its slope at $q \to 0$ and we obtained a qualitative estimate $c_s \approx 4250 \pm 150$ m/s which compares reasonably with the experimental value of $c_{s,\rm exp} \approx 4040 \pm 150$ m/s [14].

For larger q's, there is a significant difference between the $\omega_l^{\rm INS}(q)$ and the calculated AIMD $\omega_l(q)$. Moroever, as our results closely follow those obtained in CMD [7, 8] and AIMD [10] simulations, we have decided to check the accuracy of the scheme followed by Bermejo *et al* [5] to obtain their DR $\omega_l^{\rm INS}(q)$, and we applied the same procedure to out AIMD data. First, the AIMD results were used to construct the total dynamic structure factor (by weighting $S(q,\omega)$ and its self part with the corresponding coherent and incoherent neutron cross-sections of Ni). Then the total dynamic structure factor was fitted to a function comprising two Lorentzian functions for the quasielastic contributions plus a DHO model for the inelastic contribution; therefrom, the excitation frequencies of the DHO model were identified as the new calculated DR. This is also plotted in Figure 3 and it is noticed that this procedure does not reproduce the AIMD DR, $\omega_l(q)$. In fact, it yields systematically smaller frequency values which are much closer to the $\omega_l^{\rm INS}(q)$. We believe that the above results can explain the discrepancies between the INS DR, $\omega_l^{\rm INS}(q)$, and all the other DRs, either IXS or theoretical.

The transverse current spectra, $C_T(q,\omega)$, when plotted as a function of ω, may display peaks, within some q-range, which are related to propagating shear waves. From the frequencies of the peaks, a DR for the transverse modes, has been obtained and is plotted in Fig. 3. We observe two branches, a low-frequency branch whose shape is typical of simple liquid metals near melting, and a high-frequency branch, with a limited extent around q_p. The appearance of a high frequency branch is quite

 DOI: 10.1051/epjconf/201715102003

unusual in monoatomic liquids. To our knowledge, it has only been found in studies performed for liquid Li, Fe and Na at high pressure [15].

4 Conclusions.

An *ab-initio* simulation method has been used to calculate some static and dynamic properties of l-Ni near its triple point. This study has been spurred by the discrepancies between the available INS and IXS measurements.

The calculated static structure factor closely follows the corresponding experimental data, including the asymmetric shape of its second peak, which has been connected to the prevalence of icosahedral short-range order.

The calculated dynamic structure factors show side-peaks indicative of collective density excitations, and they show a good agreement with the IXS data. Moreover, the calculated dispersion relation closely follows that obtained from the IXS measurements. An analysis of the modeling scheme followed to obtain the INS dispersion relation has uncovered some shortcomings which could explain its marked disparity with the IXS dispersion relation and with other simulation results.

The transverse dispersion relation shows, besides the typical low-frequency branch, a short, high-frequency one, which has not been previously found in any other monoatomic liquid metal near melting.

BGR acknowledges the financial support of Universidad de Valladolid. LEG and DJG acknowledge the support of the MECD (FIS2014-59279-P) and JCyL (VA104A11-2).

References

[1] Y. Waseda and M. Ohtani, Phys. Stat. Sol. (b) **62**, 535 (1974); Y. Waseda, *The Structure of Non-Crystalline Materials*, (New York: McGraw-Hill, 1980).

[2] M. W. Johnson, N. H. March, B. McCoy, S. K. Mitra, D. I. Page and R. C. Perrin, Philos. Mag. **33**, 203 (1976).

[3] T. Schenk, D. Holland-Moritz, V. Simonet, R. Bellisent and D. M. Herlach, Phys. Rev. Letters **89**, 075507 (2002).

[4] M. W. Johnson, B. McCoy, N. H. March and D. I. Page, Phys. Chem. Liq. **6**, 243 (1977).

[5] F. J. Bermejo, M. L. Saboungi, D. L. Price, M. Alvarez, B. Roessli, C. Cabrillo and A. Ivanov, Phys. Rev. Letters **85**, 106 (2000).

[6] S. Cazzato, T. Scopigno, S. Hosokawa, M. Inui, W. C. Pilgrim and G. Ruocco, J. Chem Phys. **128**, 234502 (2008).

[7] M. M. G. Alemany, C. Rey and L. J. Gallego, Phys. Rev. B **58**, 685 (1998).

[8] M. D. Ruiz-Martin, M. Jimenez-Ruiz, M. Plazanet, F. J. Bermejo, R. Fernandez-Perea and C. Cabrillo, Phys. Rev. B **75**, 224202 (2007); J. Non-Cryst. Solids **353**, 3113 (2007).

[9] A. F. Voter and S. P. Chen, in *Characterization of Defects in Materials*, ed. R. W. Siegel, J. R. Weertman and R. Sinclair, MRS Symposia Proceedings No 82 (Materials Research Society, Pittsburg, 1987) p. 175

[10] N. Jakse, J. F. Wax and A. Pasturel, J. Chem. Phys. **126**, 234508 (2007).

[11] D. M. Ceperly and B. J. Alder, Phys. Rev. Letters **45**, 566 (1980); J. P. Perdew and A. Zunger, Phys. Rev. B **23**, 5048 (1981).

[12] D. Vanderbilt, Phys. Rev. B **41**, 7892 (1990).

[13] U. Balucani and M. Zoppi, *Dynamics of the Liquid State*, Clarendon, Oxford, 1994.
[14] S. Blairs, Intern. Materials Rev. **52**, 321 (2007); Phys. Chem. Liq. **45**, 399 (2007).
[15] T. Bryk, G. Ruocco, T. Scopigno and A. P. Seitsonen, J. Chem. Phys. **143**, 104502 (2015) M. Marques, L. E. Gonzalez and D. J. Gonzalez, J. Phys.: Condens. Matter **28**, 075101 (2016); Phys. Rev. B **94**, 024204 (2016).

Atomistics of self-diffusion in liquid metals

Lianwen Wang*

School of Physical Science and Technology, Lanzhou University, 730000 Lanzhou, P. R. China

Abstract. Self-diffusion in liquids is not well understood because of the lack of a clear microscopic knowledge on liquids. Here, on basis of an atomic model for liquids that we concluded from crystal melting, possible atomic details of self-diffusion in liquid metals are proposed. Accordingly, the coefficient of self-diffusion is derived for liquid elements, in reasonable agreement with experimental data.

1 Introduction

Despite intensive efforts devoted for understanding liquid metals, as signified by the series of international conferences on liquid and amorphous metals (LAM) over the past 60 years, we still have to face the fact that the liquid state is "the hardest to characterize" in comparison with the crystal and the gas states according to the presentation by Frans Spaepen given at the end of LAM-16. Such a situation is traced back through the pre-LAM ages in the early 20th century to the days of van der Waals around 1873 [1-4].

As a consequence, a paragraph on liquids written in 1967 by Norman Nachtrieb [5] can be reused today, without any changes, which is cited in the following: "The transport properties of liquids, together with structural and thermodynamic information, provide the experimental basis for theories of the liquid state. Conversely, models of the liquid state must be judged according to the success with which they account for the kinetic as well as equilibrium properties of liquids. In principle, the self-diffusion of liquid metals should provide clues to the structures and interatomic potentials. There now exist data for the self-diffusion of about 11 liquid metals, some of indifferent quality. A few studies are probably sufficiently reliable to merit serious theoretical attention at this time, but it will be obvious that improved experimental methods, applied over much wider ranges of temperature, are needed to provide better data." [5]

Here in this work, present knowledge on self-diffusion is briefly reviewed and examined in Section 2. In Section 3, a model for self-diffusion in liquid metals is developed, on basis of a microscopic liquid model that we developed in recent years, and compared with available experimental data. Finally, recent progresses on a possible crossover in self-diffusion above the melting point are discussed in Section 4. On this basis, we are trying to show that the present liquid model does help in accounting for the self-diffusion of liquid metals and, judged according to this, this liquid model is in a promising direction approaching the microscopic knowledge of liquids.

* E-mail: lwwang@lzu.edu.cn

2 Present knowledge on self-diffusion of liquid metals

Available methods for the measurement of self-diffusion coefficient of liquid metals are briefly reviewed in [6] in which the data on 12 elements, mostly reported before 1970, were also summarized. These data were measured on the ground and according to [7] and [8] the first microgravity measurement was performed in the 1970s. As a result the data could inevitably have been contaminated by convective flow in liquids [7-10] and hence show large scattering, see Figure 3 of [9] with the data on Sn as an example. Advanced techniques applying *in situ* monitoring via X-ray or neutron radiography were developed recently, which are applicable only for interdiffusion of alloys [11-13].

In comparison, under microgravity environment [7-9], at least gravity-driven convective flow could be suppressed and, although diffusion measurements were still not fully free of disturbs [7,9], the data obtained by two independent research groups showed good agreement, see Figure 2 of [9] also with Sn as an example.

Besides measurements under microgravity, quasielastic neutron scattering (QENS) [14-19] was argued to probe "the dynamics on atomic length scales and on a picosecond time scale, short enough to be undisturbed by the presence of convective flow." [18] "In order to derive the self-diffusion coefficient from QENS one has to assume that concepts developed in the framework of hydrodyanmics [20], i.e. in the limit of small wavenumber q values, are valid also at q values accessible by QENS" [18] and the self-diffusion coefficient was then related to the width of the QENS signal [18, 21]. It is however noted that the analyses in [20] were carried out by assuming that the molecules in liquids interact through "a two-body, additive, central potential" (on page 16 of [20] above Equation 2.1.1).

On page 242 of [22] E. G. D. Cohen wrote: "All kinetic theories are, what could be called, effective two-particle theories. For, in all cases, the many particle dynamics of the fluid is approximated by an effective two-particle problem" [22]. Along this line, a significant advance in the understanding of liquid state diffusion was reported by M. Dzugutov [23] that "the diffusion coefficient, reduced to a dimensionless from by scaling by the atomic collision frequency and the atomic diameter, is uniquely defined by the excess entropy (over the ideal gas [24]), a measure of the number of accessible configurations of the system" [23]. Modifications of intermolecular potentials [24] and calculating methods for the excess entropy [25, 26] seemed to lead to better applicability of this scaling law; however, indeed the issue of liquid state diffusion still remained unresolved but was just transferred to the calculation of excess entropy in terms of intermolecular potential.

From computer simulation results the influence of the type of intermolecular potential used on the final output of self-diffusion coefficient were more visible, see e.g. Figure 3 of [17].

Now, for deriving self-diffusion coefficient either from the kinetic theory or computer simulation, we have to face the fact that "establishing quantitative, predictive connections between the knowledge of the strong, complicated, and continual interactions that operate in condensed phases on the one hand, and the observable macroscopic properties on the other hand, remains a major challenge" [27].

Other theoretical approaches to self-diffusion in liquid metals simply overlooked interatomic potentials and started from e.g. the heat of vaporization [28], the Debye temperature [5] or the excess volume of a liquid over its crystal counterpart [29]. These models [5, 28, 29] represent the non-activated formalism for self-diffusion [5], see Figure 1(a), and lead the temperature T dependences of the coefficient of self-diffusion D in liquids to be $D{\sim}T^2$ [28] or $D{\sim}T$ [5]. On basis of the microgravity data, Mathiak et al [30] preferred a T^2 dependence of D.

DOI: 10.1051/epjconf/201715102004

In comparison, there is also an activated formalism [5] suggested qualitatively by J. Frenkel [31] and generally observed in measurements [6, 18, 19], see Figure 1(b), which results in ln$D$~$1/T$, i.e.

$$D = D_0 \exp\left(-\frac{Q_{SD}}{kT}\right) \tag{1}$$

where D_0 is a pre-exponential factor, Q_{SD} the activation energy of self-diffusion and k the Boltzmann's constant. At present, both D_0 and Q_{SD} are obtained by fitting the measured or simulated self-diffusion data; theoretical predictions are not successful.

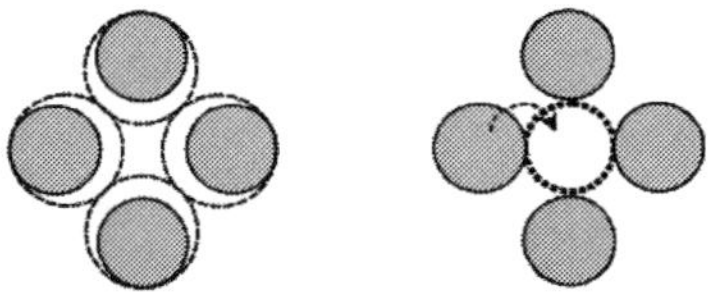

(a) (b)

Fig. 1. Two-dimension sketchy diagram of atoms in a liquid with the free volume indicated by the dashed circle in (a). Although the non-activated formalism argued that the configuration of (a) might allow self-diffusion in liquids, we tend to advocate that self-diffusion in liquids should be an activated atomic jump process towards the nearby pre-formed vacancy (dotted circle in (b)), very alike to that in the corresponding crystal.

In Section 3, the coming and details of our microscopic model in favour of Equation (1) are introduced and, after this, D_0 and Q_{SD} are estimated and compared with available reported experimental data.

3 Our microscopic model for self-diffusion in liquid metals

3.1 Starting point – estimating the energy sate of liquids

In Section 2 it is seen that the major difficulty hindering the understanding of self-diffusion in liquids is the unsuccessfulness [1] in connecting classical interatomic potentials with the properties of liquids in which many-body potentials and many-body kinetics are present. We approached this issue by considering the continuity of liquids and corresponding crystals at the instant of melting [4] and proposed to estimate the energy state of liquids by using the energy of a vacancy divided among the first shell atoms [3]. By doing so, the many-body potentials are included in the formation energy of a vacancy, with which complex calculations were avoided to know liquid properties.

3.2 Calculation of D_0

Self-diffusion in liquid metals is supposed to consist of two steps, namely, formation of a vacancy and atomic jump to the vacancy. In our melting model, the vacancy formation energy was stored in the system at the instance of crystal melting signified by squashing of the vacancy [32, 33]. On this basis we suggest that formation of a vacancy in liquids may need only the consumption of entropy but not the activation energy. After vacancy formation, the vacancy and its nearest neighbouring atoms return to a configuration similar to that before melting, see Figure 1(b), and for this reason the entropy consumption for vacancy formation in liquids is proposed to the entropy of fusion S_f. Then the probability of finding a vacancy near an atom is:

$$p_V = \exp\left(-\frac{S_f}{k}\right) \qquad (2)$$

The probability for an atomic jump to the pre-formed vacancy is suggested to be the same as that in a crystal:

$$p_m = \exp\left(\frac{S_m}{k} - \frac{E_m}{kT}\right) \qquad (3)$$

where S_m and E_m are the entropy and enthalpy of atomic migration. Then the time needed for an atomic jump in liquids is :

$$t = \frac{1}{p_V \cdot p_m \cdot f_0} \qquad (4)$$

where f_0 is an attempting frequence normally in the order of 10^{12} Hz. Finnally, with the Einstein's law for random walk [34, 35], the coefficient of self-diffusion is written as:

$$D = \frac{d^2}{6t} \qquad (5)$$

where d is the atomic distance in the liquid.

By comparing Equations (1-5) one has:

$$D_0 = \frac{d^2 f_0}{6} \exp\left(\frac{-S_f + S_m}{k}\right) \qquad (6)$$

In [36] (on page 3652) S_m was taken as $2.8k$, S_f is taken from [6] and d derived from the density of liquids reported in [6]; the obtained D_0 is compared with those reported in [6, 14, 16, 19] in Figure 2. It is seen that the calculated and the reported D_0 are comparable in order of magnitude.

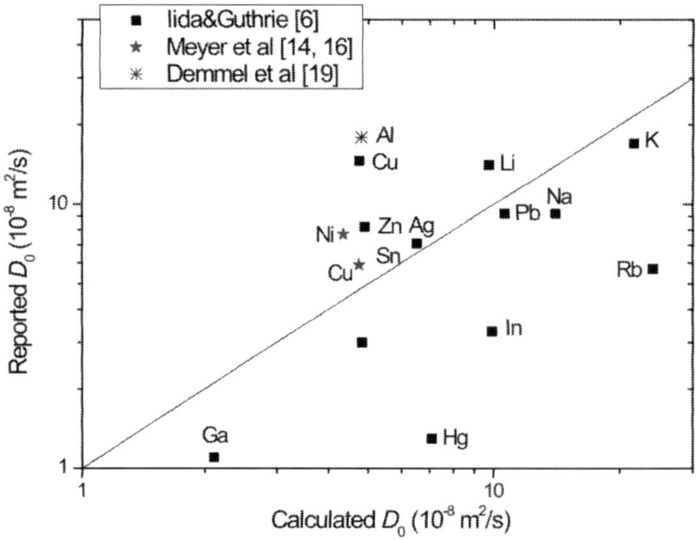

Fig. 2. Comparison of calculated D_0 by using Equation (6) with reported data.

3.3 Estimation of Q_{SD}

See Figure 3(left), at the instant of melting, a liquid nucleus was formed by the squashing of a vacancy by 2 of its N nearest neighbouring migration atoms, with the energy of the activation energy for atomic migration $2E_m$ (approximately equals to the formation energy of a vacancy) stored in the system.

Self-diffusion in liquids may require that a vacancy be pre-formed, Figure 3(middle), with the consumption of the entropy of fusion. After this, the local atomic configuration returns to that before melting, but the energy $2E_m$ remains stored; if the liquid is at the melting point, now there are 2 atoms that can migrate to the vacancy.

Of course with temperature increasing the probability for an atom to migrate to the vacancy determined by Equation (3) will increase exponentially, however the present configuration may only allow 2 atoms to migrate because simultaneous migration of more than 2 atoms may squash the pre-formed vacancy which makes atomic migration impossible.

Consequently, self-diffusion in liquids should be a superdiffusion of 2 atoms, i.e. for each activation process 2 atoms can migrate. Accordingly the activation energy Q_{SD} for self-diffusion of liquid metals is estimated to be:

$$Q_{SD} = \frac{1}{2} E_m \quad (7)$$

By comparing the E_m data compiled in [37] and Q_{SD} reported in the literature, Equation (7) is in principle verified, see Figure 4. However, the data on Ge are left out for further considerations.

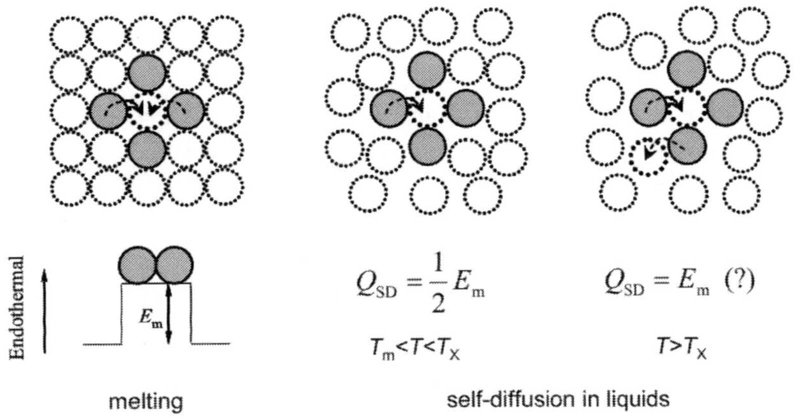

Fig. 3. (left):Two-dimension sketchy diagram of atoms in a crystal (thin dotted circle) at the instant of melting and a liquid nucleus formed by squashing of a vacancy (thick dotted circle) when there are 2 out of the N coordinating atoms of the vacancy are migratable.
(middle): For self-diffusion in liquids, after the pre-formation of the vacancy, for each activation process 2 atoms can migrate, and the activation energy of self-diffusion Q_{SD} may equal to $E_m/2$.
(right) Only for discussion. When, with temperature increasing to above the crossover T_x, two vacancies are available in the liquid nucleus, diffusion in liquids may turn to a simple activation behaviour, i.e. $Q_{SD}\sim E_m$.

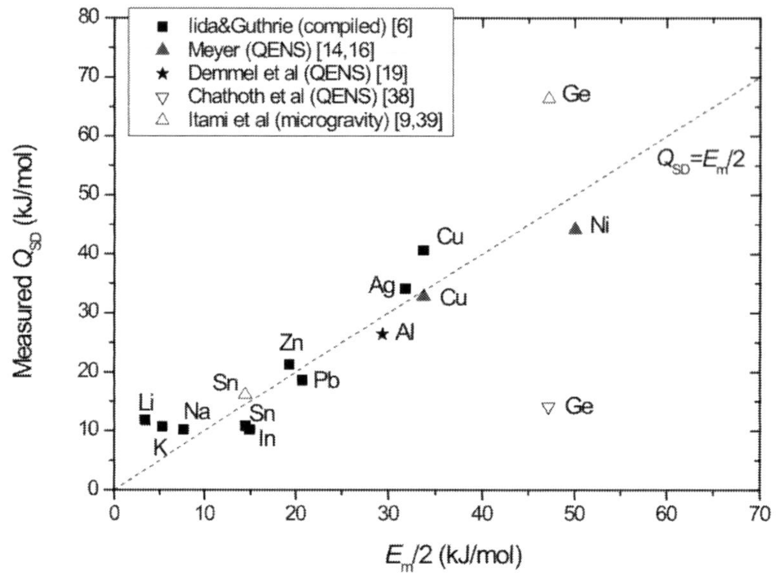

Fig. 4. Correlation between the reported activation energy for self-diffusion in liquid metals and the measured activation energy for atomic migration in crystals E_m [37]

4 Discussion

When, with temperature increasing, two vacancies are available in the liquid nucleus, diffusion in liquids may require a simultaneous formation of two vacancies. Otherwise, e.g. if only one vacancy is formed, there is still excess entropy (over the crystal counterpart) left in the liquid nucleus so that the formed vacancy will soon be squashed and atomic diffusion just could not proceed. For such a situation, self-diffusion in liquids should then turn to normal self-diffusion, leading Q_{SD} to E_m.

The scenario proposed above is very interesting in comparison with self-diffusion in crystals for which atomic jump is rare and is only possible for atoms in an activated energy state. Here, in liquids, it is seen that nearly all atoms have the energy to jump into a vacancy. However it seems that atomic migrations are too intensive so that if more than 2 atoms migrate simultaneously, the pre-formed vacancy may be squashed again. On the other hand, at high temperatures, if the excess entropy over the crystal is not completely squeezed out, a vacancy just could not be formed, which also does not permit self-diffusion.

In order to find out where the possible crossover temperature T_x may locate, an equation for the temperature dependence of the concentration of vacancies C_V in liquids is written as:

$$C_V = \exp\left(\frac{S_V}{k} + \frac{S_f}{k} - \frac{E_V}{kT}\right) \qquad (8)$$

where S_V and E_V is the entropy and enthalpy of vacancy formation. Here the entropy of fusion is added because after a vacancy is formed it will be squashed instantly by nearby migration atoms, like the process of melting. In [32] S_V is taken as $4.1k$ at the homogeneous melting point, together with S_f data from [6] and E_V data from [37], the concentration of vacancies in liquids could be calculated.

Since during melting a vacancy is squashed with its energy and volume stored in the liquids, the crossover temperature dividing Figure 3 (middle) and Figure 3(right) may be set as:

$$C_V(T_x) = 1/N \qquad (9)$$

where N is the number of first coordination atoms in the corresponding crystal. By using Equations (8) and (9) the crossover temperature T_x is calculated and plotted against the melting point T_m in Figure 5. A rough correlation between T_x and T_m is noticed namely T_x~$1.5T_m$.

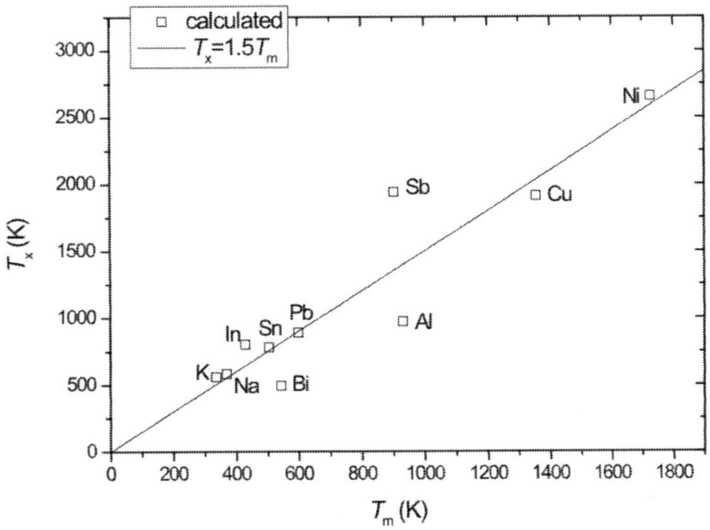

Fig. 5. Estimated crossover temperature T_x for self-diffusion in liquid metals.

Cautions are nevertheless needed for the present pure theoretical analyses and the tentative indications. Additional measurements in a wide temperature range are necessary when possible for further examination the temperature dependence of the coefficient of self-diffusion and the present analyses in comparison with others.

Anyway, in the present stage, two groups of research results are to be mentioned that may be related with the theoretical analyses carried out in this work. One is the neutron scattering evidences by Demmel et al for a dynamical crossover in liquid Rb [40], Pb [41], and Al [42] at around $1.5T_m$ which happens to coincide Figure 5 of this paper.

The other group of research results to be mentioned are the computer simulation outputs of Egami et al [43-45] for the dynamics of glass-forming liquids in which a crossover was also evidenced at about $1.5T_m$ [45].

Finally, recent *ab initio* molecular dynamic simulations of 216 atoms showed that Q_{SD} for liquid Ga increased from 8 to 16 kJ/mol at a temperature above the melting point [46].

It is less likely that the above results obtained over the years and the present theoretical analyses are not connected anyhow since all are focusing on one issue – a possible crossover in the kinetics of liquids above the melting point.

5 Conclusion

On basis of a simple microscopic model for crystal melting and for liquids, self-diffusion in liquid metals is analyzed to be an activated superdiffusion process via vacancies. The accordingly calculated prefactor and activation energy of self-diffusion are in general agreement with available experimental data. Furthermore, a possible crossover in the self-diffusion of liquid metals above the melting point is discussed in terms of the present model.

This work is supported by the Fundamental Research Funds for the Central Universities (grant number lzujbky-2015-120) and the National Natural Science Foundation of China (grant number 51101077).

References

1. J. L. Finney, Philos. Mag. **93**, 3940 (2013)
2. L. W. Wang, AIP Conf. Proc. (LAM-15) **1673**, 020013 (2015)
3. L. W. Wang, Metals **4**, 570 (2014)
4. L. W. Wang, J. Non-Cryst. Solids **383**, 59 (2014)
5. N. H. Nachtrieb, Adv. Phys. **16**, 309 (1967)
6. T. Iida, R. I. L. Guthrie, *The physical properties of liquid metals* (Oxford University Press, New York, 1993)
7. G. Mathiak, E. Plescher, R. Willnecker, Meas. Sci. Technol. **16**, 336 (2005)
8. T. Masaki, T. Fukazawa, S. Matsumoto, T. Itami, S. Yoda, Meas. Sci. Technol. **16**, 327 (2005)
9. T. Itami, T. Masaki, H. Aoki, S. Munejiri, M. Uchida, S. Matsumoto, K. Kamiyama, K. Hoshino, J. Non-Cryst. Solids **312-314**, 177 (2002)
10. L.B. Jalbert, R.M. Banish, F. Rosenberger, Phys. Rev. E **57**, 1727 (1998)
11. B. Zhang, A. Griesche, A. Meyer, Phys. Rev. Lett. **104**, 035902 (2010)
12. F. Kargl, M. Balter, Ch. Stenzel, Th. Gruhl, N. Daneke, A. Meyer, J. Phys.: Conf. Ser. **327**, 012011 (2011)
13. F. Kargl, M. Engelhardt, F. Yang, H. Weis, P. Schmakat, B. Schillinger, A. Griesche, A. Meyer, J. Phys.: Condens. Matter **23**, 254201 (2011)
14. A. Meyer, S. Stüber, D. Holland-Moritz, O. Heinen, T. Unruh, Phys. Rev. B **77**, 092201 (2008)
15. J. Horbach, R. E. Rozas, T. Unruh, A. Meyer, Phys. Rev. B **80**, 212203 (2009)
16. A. Meyer, Phys. Rev. B **81**, 012102 (2010)
17. F. Kargl, H. Weis, T. Unruh, A. Meyer, J. Phys.: Conf. Ser. **340**, 012077 (2012)
18. A. Meyer, EPJ Web Conf. **83**, 01002 (2015)
19. F. Demmel, D. Szubrin, W.-C. Pilgrim, C. Morkel, Phys. Rev. B **84**, 014307 (2011)
20. J. P. Boon, S. Yip, *Molecular hydrodynamics* (Dover Publications, New York, 1991)
21. A. G. Novikov, M. N. Ivanovskii, V. V. Savostin, A. L. Shimkevich, O. V. Sobolev, M. V. Zaezjev, J. Phys.: Condens. Matter **8**, 3525 (1996)
22. E. G. D. Cohen, Physica A **194**, 229 (1993)
23. M. Dzugutov, Nature **381**, 137 (1996)
24. J. J. Hoyt, M. Asta, B. Sadigh, Phys. Rev. Lett. **85**, 594 (2000)
25. A. Samanta, Sk. M. Ali, S. K. Ghosh, Phys. Rev. Lett. **92**, 145901 (2004)
26. N. Jakse, A. Pasturel, Sci. Rep. **6**, 20689 (2016)
27. M. Scott Shell, P. G. Debenedetti, F. H. Stillinger, J. Phys. Chem. B **108**, 6772 (2004)
28. R. A. Swalin, Acta Metall. **7**, 736 (1959)
29. M. H. Cohen, D. Turnbull, J. Chem. Phys. **31**, 1164 (1959)
30. G. Mathiak, A. Griesche, K.H. Kraatz, and G. Frohberg, J. Non-Cryst. Solids **205-207**, 412 (1996)
31. J. Frenkel, Trans. Faraday Soc. **33**, 58 (1937)
32. L. W. Wang, L. Zhang, K. Lu, Philos. Mag. Lett. **85**, 213 (2005)
33. L. W. Wang, Q. Wang, K. Q. Lu, Phlios. Mag. Lett. **87**, 19 (2007)
34. A. Einstein, Ann. Phys. **17**, 549 (1905)
35. P. Atkins, J. de Paula, *Physical chemicstry*, 8th ed. (W. H. Freeman and Company, New York, 2006), Chapter 21.12
36. L. W. Wang, Philos. Mag. **93**, 3648 (2013)

37. H. J. Wollenberger, *Point defects* in R. W. Cahn, P. Haasen eds., *Physical Metallurgy*, 4th ed (Elsevier, Amsterdam, 1996)
38. S. M. Chathoth, B. Damaschke, T. Unruh, K. Samwer, Appl. Phys. Lett. **94**, 221906 (2009)
39. T. Itami, H. Aoki, N. Higashimoto, T. Onogi, K. Sugimura, M. Kaneko, Y. Abe, Y. Arai, A. Tanji, M. Uchida, K. Iribe, H. Ando, K. Goto, N. Tateiwa, M. Koyama, T. Morita, H. Kawasaki, S. Yoda, N. Koshikawa, T. Masaki, H. Oda, T. Suzuki, T. Nakamura, Y. Nakamura, Int. J. Microgravity Sci. Appl. **16**, 79 (1999)
40. F. Demmel, A. Diepold, H. Aschauer, C. Morkel, Phys. Rev. B **73**, 104207 (2006)
41. F. Demmel, W. S. Howells, C. Morkel, J. Phys.: Condens. Matter **20**, 205106 (2008)
42. F. Demmel, A. Fraile, D. Szubrin, W.-C. Pilgrim, C. Morkel, J. Phys.: Condens. Matter **27**, 455102 (2015)
43. S.-P. Chen, T. Egami, V. Vitek, Phys. Rev. B **37**, 2440 (1988)
44. T. Tomida, T. Egami, Phys. Rev. B **52**, 3290 (1995)
45. A. Jaiswal, T. Egami, Y. Zhang, Phys. Rev. B **91**, 134204 (2015)
46. L. H. Xiong, X. D. Wang, Q. Yu, H. Zhang, F. Zhang, Y. Sun, Q. P. Cao, H. L. Xie, T. Q. Xiao, D. X. Zhang, C. Z. Wang, K. M. Ho, Y. Ren, J. Z. Jiang, Acta Mater. **128**, 304 (2017).

Analyzing the dynamic structure of liquid metals and alloys

Jean-François Wax[1,*] and *Taras* Bryk[2,3]

[1]Laboratoire de Chimie et de Physique-A2MC, Université de Lorraine-Metz, 1, boulevard Arago 57078 Metz Cedex 3, France

[2]Institute for Condensed Matter Physics, National Academy of Sciences of Ukraine, 1 Svientsitskii Street, UA-79011 Lviv, Ukraine

[3]Institute of Applied Mathematics and Fundamental Sciences, Lviv Polytechnic National University, UA-79013 Lviv, Ukraine

Abstract. Experimental and numerical improvements have stimulated a great interest in the dynamic structure of liquids during the last decades. Many unexpected features have been unveiled among which fast sound, positive dispersion and possible coupling between transverse and longitudinal excitations can be mentioned. Models used to analyze these data have to be sound and more and more rigorous. In this study, we discuss the capability of a recently proposed fitting scheme (Wax J.-F. and Bryk T. J. Phys.: Condens. Matter **25** 325104 (2013); 26 168002 (2014). Wax J.-F., Johnson M.R., and Bryk T. J. Phys.: Condens. Matter **28** 185102 (2016).) to interpret these features of the dynamic structure of liquid metals and alloys.

1 Introduction

Experimental techniques used to measure the dynamic structure of liquids, namely neutron or x-ray inelastic scattering, have made great progress during these last two decades. This allows to investigate dynamic properties with unprecedented accuracy. This is also true for simulation techniques which have benefited from the constant increase in available computing power. While the description of the interactions seems not to be a problem anymore as dynamic structure can now be obtained from ab-initio methods with a correct statistical accuracy, we are faced with a new challenge. Indeed, analyzing experimental or simulation results requires now accurate models in order to correctly interpret the observed features and not to be deluded with fitting artifacts. It is also important that parameters involved in the models have a concrete meaning in order to trace each contribution to the structure back to their origin.

In this work, a recently developed analysis scheme [1–3] inspired from the Generalized Collective Modes (GCM) approach [4] is presented and applied to simulation results of liquid metals and alloys. We focus on three features revealed by increasing accuracy of available data, namely the positive dispersion of acoustic waves in pure liquids, the so-called "fast sound" effect in mixtures and finally the possible coupling between transverse and longitudinal excitations. Beyond presenting this analysis scheme, its advantages and limitations, the purpose of this work is also to point out some aspects at which attention should be paid when fitting data.

*e-mail: jean-francois.wax@univ-lorraine.fr

After this introduction, important notions about dynamic structure will be recalled, namely how is it described and determined, as well as the concept of collective modes leading to the model that we proposed. In the third section, this model will be applied to the analysis of three topics: positive dispersion, fast sound and coupling between longitudinal and transverse excitations. Finally, we will summarize the main conclusions of this work.

2 Dynamic structure

2.1 Description and determination

In order to describe the dynamic structure of a liquid, several functions can be introduced that are all derived from the van Hove function $G(r,t)$. The intermediate scattering function

$$F(q,t) = \int G(r,t)\exp(i\mathbf{q}.\mathbf{r})\,d\mathbf{r} = \frac{1}{N}\left\langle \sum_{a=1}^{N} e^{[i\mathbf{q}.\mathbf{r}_a(0)]}.\sum_{b=1}^{N} e^{[-i\mathbf{q}.\mathbf{r}_b(t)]}\right\rangle \quad (1)$$

is its space Fourier transform; it can be computed directly from the positions obtained from molecular dynamics (MD) simulations. The dynamic structure factor

$$S(q,\omega) = \frac{1}{2\pi}\int_{-\infty}^{+\infty} F(q,t)\exp(i\omega t)dt \quad (2)$$

is the time Fourier transform of $F(q,t)$; it is also obtained directly from inelastic scattering experiments (of x-rays or neutrons). Finally, the current correlation function $J(q,\omega)$ can be deduced from $S(q,\omega)$ by simply multiplying it by ω^2/q^2.

In the case of an alloy of two species 1 and 2, we need to consider the chemical nature of the atoms. Therefore, partial functions are introduced. For a binary mixture, three partial structure factors are needed to obtain a complete description. This is also the case for the other functions. Partial functions can easily be obtained from simulation, but experiments usually only provide with a single total weighted function which, in the case of neutrons, reads

$$F_{\mathrm{INS}}(q,t) = \frac{c_1b_1^2F_{11}(q,t) + 2\sqrt{c_1c_2}b_1b_2F_{12}(q,t) + c_2b_2^2F_{22}(q,t)}{c_1b_1^2 + c_2b_2^2} \quad (3)$$

which is a linear combination of the partial ones, where b_i are neutron scattering amplitudes of corresponding atoms. The difficulty is then to disentangle the three contributions. It should also be recalled that partial functions of the Bhatia and Thornton's kinds can also be used. They describe the system in terms of atom number and concentration.

These functions have mathematical properties called sum-rules.

$$\int_{-\infty}^{+\infty} S(q,\omega)d\omega = F(q,0) = S(q), \quad (4)$$

$$\int_{-\infty}^{+\infty} \omega S(q,\omega)d\omega = i\left.\frac{\partial F(q,t)}{\partial t}\right|_{t=0} = 0, \quad (5)$$

$$\int_{-\infty}^{+\infty} \omega^2 S(q,\omega)d\omega = -\left.\frac{\partial^2 F(q,t)}{\partial t^2}\right|_{t=0} = q^2k_BT/m, \quad (6)$$

$$\int_{-\infty}^{+\infty} \omega^3 S(q,\omega)d\omega = -i\left.\frac{\partial^3 F(q,t)}{\partial t^3}\right|_{t=0} = 0 \quad (7)$$

They are important since they have consequences on their asymptotic behavior. For instance, the second relation implies that $F(q,t)$ must have zero slope as t tends to zero, but also that

$J(q, \omega)$ must tend to zero as ω tends to infinity. Models used to interpret the data should obey as many of these sum rules as possible. Of course, similar relations also exist in the case of alloys.

In order to determine these functions, on the experimental side, dynamic structure can be probed at the atomic scale by inelastic neutron or x-rays scattering. This last technique has benefited from the advent of high flux synchrotron sources which has strongly boosted the study of dynamic structure over the last two decades. The function that is obtained is the dynamic structure factor $S(q, \omega)$. On the simulation side, usual limitations are of course recovered in terms of accuracy of the interactions for classical MD, and of statistical accuracy for ab-initio simulations. However, the results do not need to be corrected from experimental side-effects. Moreover, partial functions are easily computed in the case of alloys, as well as transverse dynamics ones. The function usually computed is the intermediate scattering function $F(q, t)$.

2.2 Collective modes

A central concept used to analyze the dynamic structure of a liquid is that of collective modes. It can be understood looking at the shape of the dynamic structure factor that is usually composed of several peaks. Each peak corresponds to one or more modes, so that the dynamic structure stems from sum of contributions from several modes. They can be of relaxing or of propagating kind and can easily be distinguished. Propagating modes usually contribute to side peaks in $S(q, \omega)$ while relaxing modes contribute to the central peak. Looking at $F(q, t)$, the propagating ones appear as oscillations while the relaxing ones contribute as exponential decays [3].

There is one (ω, q) domain in which the leading contributions to $S(q, \omega)$ are well-understood: the macroscopic limit which corresponds to the hydrodynamic model. In this length and time ranges, conservation rules of particle density, momentum and energy lead to slow long-lived collective modes which give rise to well-defined sharp peaks. The central one, called Rayleigh peak, is connected to thermal relaxations, because macroscopic adiabatic propagation of sound causes small differences in local temperature for compressed and uncompressed regions. The pair of side peaks, called Brillouin peaks, correspond to density waves related to sound propagation. The side-peak positions are connected with the sound velocity and the dispersion curve is linear in this range. In the case of alloys, another quantity is conserved, namely the chemical composition which leads to a second contribution to the central peak from relaxation connected with mutual diffusivity of species.

At interatomic distances, we enter the so-called kinetic regime and additional modes contribute to the dynamic structure measured in x-rays or neutrons experiments. Kinetic modes include structural relaxation and thermal waves, for instance. It is very easy to distinguish between hydrodynamic and kinetic modes. In order to propagate over macroscopic distances, the damping of hydrodynamic modes must tend to zero with q. Conversely, kinetic mode become evanescent on long distances and their damping tend to a non-zero value in the same limit. It is also useful to notice that, at a given q-value, two coupled relaxation modes with comparable relaxation times can mix into a propagation mode and vice-versa.

2.3 Model used

Whatever the model or theory used to analyze the dynamic structure of a liquid, it is usually considered as the sum of mode contributions. Among all the available approaches, the Generalized Collective Mode model developed by Bryk and co-workers [4] is of particular interest. Indeed, beyond proposing tractable expressions to fit the data, it also provides with a consistent picture as to the origin of the various modes involved. This approach adds kinetic modes to the hydrodynamic description by taking into account non-hydrodynamic variables. By determining the eigen-functions and eigen-values of the collective variables obtained from MD simulations, it can determine the collective modes and their contributions to the dynamic structure functions. We shall call "full-GCM" this method which allows to understand quite completely the nature of the mechanisms involved in the dynamic structure. Of course, it cannot be applied directly to experimental data. But, interestingly, it leads to expressions that are rather simple and that we can easily use to fit experimental or simulation results.

Therefore, we consider the following expression:

$$F(q,t) = \sum_{i=1}^{N_r} A_i \exp\left[-\alpha_i.t\right] + \sum_{j=1}^{N_p} B_j \exp\left[-\beta_j.t\right] \cos\left[\omega_j.t + \varphi_j\right]. \tag{8}$$

Relaxing modes are depicted by exponential decays, while propagating modes are taken into account by damped cosine functions. $S(q,\omega)$ and $J(q,\omega)$ also have analytic expressions. Fitting parameters A_i, α_i, B_j, β_j, ω_j, and φ_j are q dependent quantities. The number of modes depends on the system considered, but anyway, we are faced with the question of the number of free parameters. Indeed, considering just one relaxing mode and one propagating mode would lead to six independent parameters in a pure liquid.

In order to reduce this number, we found it useful to consider the sum-rules mentioned before. For instance, introducing the first three relations, we were able to reduce the number of independent parameters down to three in the case of two modes for pure liquids [1]. But the most important is that this guarantees a correct asymptotic behavior of dynamic structure functions and especially of $J(q,\omega)$ (see Ref. [2] for a comparison with Damped Harmonic Oscillator (DHO) model which leads to a wrong high frequency asymptote of $J(q,\omega)$ when combined with a central lorentzian due to wrong behavior of this latter). This model appeared to be efficient whatever the q-value (before, after, or at the position of the first peak of $S(q)$), in the case of both pure liquids and alloys.

3 Results and discussion

We turn now to some topics of interest that can be highlighted using our model. We will discuss first the so-called positive dispersion, then the fast sound, and finally, the possible coupling between transverse and longitudinal modes. These topics show increasing complexity.

3.1 Positive dispersion

Considering the peaks of $S(q,\omega)$, plotting the corresponding frequencies versus wave-vector, we obtain the dispersion curve. As q tends to zero, the slope should correspond to the adiabatic speed of sound in agreement with the hydrodynamic theory. However, it is usually observed that the dispersion curve lies above this asymptotic limit, so that the slope of the dispersion curve is locally higher than this low-q limit. The positive sound dispersion is caused by viscoelastic transition in liquids, which consists in difference between macroscopic bulk modulus, which is defined by adiabatic compressibility of the system, and high-frequency elastic moduli, which depend on microscopic forces acting on particles. The estimation of positive sound dispersion from experiments or simulations may depend on the fitting model used (or even on the function, J or S if one simply plots the position of the maximum). The positive dispersion is a largely observed phenomenon [5] that can be rather high and reach 15% in the case of silicon.

DOI: 10.1051/epjconf/201715102005

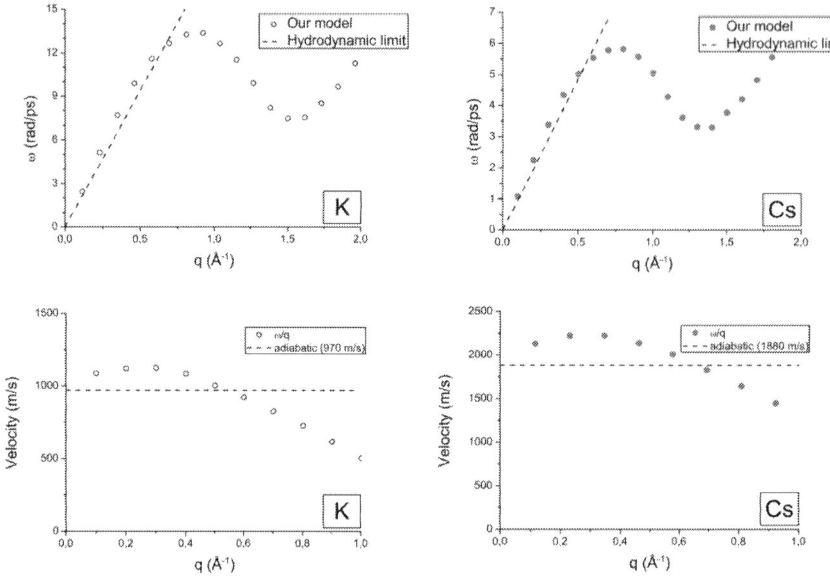

Figure 1. Dispersion curves (first line) and apparent speed of sound ω/q (second line) for liquid K and Cs. Asymptotic hydrodynamic limit is also drawn.

In Figure 1, we have plotted the dispersion curves corresponding to our studies of K [2] and Cs [3]. These results were obtained from simulations with 2048 particles in a thermodynamic state close to the melting point. The results were analyzed using our model involving one relaxation mode and one propagation mode and three sum-rules were fulfilled. Complete details about the procedure can be found in Refs. [1, 2]. The hydrodynamic asymptote is also plotted. In order to illustrate quantitatively the positive dispersion, it is convenient to consider the quantity ω/q which dimension is velocity. In these curves, the positive dispersion is clearly visible and reaches 18 % for potassium and 16 % for cesium.

As a conclusion about this point, it should pointed out that: (i) the phenomenon is clearly visible and confirmed by our model; (ii) the parameters obtained smoothly vary with q. Trying to add additional modes leads to scattered data for this mode; (iii) a fit remains a fit and we can not determine the origin of the phenomenon from it alone. A deeper analysis is required. For instance, using full-GCM approach [6], it has been possible to connect the existence of positive dispersion to a coupling between sound excitation and non-hydrodynamic structural relaxation. The hypothesis of a coupling of sound propagation with hydrodynamic thermal relaxation (leading to negative dispersion) has been ruled out by this study.

3.2 Fast sound

This phenomenon is observed in the case of alloys. It was first predicted by simulation [7, 8], then observed experimentally in the case of Helium-Neon mixture [9]. To cut the long story short, two pairs of side peaks, associated with two dispersion branches are observed in the dynamic structure factor of alloys, when the mass-ratio between components is high enough.

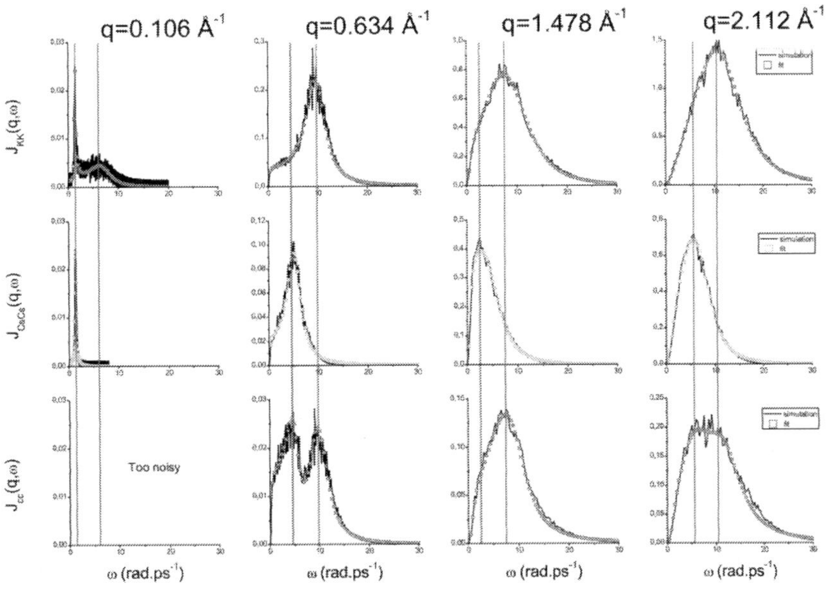

Figure 2. Partial $J(q,w)$ functions, namely K-K, Cs-Cs, and Bhatia-Thornton $C-C$ for K-Cs alloy. Solid lines are raw simulation data while open squares are the curves fitted using our model. The vertical red lines are only guide for the eyes corresponding to positions of the maximum.

As the second branch appears to have a slope higher than the branch corresponding to the adiabatic speed of sound, this phenomenon has been called "fast sound". However, only one such branch survives in the hydrodynamic limit and a lively debate followed about the exact nature and origin of this phenomenon. For instance, it was believed for a long time that it could only exist in mixtures with high mass-ratio between its components.

It could never have been understood without simulation. Indeed, the answer lies in the partial structure functions which are out of reach of experiment. On a technical point of view, it is also necessary to define a correct fitting function, taking care to the fact that the number of independent parameters remains reasonable in order to remain meaningful.

Results for liquid K-Cs alloy at equiatomic composition [3] are presented in Figure 2. The partial dynamic structure functions were fitted considering two relaxation and two propagation modes. The first two sum-rules were applied. In this alloy, the mass-ratio is about 3.4 and only one dispersion branch had been observed experimentally [10].

As in the case of pure liquids, the model is able to accurately account for the behavior of these functions. We can also see that each partial function K-K and Cs-Cs has a peak at a different positions for q values not too small. These two peaks are nicely exhibited by J_{CC} at the second q-value. Moreover, it clearly appears that, above a given q-value, potassium atoms convey the high frequency mode, while Cs atoms convey the low frequency one. This corresponds to the so-called kinetic regime. Looking now at what happens in the low-q limit, we observe a transition as both species tends to vibrate at the same low frequency. The high frequency mode contribution vanishes as it is expected in order to recover the hydrodynamics predictions.

 DOI: 10.1051/epjconf/201715102005

Looking at the parameters of these modes (not shown here, please refer to [3]), considering first their frequencies, the low frequency is very similar to the corresponding pure Cs branch. On the other hand, the high frequency branch has only remote resemblance with the pure potassium branch and, notably, it tends to a non-zero limit as q tends to zero. Looking at the damping, this mostly indicates that the low-frequency mode is of hydrodynamic nature, while the high-frequency one is kinetic. Of course, what we have been able to highlight in the case of K-Cs with a rather low mass-ratio can also be observed in the case of mixtures with high ratio like Li-Tl or Li-Bi, for instance.

To conclude this part, two aspects should be mentioned. The first one is the importance of considering the partial functions and their q-dependent contributions to the total correlation functions, and thus, the strategic position of numerical simulation. The second one, is that fitting such an intricate wealth of data requires a well established model which should be as rigorous as possible. Otherwise, the same data set could lead to very different sets of results.

3.3 Longitudinal/Transverse coupling

Transverse excitations should not be observed in the first Brillouin zone in inelastic scattering experiments. However, a discussion has been opened a few years ago when Hosokawa and co-workers [11] claimed anomalies in their measurements of Ga. This anomaly for wave numbers inside the first pseudo-Brillouin zone has also been reported in Na [12], Sn [13], Cu, Fe [14], and Zn [15]. On the basis of a DHO fit, they noticed the coincidence between the anomaly and the transverse excitations frequencies obtained by simulation. Thus, it was proposed that these anomalies could be attributed to transverse modes. But up to now, no theoretical model predicts the existence of such transverse mode contribution. Moreover, MD simulations performed for several liquid metals do not show evidences of identical peak frequencies in longitudinal and transverse current correlation functions when the wave numbers are within the first pseudo-Brillouin zone [16, 17].

As fit is sensitive to the model used, we search for another possible explanation [17]. In order to discuss the possible origins of the anomaly detected in the data of several metals, we performed ab-initio simulation of sodium. To analyze the data, we considered several methods. We computed the longitudinal structure functions and fitted the data using two propagating modes. We first analyzed these modes using two DHO functions fulfilling one sum-rule, then using two GCM functions (see Eq.8) fulfilling three sum-rules. We also computed the transverse waves that propagate in the kinetic range and performed a full-GCM analysis of the data.

The results are plotted in Figure 3. The acoustic branch obtained from peak positions of $J(q, \omega)$ agrees favorably with experimental data. Additional modes obtained considering two DHO's and two GCM contributions are different, illustrating the relevance of fulfilling as many sum-rules as possible. This also demonstrates that both approaches confirm the existence of a minor second mode. If we now consider the transverse excitations, they are located between both fits. This proximity is the main argument in favor of the coupling between longitudinal and transverse modes. But, if we consider full-GCM results, this theory predicts that thermal waves propagate at about the same frequencies as those by the two GCM fit at wave-vectors lower than the first peak position. The advantage of this explanation is that it is consistent with existing theoretical model. Of course, in the first peak region, the huge structural relaxation contribution prevents to get accurate data about the minor second mode by fitting procedures.

So, to summarize, a weak second propagating mode can be observed. Provided data are accurate enough, its quantitative description is possible by fitting methods. However, it is helpful to take sum-rules into account. The more is the better (notwithstanding fitting

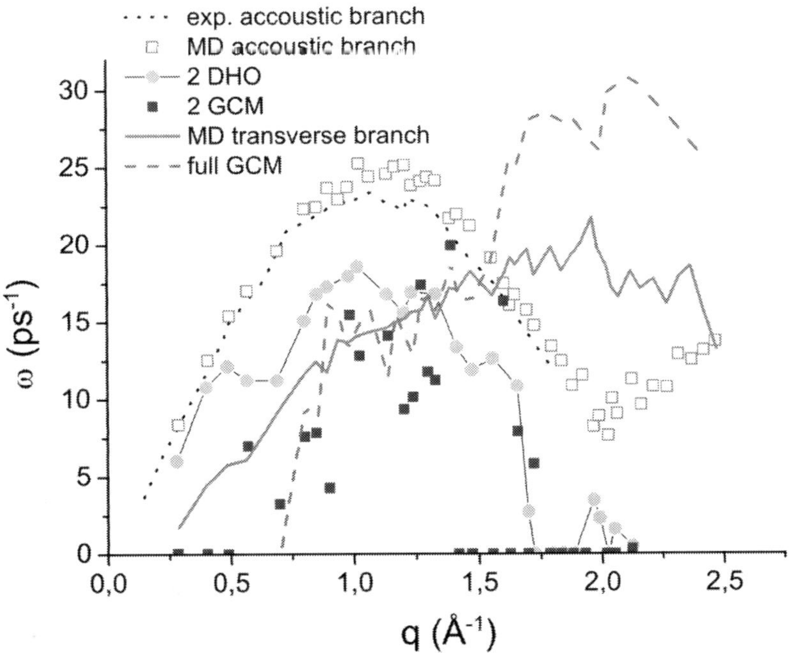

Figure 3. Frequencies of the collective excitations. Experimental and computed acoustic waves are plotted as dotted lines and open red squares, respectively. Additional modes frequencies obtained using two DHO and two GCM functions are plotted as full circles and full squares, respectively. Transverse mode and heat waves frequencies are drawn as solid and dashed lines, respectively.

complexities). Indeed, between two DHO's and two GCM, important differences appear. Finally, instead of coupling with transverse excitations, we propose another explanation about the nature of this propagating mode that could be heat waves.

4 Conclusion

Considering three different phenomena observed in the dynamic structure of liquid metals and alloys, we have illustrated the capability of a recently proposed fitting model derived from GCM approach. This model can be applied to either simulation or experimental results. The parameters of the collective modes contributing to the functions obtained can be understood in the light of full-GCM results.

However, the following point should be remembered: a fit remains a fit and the results obtained depend on what one put in the model. Tracking additional modes, for instance, leads to additional parameters and very scattered results if the modes are weakly contributing to the data. Available safety barriers like sum-rules or considering simultaneously several functions (F, S, and J) should not be neglected. As the overall accuracy of both experimental and simulation determinations of the dynamic structure keeps increasing, we hope that this model will help improving our understanding of the dynamic structure of liquids and alloys.

Acknowledgments The PMMS (Pôle Messin de Modélisation et de Simulation) is greatly acknowledged for providing us with computer time.

References

[1] Wax J.-F. and Bryk T., J. Phys.: Condens. Matter **25**, 325104 (2013); **26** 168002 (2014).
[2] Wax J.-F. and Bryk T., AIP Conf. Proc. **1673**, 020010 (2015).
[3] Wax J.-F., Johnson M.R., and Bryk T., J. Phys.: Condens. Matter **28**, 185102 (2016).
[4] Bryk T. and Chushak Y., J. Phys.: Condens. Matter **9**, 3329 (1997).
[5] Scopigno T., Ruocco G., and Sette F., Rev. Mod. Phys. **77**, 881 (2005).
[6] Bryk T., Mryglod I., Scopigno T., Ruocco G., Gorelli F., and Santoro M., J. Chem. Phys. **133**, 024502 (2010).
[7] Jacucci G., Ronchetti M., and Schirmacher W., J. Phys. (Paris) Coll. **C8**, 385 (1984).
[8] Bosse J., Jacucci G., Ronchetti M., and Schirmacher W., Phys. Rev. Letters **57**, 3277 (1986).
[9] Montfrooij W., Westerhuijs P., de Haan V.O., and de Schepper I.M., Phys. Rev. Letters **63**, 544 (1989).
[10] Bove L.E., Sacchetti F., Petrillo C., and Dorner B., Phys. Rev. Letters **85**, 5352 (2000).
[11] Hosokawa S., Inui M., Kajihara Y., Matsuda K., Ichitsubo T., Pilgrim W.-C., Sinn H., González L.E., González D.J., Tsutsui S., and Baron A.Q.R., Phys. Rev. Letters **102**, 105502 (2009).
[12] Giordano V.M. and Monaco G., Proc. Natl. Acad. Sci. U.S.A. **107**, 21985 (2010).
[13] Hosokawa S., Munejiri S., Inui M., Kajihara Y., Pilgrim W.-C., Ohmasa Y., Tsutsui S., Baron A.Q.R., Shimojo F., Hoshino K., J.Phys.: Condens. Matter **25**, 112101 (2013).
[14] Hosokawa S., Inui M., Kajihara Y., Tsutsui S., Baron A.Q.R., J.Phys.: Condens. Matter **27**, 194104 (2015).
[15] Zanatta M., Sacchetti F., Guarini E., Orecchini A., Paciaroni A., Sani L., and Petrillo C., Phys. Rev. Letters **114**, 187801 (2015).
[16] Bryk T., Ruocco G., Scopigno T., and Seitsonen A.P., J. Chem. Phys., **143**, 110204 (2015).
[17] Bryk T. and Wax J.-F., J. Chem. Phys. **144**, 194501 (2016); **144**, 219903 (2016).

Orbital free ab initio study of static and dynamic properties of some liquid transition metals

G. M. Bhuiyan[1,*], *Mohammad Riazuddin* Molla[1,2,**], *A. Z. Ziauddin* Ahmed[1,***], *L. E.* Gonzalez[3], and *D. J.* Gonzalez[3]

[1]Department of Theoretical Physics, University of Dhaka, Dhaka
[2]Department of Mathematics, University of Dhaka, Dhaka
[3]Departamento de Física Teórica, Universidad de Valladolid, Valladolid, Spain

Abstract. Several static and dynamic properties of liquid transition metals Cr, Mn and Co are studied for the first time using the orbital free ab-initio molecular dynamics simulation (OF-AIMD). This method is based on the density functional theory (DFT) which accounts for the electronic energy of the system whereas the interionic forces are derived from the electronic energy via the Hellman-Feynman theorem. The external energy functional is treated with a local pseudopotential. Results are reported for static structure factors, isothermal compressibility, diffusion coefficients, sound velocity and viscosity and comparison is performed with the available experimental data and other theoretical calculations.

1 Introduction

We have studied some static and dynamic properties of three liquid transition metals namely, *l*-Cr, *l*-Mn and *l*-Co by using the Orbital Free Ab-initio Molecular Dynamics (OF-AIMD) simulation method. These elements are lying in the middle of the 3*d* series and share some common features with the other transition metals. For example, they show strong *d*-band scattering and *sp-d* hybridization effect. The *d*-electrons are neither completely free to be treated by the extended band approach nor tightly bound to use the tight binding theory. In addition, physical properties of transition metals lying in the middle of the transition series are strongly affected by the interplay between the bonding and antibonding states [1, 2]. Glasses based on transition metals, being strong, are used in modern technological development. In order to understand physical properties of glasses, in particular the glass transition temperature, we often require the knowledge of diffusion coefficient, D, in the liquid phase. Because of the high melting temperatures of *l*-Cr, *l*-Mn and *l*-Co, the measurement of their diffusion coefficients is very difficult and, in fact, no experimental data for D are available for these liquid metals. So, in order to study the static and dynamic properties of the aforesaid metals a very useful option is computer simulation. For the last three decades, computer simulation has been a powerful tool to study different properties of condensed matter. Most ab-initio molecular dynamics simulation methods (AIMD) are based on density functional theory (DFT) [3, 4].

*e-mail: gbhuiyan@du.ac.bd
**e-mail: mriazmath@gmail.com
***e-mail: abuzafur@yahoo.com

Although the AIMD methods based on Kohn-Sham orbital version of DFT (KS-AIMD methods) are, in principle, more accurate, however, they imply heavy computational demands that limit both the size of simulation box and the simulation time. This limitation becomes worse if d-electrons are included in the calculation. However, some of the above problems can be partly alleviated by employing the Hohenberg-Kohn (orbital-free) version of the DFT (OF-AIMD method). The OF-AIMD simulation method allows to simulate larger systems, up to a few thousand particles. So, for our metals lying in the middle of the $3d$-series the application of OF-AIMD is quite justificatory. We stress that, to our knowledge, this is the first AIMD simulation study for l-Cr, l-Mn and l-Co.

The layout of this report is the following. Section 2 is devoted to the theories relevant to the OF-AIMD study for static and dynamic properties. We present results and discussion in section 3. This article is concluded in section 4.

2 Theory

The liquid metal is modelled as a set of N ions, each of valence Z, enclosed in a volume V, and interacting with $N_e = NZ$ electrons through an effective potential $v(r)$. According to DFT, the ground state electronic energy functional can be written as [3, 4]

$$E[n(\vec{r})] = T_s[n(\vec{r})] + E_H[n(\vec{r})] + E_{\rm xc}[n(\vec{r})] + E_{\rm ext}[n(\vec{r})] \tag{1}$$

where $T_s[n]$, $E_{\rm H}[n]$, $E_{\rm xc}[n]$ and $E_{\rm ext}[n]$ represent the electronic kinetic energy of non-interacting electrons, the Hartree energy, the exchange-correlation energy and the energy of interaction with the external potential, respectively. Additional details can be found in Refs. [5, 6]. The electronic exchange-correlation energy functional is treated with the local density approximation (LDA). As $E_{\rm ext}[n] = \int d\vec{r}\, n(\vec{r})\, v_{ext}(r)$, the external potential is modelled by a local ionic pseudopotential, $v_{ps}(r)$ [6]. The $v_{ps}(r)$ (see Ref. 5) has some parameters, (A, R_C and a), which have been chosen so that the OF-AIMD simulation reproduces the experimental static structure factors. To include the sd-hybridization effects we follow Wills and Harrison [7] and choose effective valence $Z = 1.5$ for all systems. More specifically we have used the following values for A, R_c and a (in atomic units): 0.050, 1.37, 0.65 (l-Cr), 0.005, 1.40, 0.75 (l-Mn) and 0.005, 1.39, 0.85 (l-Co) respectively.

3 Results and Discussions

The OF-AIMD simulation was performed by using 500 particles in a cubic simulation cell whose size was chosen to reproduce the experimental ionic number density. Given the ionic positions at time t, the energy functional is minimized with respect to the total electron density represented as $n(\vec{r}) = |\psi(\vec{r})|^2$, where $\psi(\vec{r})$ stands for an effective orbital. This is expanded in a plane wave basis set whose cutoff energies $E_{\rm cut} = 14, 11, 14$ Ryd were found adequate for l-Cr, l-Mn, and l-Co, respectively. Forces on the ions are derived from Hellmann-Feynman theorem. The ionic positions and velocities are updated by solving the Newton's equation of motion where the Verlet leapfrog algorithm is used with a time step of 7.6×10^{-3} ps for the three systems.

3.1 Static Properties

Figure 1(a) shows the form factor of non-Coulombic part of the v_{ps} for l-Cr, l-Mn and l-Co. It has the largest value at $q \to 0$ and quicky decreases before exhibiting a weak oscillatory behavior. The simulation allows a direct computation of both $S(q)$ and $g(r)$, and figure 1(b)

 DOI: 10.1051/epjconf/201715103001

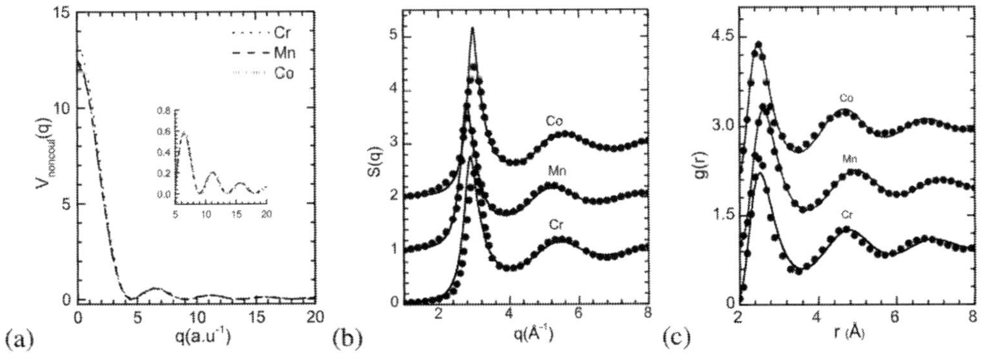

Figure 1. (a) Non-Coulombic potential, (b) static structure factor, $S(q)$ and (c) pair correlation function, $g(r)$, for *l*-Cr, *l*-Mn and *l*-Co at 2173, 1533 and 1823 K, respectively. The solid lines stand for OF-AIMD results and the closed circles for XR diffraction data.

shows the obtained $S(q)$ along with the respective x-ray data [8] for *l*-Cr, *l*-Mn and *l*-Co at T= 2173, 1533 and 1823 K, respectively. The main peak is overestimated in all cases but its position at q_p = 2.90, 2.76 and 2.97 Å^{-1} agrees with experiment. The positions and amplitudes of the subsequent peaks show a reasonable agreement with the respective experimental data. In the long wavelength limit the static structure factor is connected to thermodynamics via the relation $S(q \to 0) = \rho k_B T \chi_T$, where χ_T, ρ, k_B and T denote the isothermal compressibility, the average ionic number density, Boltzmann's constant and temperature, respectively. We have estimated $S(q \to 0)$ by fitting our calculated $S(q)$ with a quadratic function, within the range $q < 0.7$Å^{-1}, and we found $\chi_T = 0.79 \pm 0.01$, 1.27 ± 0.03 and 0.69 ± 0.01 (in 10^{-12}cm^2 dyne^{-1} units) for *l*-Cr, *l*-Mn and *l*-Co, respectively. Values of χ_T, in the same units, for the corresponding solid phase at room temperature are 0.60, 0.83 and 0.55, respectively. The enhancement of compressiblity in the liquid phase near melting temperature is found to be 30% to 40% larger than parent solids. To our knowledge, no experimental data are available for the liquid phase, but an estimation from the low-*q* experimental $S(q)$ values of Waseda [8] has yielded $\chi_T \approx 0.82$, 1.73 and 0.89 (same units) for *l*-Cr, *l*-Mn and *l*-Co, respectively. The pair correlation function $g(r)$ provides information about the short range order in the liquid state. Figure 1(c) depicts our calculated $g(r)$ for *l*-Cr, *l*-Mn and *l*-Co and they are compared with the x-ray data of Waseda [8]. We observe an overall agreement for the positions and amplitudes of the peaks. The coordination numbers were obtained by integrating $4\pi r^2 \rho\, g(r)$ up to the first minimum of the pair distribution function and the obatined results were 12.08, 12 and 12.43 for *l*-Cr, *l*-Mn and *l*-Co respectively. For comparison, we note that a similar calculation using the experimental pair distribution functions [8] yielded 11.68, 11.69, 11.75 for *l*-Cr, *l*-Mn and *l*-Co, respectively.

3.2 Dynamic properties: Single Particle Dynamics

The self-intermediate scattering function

$$F_s(q,t) = \frac{1}{N}\langle \sum_{j=1}^{N} \exp[i\vec{q} \cdot \vec{R}_j(t+t_0)] \exp[-i\vec{q} \cdot \vec{R}_j(t_0)] \rangle, \qquad (2)$$

provides information about the single particle dynamics over different length scales going from hydrodynamic ($q \to 0$) to free particle ($q \to \infty$) limit. Here $\langle ... \rangle$ denotes the average

DOI: 10.1051/epjconf/201715103001

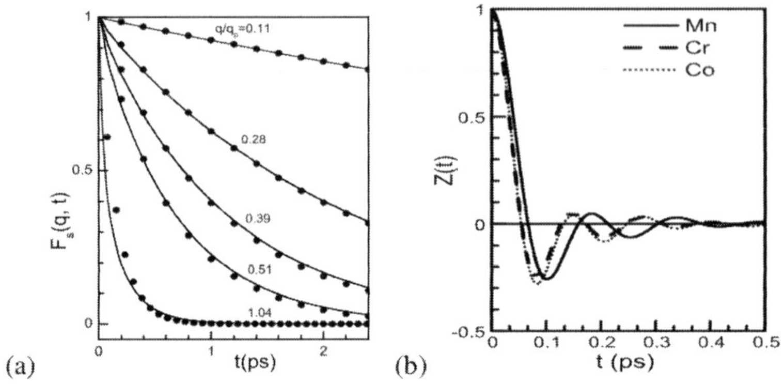

Figure 2. (a) $F_s(q,t)$ of l-Cr for several q. Full lines: OF-AIMD results. Full circles: Gaussian approximation (b) Normalized velocity autocorrelation functions of *l*-Cr , *l*-Mn, and *l*-Co.

over time origin and wave vectors with the same modulus. The velocity autocorrelation function (VACF) of a tagged ion in the fluid, $Z(t)$, is related to the time derivative of $F_s(q,t)$ in the hydrodynamic limit [9] but we have evaluated it from its definition

$$Z(t) = \langle \vec{v}_1(t)\vec{v}_1(0)\rangle / \langle v_1^2 \rangle. \tag{3}$$

The self diffusion coefficient, D, is related to the time integral of $Z(t)$ via the Green-Kubo relation

$$D = \frac{k_B T}{m} \int Z(t)dt. \tag{4}$$

It is interesting to note that, $F_s(q,t)$ may be expressed within the Gaussian approximation (GA) as $F_s(q,t) = \exp[-\frac{1}{6}q^2 \delta R^2(t)]$, where $\delta R^2(t)$ stands for the mean square displacement. Figure 2(a) shows, for l-Cr, the calculated $F_s(q,t)$, for several values of q. and it is compared with the result from the Gaussian approximation. $F_s(q,t)$ displays the typical monotonic non-linear decrease with time which becomes faster with increasing q/q_p values; in fact this trend is similar to that of the simple liquid metals near their respective triple points. It is noticed from the figure that the GA completely merge with $F_s(q,t)$ for low q values, but deviates somewhat at small t and large q. This clearly reveals the shortcomings of the GA. We note here that similar features have also been observed for *l*-Mn and *l*-Co.

The velocity autocorrelation function, $Z(t)$, is depicted in figure 2(b) for *l*-Cr, *l*-Mn and *l*-Co. The negative value of $Z(t)$ indicates backscattering of ions due to cage effects produced by the shell formed by surrounding neighbours. The self-diffusion coefficient, D, has been calculated from $Z(t)$ and we have obtained $D = 0.71 \pm 0.01$, 0.53 ± 0.01 and 0.41 ± 0.02 Å^2/ps for *l*-Cr, *l*-Mn and *l*-Co, respectively. To our knowledge, no experimental data are available, therefore we have compared with other theoretical estimates. The scaling law of Dzugutov yields a value of D=0.712 Å^2/ps for *l*-Cr. For *l*-Mn other calculations have yielded D= 0.52 [10] and 0.42 [11] whereas for *l*-Co another theoretical calculation obtained 0.41 Å^2/ps [12].

3.3 Dynamic properties: Collective dynamics

The intermediate scattering function, $F(q,t)$ (see Ref.11), provides information about the collective dynamics of density fluctuations and its time Fourier transform yields the dynamic

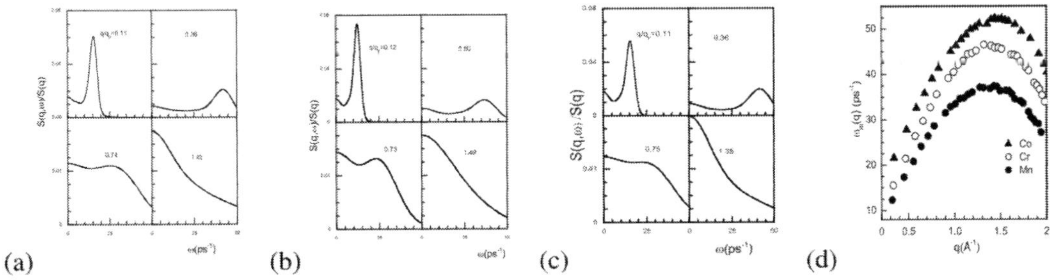

Figure 3. $S(q,\omega)$ for (a) *l*-Cr, (b) *l*-Mn, (c) *l*-Co and (d) their dispersion curves.

Table 1. Transport coefficients for *l*-Cr, *l*-Mn and *l*-Co.

Sys-tem	D(Å^2/ps)		c_S(m/s)		η(mPas)	
	OF-AIMD	Theo.	OF-AIMD	Exp.	OF-AIMD	Exp. (Theo.)
Cr	0.71	0.71	4586	4520	2.96	(2.96)
Mn	0.53	0.52, 0.42	3790	3710	2.74	(2.73)
Co	0.41	0.37, 0.41	4547	4090	4.15	4.1-5.3, 3.9

structure factor $S(q,\omega)$. Figures 2(a-c) show the calculated $S(q,\omega)$ for *l*-Cr, *l*-Mn and *l*-Co for several q values. It is noticed that up to $q/q_p \approx 0.73$, 0.77 and 0.75 (for *l*-Cr, *l*-Mn and *l*-Co, respectively) the obtained $S(q,\omega)$ exhibit well defined side peaks indicating the existence of collective density excitations. At $q/q_p > 0.73, 0.77$ and 0.75 a shoulder instead of a peak is observed first, and therefrom the $S(q,\omega)$ decrease monotonically. From the positions of the side peaks, $\omega_m(q)$, a dispersion relation of the density fluctuations has been obtained (see figure 3(d)). The slope of the dispersion curves evaluated at $q \to 0$ provides an estimation of adiabatic sound velocity c_s. It is interesting to note that c_s is connected to the ratio of specific heat γ which inturn is related to the Landau-Placzek ratio. The values of γ obtained from c_s are found to be 1.05, 1.09 and 1.09 for *l*-Cr, *l*-Mn and *l*-Co, respectively. Table 1 shows that the calculated c_s are in good agreement with the respective experimental values. Another interesting dynamical magnitude is provided by the current

$$\vec{j}(q,t) = \sum_{j=1}^{N} \vec{v}_j \exp[i\vec{q}\cdot\vec{R}_j(t)], \tag{5}$$

associated with the overall motion of the particles. This is usually split into a longitudinal component, $\vec{j}_l(q,t)$, parallel to $\vec{q}$ and, a transverse component, $\vec{j}_t(q,t)$, perpendicular to $\vec{q}$. The corresponding autocorrelation functions are the longitudinal, $C_l(q,t)$, and transverse, $C_t(q,t)$, correlation functions. $C_l(q,t)$ is basically the second time derivative of $F(q,t)$, and it does not bear any new physics. On the other hand, the transverse current correlation function $C_t(q,t)$ is related to the relaxation of shear stresses, and in the hydrodynamic limit yields the shear viscosity coefficient, η, through the memory function representation of $C_t(q,t)$. Values of the shear viscosity calculated for *l*-Cr, *l*-Mn and *l*-Co are presented in table 1 and we observe excellent agreement with experimental data [13] for *l*-Co. For *l*-Cr and *l*-Mn we are not aware of any experimental data, but other theoretical estimates are very close to the present ones.

4 Conclusion

Static and dynamic properties of *l*-Cr, *l*-Mn and *l*-Co transition metals near their respective triple points, have been studied near for the first time by using an AIMD simulation method. The calculated static structure agrees well with diffraction data and the calculated diffusion coefficients show a qualitative agreement with other theoretical estimates. Since experimental data for D do not exist, the present OF-AIMD values may be reliably used in other studies. The adiabatic sound velocity, c_S, obtained from our simulation data for *l*-Cr and *l*-Mn agree very well with experiment, but for *l*-Co agreement is not so good. In the case of shear viscosity, η, agreement between OF-AIMD and experiment is excellent for *l*-Co. Experimental data for η are not available for *l*-Cr and *l*-Mn , but other theoretical values are very close to ours. This agreement provides additional confidence on our obtained simulation values. Finally, we conclude that the OF-AIMD simulation method describes static and dynamic properties of liquid transion metals rather well and we expect that application of the KS-AIMD simulation method for these systems might improve results further.

GMB thanks AvH foundation for the grant subsidy to attend the LAM-16 conference, MRM acknowledges Ministry of Science and Technology, Bangladesh for fellowship under the project of Bangabandhu Fellowship on science and ICT, LEG and DJG acknowlegde the support of MECD (FIS2014-59279-P) and JCyL (VA104A11-2).

References

[1] K. A. Gschneider, Solid State Physics **16**, 275 (1974).

[2] J. Friedel, The physics of Metals 1. Electrons, ed. J.M. Ziman, Cambridge University Press(1971).

[3] P. Hohenberg, and W. Kohn, Phys. Rev. B **136**, 864 (1964).

[4] W. Kohn, and L. J. Sham, Phys. Rev. A **140**, 1133 (1965).

[5] G. M. Bhuiyan, L. E. González, and D. J. González, Cond. Matt. Phys. **15**, 33604, 1 (2012).

[6] G. M. Bhuiyan ,L. E. González, and D. J. González, Euro. Phys. J. Web of Conferences, **15**, 011011, (2011).

[7] J. M. Wills and W. A. Harrison, Phys. Rev. B,**28**,3199 (1988).

[8] Y. Waseda, The Structure of Non-Crystalline Materials. McGraw-Hill, New York, 1980.

[9] U. Balucani, and M. Zoppi, *Dynamics of the Liquid State*,Clarendon, Oxford, 1994.

[10] P. B. Thakor, Y. A. Sonvane, and A. R. Jani, AIP Conf. Proc. **1447**,915 (2012).

[11] R. C. Gosh, M. R. Amin, and G. M. Bhuiyan, J. Mole. liquids **188**,148 (2013).

[12] I. Yokoyama, Physica B **291**,145 (2000).

[13] G. Kaptay, Z. Metall.**1**,96 (2005); I. Egry, Scr. Metall. Mater. **28**,1273 (1993).

EPJ Web of Conferences **151**, 03002 (2017) DOI: 10.1051/epjconf/201715103002
LAM-16

Ab-initio study of several static and dynamic properties of liquid palladium and platinum.

L. E. González[1,*], *D. J.* González[1], *Mohammad Riazuddin* Molla[2,3], *A. Z. Ziauddin* Ahmed[2], and *G. M.* Bhuiyan[2]

[1]Departamento de Física Teórica, Universidad de Valladolid, Valladolid, Spain
[2]Department of Theoretical Physics, University of Dhaka, Dhaka
[3]Department of Mathematics, University of Dhaka, Dhaka

Abstract. We report a study on several static and dynamic properties of liquid Pd and Pt metals at thermodynamic conditions near their respective triple points. The calculations have been carried out by an *ab initio* molecular dynamics simulation technique. Results are reported for several static structural magnitudes which are compared with the available X-ray diffraction. As for the dynamic properties, results have been obtained for both single and collective dynamical magnitudes as well as for some transport coefficients which are compared with the corresponding experimental data.

1 Introduction.

Ab-initio molecular dynamics (AIMD) methods based on density functional theory [1] (DFT) are a usual technique for the study of a wide range of condensed matter systems. Most AIMD methods are based on the Kohn-Sham (KS) orbital representation of Density Functional Theory (DFT), whose application poses heavy computational demands and therefore somehow limits the size of the systems under study as well as the simulation times. Nevertheless, this drawback is offset by the very good accuracy provided by these *ab-initio* methods.

This paper reports an AIMD simulation study on several static and dynamic properties of liquid Palladium (l-Pd) and Platinum (l-Pt) at thermodynamic states near their respective triple point. Oddly enough, these two transition metals have not attracted as much theoretical and/or experimental work as most of the other transition metals and, to our knowledge, only a few classical molecular dynamics (CMD) simulations have been carried out on these two liquid metals. Alemany *et al* [2] have used CMD simulations, combined with interatomic potentials constructed within the framework of the embedded atom model, in order to evaluate the static structure factor, $S(q)$, and some transport coefficients of l-Pd and l-Pt. A similar calculation was also performed by Kart *et al* [3], but using a different interatomic potential. On the experimental side, we are only aware of the X-ray diffraction (XD) measurements of Waseda [4], who determined the $S(q)$ of l-Pd and l-Pt near their respective triple points. As for the dynamical structure, no measurements have been performed yet.

*e-mail: luisen@metodos.fam.cie.uva.es

This dearth of information has prompted us to perform AIMD simulations in order to evaluate several structural and dynamical properties as well as some transport coefficients. We emphasize that this is the first *ab initio* study on these two metals and, given the superior accuracy of the AIMD versus CMD simulations, we expect these results may stimulate further experimental and/or theoretical work.

In section 2 we briefly describe the theory underlying the present AIMD simulations, and in section 3 we report and discuss some structural and dynamical results which are compared with other previous studies and experimental data. Finally some conclusions are drawn.

2 Computational method.

The total potential energy of N ions with valence Z, enclosed in a volume Ω, and interacting with $N_e = NZ$ valence electrons is written, within the Born-Oppenheimer (BO) approximation, as the sum of the direct ion-ion coulombic interaction energy, $E_{i-i}[\{\vec{R}_l\}]$, plus the ground state energy of the electronic system subjected to the ionic external potential created by the ions, $V_{\rm ext}(\{\vec{R}_l\})$,

$$E(\{\vec{R}_l\}) = E_{i-i}[\vec{R}_l] + E_g[n_g(\vec{r}), V_{\rm ext}(\{\vec{R}_l\})] \,, \tag{1}$$

where $n_g(\vec{r})$ is the ground state electronic density and $\vec{R}_l$ are the ionic positions. According to DFT, the ground state electronic density, $n_g(\vec{r})$, minimizes the energy functional

$$E[n(\vec{r})] = T_s[n] + E_{\rm ext}[n] + E_H[n] + E_{\rm xc}[n] \,, \tag{2}$$

where $T_s[n]$ is the non-interacting electronic kinetic energy, $E_{\rm ext}[n]$ is the electron-ion interaction energy, $E_H[n]$ is the electrostatic Hartree energy and $E_{\rm xc}[n]$ is the electronic exchange-correlation energy for which we have used the local density approximation [5]. The ion-electron interaction has been described by an ultrasoft pseudopotential [6].

We have performed Kohn-Sham AIMD calculations for l-Pd and l-Pt at thermodynamic conditions near their respective triple point. Specifically, for l-Pd we have studied a thermodynamic state characterized by an ionic number density $\rho = 0.0594$ Å^{-3} and a temperature $T = 1873$ K whereas for l-Pt we used $\rho = 0.0577$ Å^{-3} and $T = 2053$ K. For each system, the AIMD simulations were performed by using 120 atoms (i.e. $120\times(9d{+}1sp) = 1200$ valence electrons) in a cubic supercell. Both systems were thermalized during 15.0 ps and therefrom, microcanonical AIMD simulations were performed over 21000 (18500) time steps in l-Pd (l-Pt). As the time step was 0.0055 ps for both systems, the total simulation time amounted to 115.5 ps (l-Pd) and 101.75 ps (l-Pt).

The configurations were generated under the BO approximation solving the Kohn-Sham equations on a plane-wave basis set for a given configuration of ions, calculating the forces on the ions using the Hellmann-Feynman theorem, and solving the corresponding Newton's equations for constant total energy, as implemented in the Quantum Espresso package [7]. We have used a plane-wave representation with an energy cutoff of 25 Ryd and the single Γ point was used in sampling the Brillouin zone. Finally, the 21000 configurations generated for l-Pd (18500 configurations for l-Pt) were used in the evaluation of the corresponding static and dynamic properties.

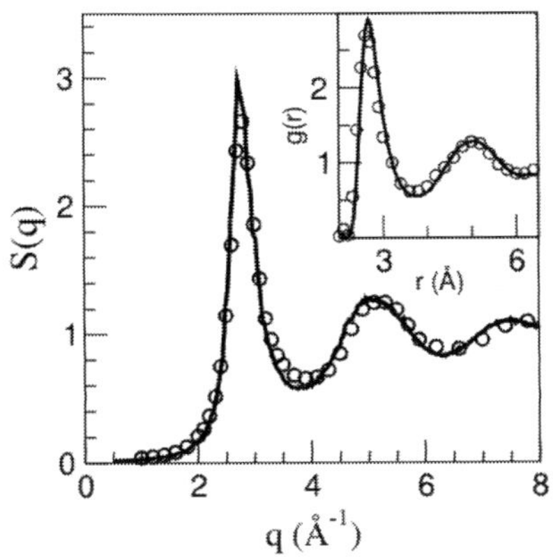

Figure 1. Static structure factor, $S(q)$, of l-Pd at $T = 1873$ K. Full line: AIMD calculations. Open circles: XD data from Waseda [4]. The inset shows the pair distribution function $g(r)$.

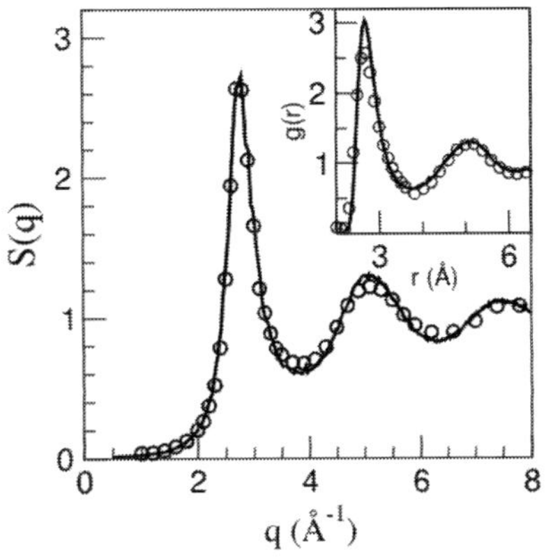

Figure 2. Same as the previous figure, but for l-Pt at $T = 2053$ K.

3 Results and discussion.

3.1 Static properties

The AIMD simulations provide a direct evaluation of both the static structure factor, $S(q)$, and the pair distribution function, $g(r)$. Figure 1 shows the AIMD result for the $S(q)$ of l-Pd along with the corresponding XD data [4]. The calculated $S(q)$ has a symmetric main peak at $q_p = 2.79$ Å^{-1}, whose shape and amplitude practically coincide with experiment. As for the subsequent oscillations, we observe a very small mismatch. The AIMD obtained $S(q)$ of l-Pt is depicted in Figure 2 and shows an excellent agreement with the XD data [4]. There is a symmetric main peak at $q_p = 2.77$ Å^{-1} and the following oscillations closely follow the experimental ones.

A similar good agreement is found for the AIMD calculated $g(r)$. The coordination number (CN) has been evaluated by integrating the $g(r)$ up to the position of its first minimum, which for both l-Pd and l-Pt stands at the same value $R_{\text{min}} = 3.77$ Å, yielding an identical result of CN ≈ 12.8 atoms, which is a typical value of the simple liquid metals near their triple point [8].

We have estimated the isothermal compressibility, κ_T, by resorting to the relation $S(q \to 0) = \rho k_B T \kappa_T$ where k_B is Boltzmann's constant. The AIMD $S(q)$ has been extrapolated to $q \to 0$, using $S(q) = s_0 + s_2 q^2$, yielding $S(q \to 0) = 0.013 \pm 0.001$ for l-Pd and 0.0095 ± 0.0010 for l-Pt (the uncertainties have been estimated from the results obtained using different q-ranges in the fits). This leads to the values $\kappa_T = 4.70 \pm 0.20$ (l-Pd) and 1.80 ± 0.15 (l-Pt) (in units of 10^{11} m^2 N^{-1}). Comparison with experiment, can only be performed for l-Pt for which there is a experimental data, $\kappa_T \approx 0.84 \pm 0.02$ [9].

DOI: 10.1051/epjconf/201715103002

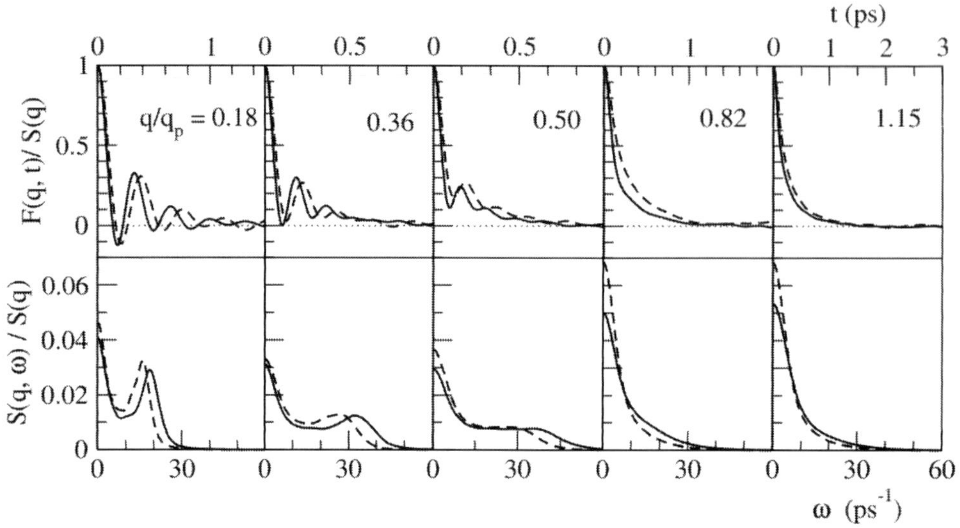

Figure 3. Upper part: Intermediate scattering functions, $F(q,t)$, of l-Pd at $T = 1873$ K (full lines) and l-Pt at $T = 2053$ K (dashed lines) at several values of q/q_p. Lower part: Same as before but for the dynamic structure factors, $S(q,\omega)$

3.2 Dynamic properties

3.2.1 Collective dynamics

The collective dynamics of density fluctuations in a liquid is described by the intermediate scattering function, $F(q,t)$, defined as

$$F(q,t) = \frac{1}{N}\left\langle \left(\sum_{j=1}^{N} e^{-i\vec{q}\cdot\vec{R}_j(t+t_0)}\right)\left(\sum_{l=1}^{N} e^{i\vec{q}\cdot\vec{R}_l(t_0)}\right)\right\rangle , \tag{3}$$

where $\langle ... \rangle$ stands for the ensemble average. Fourier Transform (FT) of the $F(q,t)$ into the frequency domain leads to the dynamic structure factor, $S(q,\omega)$. Another magnitude associated with the density fluctuations is the current due to the overall motion of the particles, i.e.

$$\vec{j}(q,t) = \sum_{j=1}^{N} \vec{v}_j(t)\ \exp[i\vec{q}\cdot\vec{R}_j(t)] , \tag{4}$$

where $\vec{v}_j(t)$ is the velocity of particle j at time t. The current is usually split into longitudinal ($\vec{j}_L(q,t)$) and transverse ($\vec{j}_T(q,t)$) components with respect to $\vec{q}$. Therefrom, the longitudinal, $C_L(q,t)$, and transverse $C_T(q,t)$, current correlation functions are obtained as

$$C_L(q,t) = \frac{1}{N}\langle \vec{j}_L(q,t)\cdot\vec{j}_L^*(q,0)\rangle \qquad C_T(q,t) = \frac{1}{2N}\langle \vec{j}_T(q,t)\cdot\vec{j}_T^*(q,0)\rangle \tag{5}$$

and the time FT gives the associated spectra, $C_L(q,\omega)$ and $C_T(q,\omega)$ respectively.

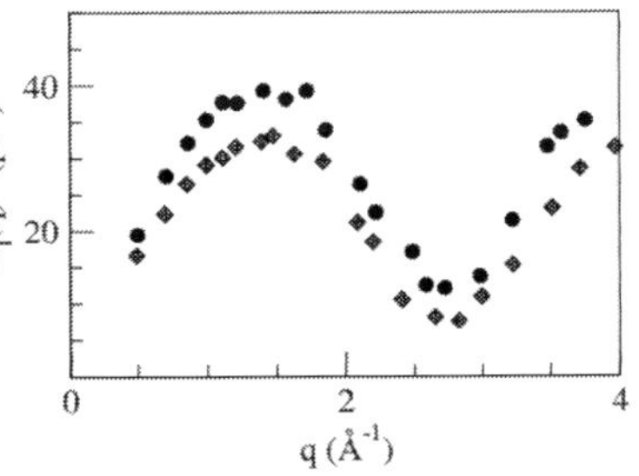

Figure 4. Longitudinal dispersion relation for l-Pd at $T = 1873$ K (circles) and for l-Pt at $T = 2053$ K (diamonds).

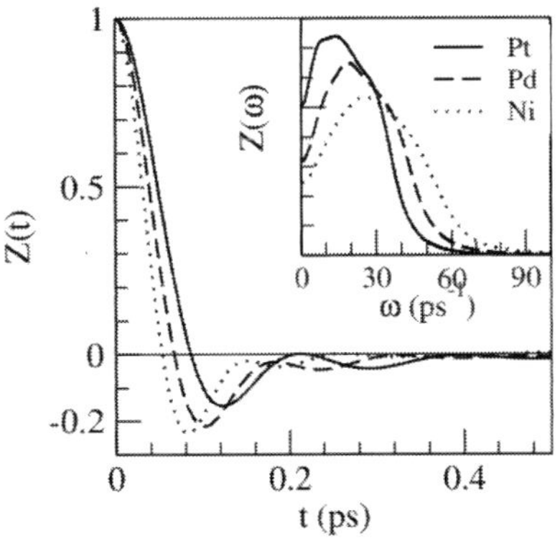

Figure 5. Normalized AIMD calculated velocity autocorrelation function of l-Pd, l-Pt and l-Ni near their respective melting points. The inset represents the corresponding power spectrum $Z(\omega)$.

Figure 3 depicts, for some q/q_p-values, the AIMD results for $F(q,t)$ and $S(q,\omega)$ in both l-Pd and l-Pt. For small q-values, the $F(q,t)$ show the typical oscillatory behaviour which lasts up to $q \approx (3/5)\ q_p$, with the amplitude of the oscillations becoming weaker with increasing q-values. Moreover, these oscillations are superposed on a discernible diffusive component. At $q \approx q_p$, the $F(q,t)$ exhibit the typical slow decay, known as "de Gennes narrowing".

The associatted $S(q,\omega)$ show side-peaks, indicative of collective density excitations, up to $q \approx (3/5)q_p$; therefrom the side-peaks evolve into shoulders which last until $q \approx (4/5)q_p$ and for greater q's the corresponding $S(q,\omega)$ show a monotonic decreasing behavior.

We have used Eqs. 4-5 to evaluate the current correlation functions $C_{L,T}(q,t)$ and their spectra $C_{L,T}(q,\omega)$. Our AIMD calculated $C_L(q,\omega)$ show a peak for each q-value and from the frequencies of those peaks, the dispersion relation for the longitudinal modes, $\omega_l(q)$, has been obtained and is plotted in Figure 4.

In the low-q region, the slope of the $\omega_l(q)$ curve gives a q-dependent adiabatic sound velocity that in the limit $q \to 0$ reduces to the bulk adiabatic sound velocity, c_s. Using the $q \leq 1.2$ Å^{-1} values of our calculated $\omega_l(q)$, we have evaluated its slope at $q \to 0$ and obtained $c_s \approx 3450 \pm 150$ m/s (l-Pd) and $c_s \approx 3000 \pm 150$ m/s (l-Pt). The AIMD result for l-Pt compares well with the experimental data, $c_{s,\exp} \approx 3053 \pm 150$ m/s [9].

The transverse current correlation function, $C_T(q,t)$, provides information on the shear modes. Its spectrum, $C_T(q,\omega)$, when plotted as a function of ω, may display peaks, within some q-range, which are related to propagating shear waves. For both l-Pd and l-Pt, we have found that the $C_T(q,\omega)$ already shows a peak for the smallest attainable q-value, i.e. $q_{min} = 0.18\ q_p$ and it lasts up to $q \leq 3q_p$, which is typical of the simple liquid metals [8].

From the AIMD results for the $C_T(q,t)$, the shear viscosity coefficient, η, has been calculated as follows [8, 10, 11]. The memory function representation of the $C_T(q,t)$

$$\tilde{C}_t(q,z) = \frac{1}{\beta m}\left[z + \frac{q^2}{\rho m}\,\tilde{\eta}(q,z)\right]^{-1} , \tag{6}$$

where the tilde denotes the Laplace transform, introduces a generalized shear viscosity coefficient, $\tilde{\eta}(q,z)$. The area under the normalized $C_T(q,t)$ gives $\beta m\ \tilde{C}_T(q,z=0)$, wherefrom values for $\tilde{\eta}(q,z=0)$ can be obtained and when extrapolated to $q=0$ give the usual shear viscosity coefficient η; additional details are given in References [8, 10, 11]. For l-Pd, we have obtained $\eta_{\mathrm{AIMD}} \approx 4.57 \pm 0.15$ GPa ps which compares well with the experimental value of $\eta_{\mathrm{exp}} = 4.22 \pm 0.15$ GPa ps [12]; moreover, the CMD simulation studies have yielded smaller values, i.e. $\eta \approx 3.68 \pm 0.50$ GPa ps [2] and $\approx 2.91 \pm 0.86$ GPa ps [3]. As for l-Pt, our result $\eta_{\mathrm{AIMD}} \approx 4.90 \pm 0.25$ GPa ps is close to the smaller of the two available experimental data $\eta_{\mathrm{exp}} = 4.82$ and 6.74 GPa ps [12]; moreover, the latter one is closer to the CMD simulation result $\eta \approx 6.08 \pm 0.95$ GPa ps [2].

3.2.2 Single particle dynamics

An important magnitude is the normalized velocity autocorrelation function (VACF) of a tagged ion in the fluid, $Z(t)$, defined as

$$Z(t) = \langle \vec{v}_1(t) \cdot \vec{v}_1(0)\rangle / \langle v_1^2 \rangle \tag{7}$$

with $\vec{v}_1(t)$ being the velocity of a tagged ion in the fluid at time t.

Figure 5 depicts the calculated $Z(t)$ along with the respective power spectrum (for the sake of comparison we have also included liquid Ni [13]). The $Z(t)$ exhibit the typical backscattering behavior with a first minimum followed by rather weak oscillations and the $Z(t)$ always remaning negative. Notice that the $Z(t)$ are qualitatively very similar and the minor differences are related to the increasing ionic mass when going from Ni to Pt.

The self-diffusion coefficient, D, can be found by either the time integral of $Z(t)$ or from the slope of the mean square displacement $\delta R^2(t) \equiv \langle |\vec{R}_1(t) - \vec{R}_1(0)|^2 \rangle$. Both routes have yielded the same value which, moreover, coincides for both metals, i.e. $D_{\mathrm{AIMD}}(Pd) = D_{\mathrm{AIMD}}(Pt) = 0.27 \pm 0.02$ Å^2/ps. Unfortunately, the lack of experimental data implies that comparison can only be performed with the avaliable CMD simulations; thus Alemany *et al* [2] have obtained $D_{\mathrm{CMD}}(\mathrm{Pd}) = 0.403 \pm 0.003$ Å^2/ps and $D_{\mathrm{CMD}}(\mathrm{Pt}) = 0.281 \pm 0.003$ Å^2/ps. On the other hand, the CMD study of Kart *et al* [3] gave $D_{\mathrm{CMD}}(Pd) = 0.496 \pm 0.006$ Å^2/ps.

4 Conclusions.

An *ab-initio* molecular dynamics simulation method has been used to calculate some static and dynamic properties of l-Pd and l-Pt near their respective triple point. These two metals have not attracted much attention and therefore there is not much theoretical and/or experimental information to compare with. Concerning the static structure, the *ab initio* results for the pair distribution function, $g(r)$, and the static structure factor, $S(q)$, are in good agreement with the experimental data. The calculations account for the shape and position of the main peak of $S(q)$ as well as the subsequent peaks. The obtained coordination numbers are typical of the simple liquid metals.

The intermediate scattering functions, $F(q,t)$, show at small q-values, strong oscillations superposed on a distinct diffusive component. The associated dynamic structure factors, $S(q,\omega)$, show side-peaks which are indicative of collective density excitations. The calculated transport coeffcients show good agrement with the available experimental data.

LEG and DJG acknowledge the support of MECD (FIS2014-59279-P) and JCyL (VA104A11-2). MRM acknowledges the financial support of the Ministry of Science and Technology (Bangladesh) under the Bangabandhu Fellowship on Science and ICT.

References

[1] P. Hohenberg and W. Kohn, Phys. Rev. **136**, 864 (1964); W. Kohn and L.J. Sham, Phys. Rev. **140**, A1133 (1965).
[2] M.M.G. Alemany, C. Rey and L.J. Gallego, J. Chem. Phys. **109**, 5175 (1998); Phys. Rev. B **60**, 9208 (1999)
[3] S.O. Kart, M. Tomak, M. Uludogan and T. Cagin, J. Non-Cryst. Solids, **337**, 101 (2004).
[4] Y. Waseda, *The Structure of Non-Crystalline Materials*, (New York: McGraw-Hill, 1980).
[5] D.M. Ceperley and B.J. Alder, Phys. Rev. Letters **45**, 566 (1980); J.P. Perdew and A. Zunger, Phys. Rev. B **23**, 5048 (1981).
[6] D. Vanderbilt, Phys. Rev. B **41**, 7892 (1990)
[7] S. Baroni, A. Dal Corso, S. de Gironcoli, P. Giannozzi, C. Cavazzoni, G. Ballabio, S. Scandolo, G. Chiarotti, P. Focher, A. Pasquarello, K. Laasonen, A. Trave, R. Car, N. Marzari, A. Kokalj, http://www.pwscf.org/
[8] U. Balucani and M. Zoppi, *Dynamics of the Liquid State*, Clarendon, Oxford, 1994, pag. 261
[9] S. Blairs, Intern. Materials Rev. **52**, 321 (2007); Phys. Chem. Liq. **45**, 399 (2007)
[10] B.J. Palmer, Phys. Rev. E **49**, 359 (1994)
[11] U. Balucani, J.P. Brodholt, P. Jedlovszky and R. Vallauri, Phys. Rev. E **62**, 2971 (2000).
[12] T. Iida and R.I.L. Guthrie, *The Thermophysical Properties of Metallic Liquids* (Oxford Univ. Press, 2015)
[13] B. G. del Rio, L. E. González and D. J. González, *these proceedings.*

Orbital-free ab initio molecular dynamics study of the static structure and dynamic properties of the free liquid surface of Sn.

B. G. del Rio[1,*] and *L. E.* González[1]

[1]Departamento de Fisica Teórica, Facultad de Ciencias, Universidad de Valladolid, Valladolid, SPAIN

Abstract. We report results of an orbital-free ab initio molecular dynamics (OF-AIMD) study of the free liquid surface (FLS) of Sn at 1000 K and 600 K. A key ingredient in the OF-AIMD method is the local pseudopotential describing the ions-valence electrons interaction. We have used a force-matching method to derive a local pseudopotential suitable to account for the variation of the forces from the bulk to the FLS. We obtain very good results for structural properties, such as the reflectivity, including the characteristic shoulder it presents in x-ray experiments. Moreover we have been able to study ab initio for the first time the evolution in some dynamical properties as we move from the central region, where the system behaves like the bulk liquid, to the FLS.

1 Introduction.

Molecular dynamics (MD) simulations have become an essential technique in the study of a wide variety of condensed matter systems, and during the last three decades there has been an increasing amount of systems studied by using ab initio molecular dynamics (AIMD) methods based on density functional theory (DFT) [1]. Most AIMD methods are based on the Kohn-Sham (KS) orbital representation of DFT (KS-AIMD methods) which, although very accurate, imposes high computational demands that severely limit the size of the systems to be studied as well as the extent of the simulation times. Nevertheless, these constraints may be drastically reduced by the orbital-free ab-initio molecular dynamics (OF-AIMD) simulation method [2], whose basic variable is the total valence electronic density instead of the electronic orbitals of the KS-AIMD formulation. The OF-AIMD method substantially reduces the number of variables needed to describe the electronic state and therefore allows to study larger systems for longer simulation times. However, this improvement is achieved at the price of resorting to an approximate description of the electronic kinetic energy, and then, in principle, the OF-AIMD approach is less accurate than the KS-AIMD method. Nevertheless, despite this caveat, the OF-AIMD has provided good descriptions for several static and dynamic properties of a variety of bulk simple liquid metals [3, 4] and free liquid surfaces (FLS) [5].

*e-mail: beatriz@metodos.fam.cie.uva.es

DOI: 10.1051/epjconf/201715103003

In 1981 Rice and co-workers predicted, from theoretical considerations and Monte Carlo simulations, that the atomic density near the free surface of a metallic liquid would change in a nonmonotonic way, showing a stratification in layers parallel to the interface [6]. The experimental confirmation of surface layering was first observed, several years later, in the x-ray reflectivity of liquid Hg [7]. Since then, the reflectivities of other five elemental liquid metal surfaces have been measured, namely, Ga, In, K, Sn and Bi [8], all of them displaying (or suggesting, in the case of K) the expected layering peak. Moreover, Sn and Bi have shown a weak wide shoulder at a lower q_z that is not present in the other systems. The data obtained for these liquid metals and several binary alloys [9] have shown that the FLS exhibits an oscillatory surface-normal density profile (DP) which extends several atomic diameters into the bulk liquid. In these works the shape of the DPs were deduced by fitting the measured reflectivities to those predicted within a specific profile model, namely, the distorted crystal model, or variations thereof [8]. Although it would be desirable to obtain the DP in a model-independent way, the lack of phase information in the measurements renders this impossible.

In this paper we report a study of the surface structure of liquid Sn (l-Sn), performed through OF-AIMD simulations. Moreover, we have also analyzed the collective dynamics of the system at different depths from the FLS. To date, we are awere of only one experimental study of this topic using inelastic x-ray scattering in the FLS of indium [10]. This study showed that the effect of capillary waves is not only present on the surface, but also at very deep regions below the FLS. To our knowledge, the present paper reports the first ab initio study on the collective dynamics of a free liquid surface.

2 Computational method.

The total potential energy of N ions with valence Z, enclosed in a volume Ω, and interacting with $N_e = NZ$ valence electrons is written, within the Born-Oppenheimer approximation, as the sum of the direct ion-ion coulombic interaction energy and the ground state energy of the electronic system under the external potential created by the ions.

According to DFT, the ground state electronic density, $n_g(r)$, minimizes an energy functional given by the sum of the non-interacting electronic kinetic energy, $T_s[n]$, the electron-ion interaction energy, $E_{\text{ext}}[n]$, the classical electrostatic Hartree energy, $E_H[n]$, and the exchange-correlation energy, $E_{xc}[n]$, for which we have used the local density approximation, as parametrized by Perdew and Zunger [11]. Different expressions have been proposed for $T_s[n]$ and in this work we have used an average density model [3, 4].

The ion-electron interaction is described by means of pseudopotentials. In the OF-AIMD method the valence electron density is the basic variable, and non-local pseudopotentials cannot be used. The construction of the local pseudopotentials for our system has two steps. We start from an initial local pseudopotential proposed for Sn by Bhuiyan *et al* [12] which, in combination with the OF-AIMD method, has provided a good description of the static properties of liquid Sn at thermodynamic conditions near its triple point. However, the method used to construct the pseudopotential is not completely optimal, since it only tries to represent the static behavior but not the dynamical one, where the interaction between the different particles is taken into account at different time and length scales. A more realistic method should explicitly account for the presence of atoms around the central one and their effect on the force it feels. For this purpose, we have used a force matching method developed recently [13] to obtain a more accurate LPS. In figure 1 we have plotted the local pseudopotential before and after the force-matching process. We can observe how, in order to better represent the interactions, the pseudopotential has to adopt a higher value in reciprocal space near $q = 0$. Using as benchmark the results of a KS-AIMD calculation for l-Sn [14], in figure 2 we have plotted the differences in the x and z components of the forces felt by the atoms as obtained in the OF and in the KS calculations at a particular liquid configuration before and after the force matching process. We can observe how with the initial local pseudopotential we can recover reasonably well the forces in the bulk region, but not at the surface (particles are sorted according to their z coordinate). However, after the force matching process, these forces are recovered and the ones in the bulk are improved.

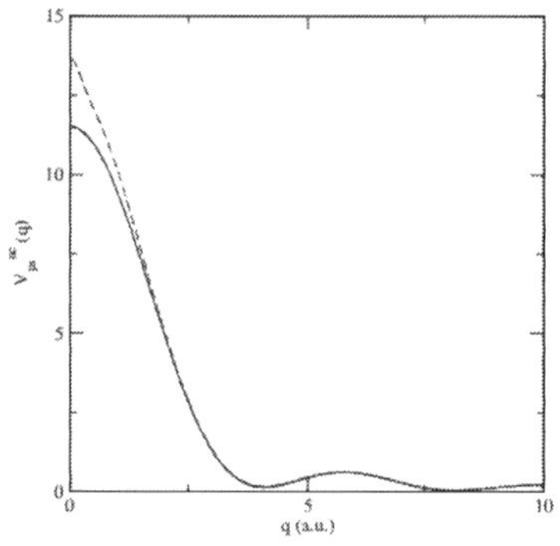

Figure 1. Non coulombic part of the local pseudopotential before (full curve) and after (dashed curve) force-matching.

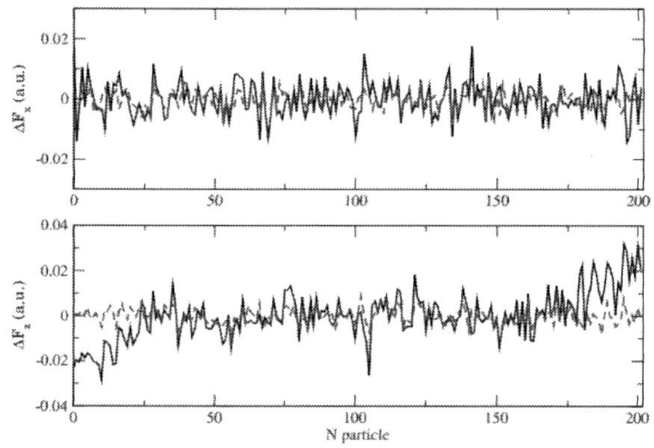

Figure 2. Difference between the x and z components of the forces as computed with KS-AIMD simulations [14] and OF-AIMD calculations, before (solid line) and after (dashed line) force-matching.

3 Results and discussion.

The simulated sample contains 2000 Sn atoms, enclosed in a tetragonal box with $L_x = L_y = 34.25$ Å, and $L_z = 66.0$ Å in the case of 1000 K. At the lower temperature of 600 K, the dimensions of the tetragonal box are $L_x = L_y = 33.80$ Å, and $L_z = 66.0$ Å. In both cases the atoms formed a slab of initial width (approximately 50 Å) corresponding to the experimental number density (0.035 and 0.337 Å^{-3} at 600 and 1000 K respectively) with 8 Å of vacuum at each side. Periodic boundary conditions are applied in all directions, meaning that each surface has a separation of 16 Å with the adjacent image. This distance has proven to be enough for the different images not to interact with each other [15, 16]. The system was first allowed to evolve at the desired temperature until the slab width accomodated the requirement of null pressure and equilibrium was reached, and after that, 20000 configurations were generated using a time step of 4 fs. These were used for the analysis of the properties of the slabs. The slab densities after equilibration were 0.0340 and 0.0332 Å^{-3} at 600 and 1000 K, in good agreement with the experimental ones.

DOI: 10.1051/epjconf/201715103003

3.1 Static structure

The averange ionic DP was computed from a histogram of the particle positions relative to the slab's center of mass, so that both halves of the slab are being averaged. The obtained results are shown in figure 3, where we observe stratification for at least four layers into the bulk liquid. The spacing between the outermost layer and the first inner one is $\lambda_{\mathrm{out}} = 2.83$ Å, whereas the inner interlayer distances are somewhat larger, namely, $\lambda_{\mathrm{in}} = 2.97$ Å at 1000 K. It should be noted that the average DP obtained through the KS-AIMD [14] led to equal interlayer distances of 3.1 Å, which highlights the dependence of the average DP on the conditions of the simulation, in particular the lateral size of the box. At the lower temperature of 600 K the interlayer distances obtained are essentially all equal, with $\lambda_{\mathrm{out}} \approx \lambda_{\mathrm{in}} = 2.86$ Å.

In order to compute the reflectivity and compare with the experimental reported results, capillary waves have to be substracted from the system. The most convenient method is to obtain the intrinsic density profile through the calculation of the separation surface between the liquid and vapour phases, which is smoothed afterwards, following the procedure of Chacon and Tarazona [17]. In figure 3 we show the intrinsic DP along with the average DP for the two temperatures studied. The layered structure of the interface is more marked in the intrinsic profiles. Moreover, the obtained interlayer distances are modified from those in the average DP. The outermost interlayer distance is $\lambda_{\mathrm{out}} = 2.96$ Å, with the inner interlayer distances being $\lambda_{\mathrm{in}} = 3.02$ Å at 1000 K. At the lower temperature of 600 K the difference between the outermost and inner interlayer distances is more emphasized, with $\lambda_{\mathrm{out}} = 2.83$ Å and $\lambda_{\mathrm{in}} = 3.01$ Å. It is important to note that, despite the differences between the average DPs obtained in this work and the one obtained in the KS-AIMD study, the corresponding intrinsic DPs practically coincide (see figure 3).

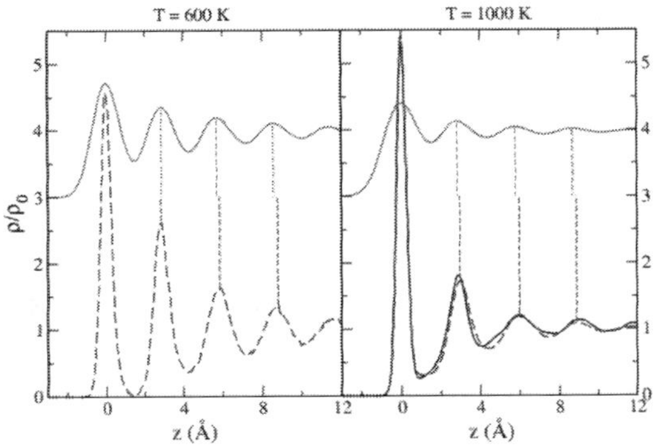

Figure 3. Average ionic DP (full curve, shifted upwards by 3 units) and intrinsic ionic DP (dashed curve) at T=600 K and 1000 K. At 1000 K the lower full curve corresponds to the intrinsic ionic DP obtained using the simulation data from [14].

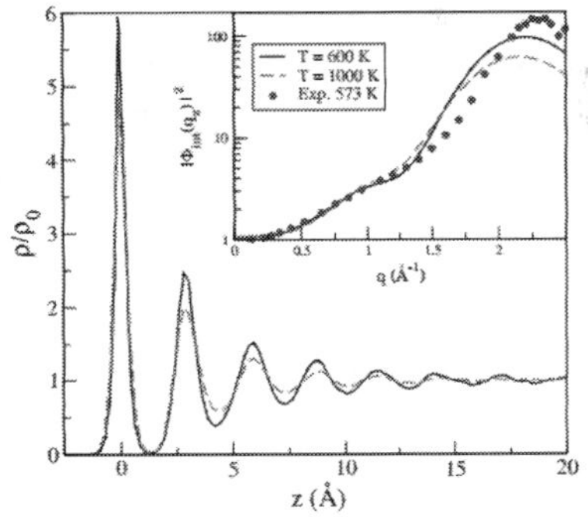

Figure 4. Comparison of the intrinsic electronic density profiles at 600 K (full line) and 1000 K (dashed line). Inset: comparison of reflectivities at 600 K (full line) and 1000 K (dashed line) with the experimental reflectivity (circles) [18].

The x-ray reflectivity, which is the property amenable to experimental observation, is related to the electronic DP. We have calculated the intrinsic electron DP, as obtained from a superposition of atomic electron densities centered at each ionic position, using the smoothed separation surface. The intrinsic reflectivity is calculated as $R_{\rm int}(q_z) = |\Phi_{\rm int}(q_z)|^2$, where

$$\Phi_{\rm int}(q_z) = \frac{1}{\rho_{e0}} \int_0^\infty dz \exp[-iq_z z] \frac{d\rho_{e;{\rm int}}(z)}{dz} \tag{1}$$

where ρ_{e0} is the average bulk electron density. The results obtained are shown in figure 4, where we see a characteristic peak located around $q_z \simeq 2.2$ Å^{-1}, which is close to $2\pi/\lambda_{\rm in,600K} \simeq 2.1$ Å^{-1}, and is related to the layering of the DP. Morever, at around $q_z \simeq 0.8$ Å^{-1} we obtain a shoulder, more emphasized at 600 K than at 1000 K, which also appears in the experimental measurements perfomed by Shpyrko *et al* [18]. The experimentalists ascribed the origin of the shoulder to an increased density in the outermost layer related to its smaller interlayer distance. However this interpretation may be biased by the model that was used and a deeper analysis would be valuable. This will be reported elsewhere.

3.2 Dynamic properties

We want to study the collective dynamics in the slab for different regions at different depths from the interface. To this end we define the collective properties (number of atoms, particle density, particle current) corresponding to a given region by multiplying the usual definition by a window function that is one inside the region and falls smoothly to zero outside.

From these we compute the corresponding time correlation functions. For instance, the intermediate scattering function, $F(q,t)$ is obtained as

$$F(\vec{q},t) = \langle \rho(-\vec{q}, t+t_0)\rho(\vec{q},t_0)\rangle \ , \tag{2}$$

and its FT into the frequency domain is the dynamic structure factor, $S(q,\omega)$.

The symmetry of the system defines a special direction, the z-axis, around which the properties must show cylindrical symmetry. We are particularly interested in waves that propagate parallel to the interface, and therefore among the $\vec{q}$ vectors allowed by the periodic boundary conditions, only those that are perperdicular to the z-axis have been used. In the averages like those of equation 2 we consider all vector orientations that fulfill this condition.

The calculated $S(q,\omega)$ presented in figure 5 at different depths in the slab show side-peaks, indicative of collective density excitations, up to $q \approx 1.23$ Å^{-1}. Moreover, as one approaches the interface region coming from the bulk, the frequency of the side peaks increases, while the corresponding widths also increase, making the peaks less marked. From the position of the side peaks with respect to q we can obtain the dispersion relation (DR) of the dynamic structure factor, depicted in figure 6. Evaluating the slope of the curve for small q it is possible to obtain an estimate of the adiabatic sound velocity at the corresponding layer. In this case we obtain an increase in the sound velocity from 2700 ± 150 m/s at the bulk region to 3600 ± 200 m/s in the outermost layer. This increase suggests a higher ionic density and a lower compressibility in the outermost layer with respect to the bulk region.

We have also computed the longitudinal and transverse components of the current due to the overall motion of the particles. For the transverse current, two independent components appear due to symmetry, and here we are only considering the component associated to the motion of the particles along the z axis. In figures 7 and 8 we have depicted the DR of the spectra of both currents. In the case of ω_L, we clearly

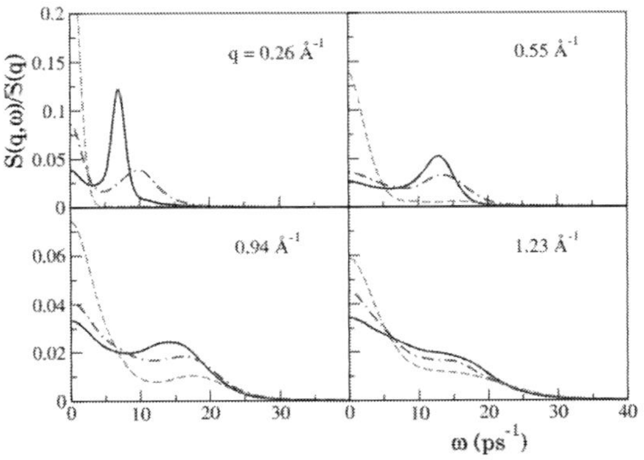

Figure 5. Dynamic structure factor at different q values and depths: outmost layer (dashed curve), first inner layer (dot-dashed curve) and bulk region (full curve).

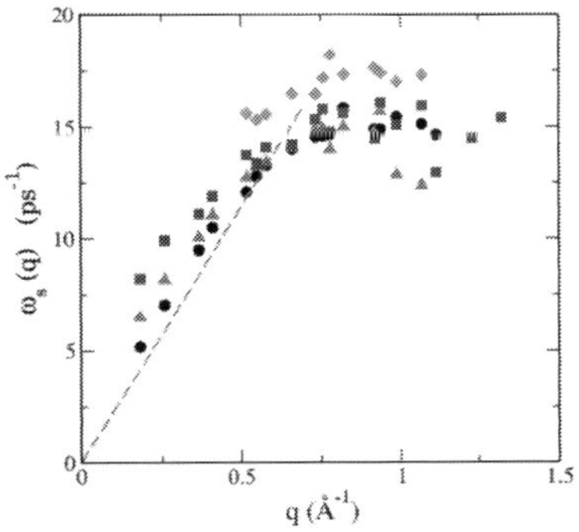

Figure 6. Dispersion relation of $S(q,\omega)$ at different depths: outmost layer (diamonds), first inner layer (squares), second inner layer (triangles), and bulk region (dots). Dashed curve: Experimental value of the adiabatic sound velocity at 1273 K [19] (2461 m/s).

observe, apart from the usual sound propagation mode, a second contribution at lower frequencies that only appear at small q. This second contribution relates to the DR of capillary waves described by [8], $\omega \cong \sqrt{\gamma/\rho}\, q^{3/2}$. Note that the effect of the capillary waves not only appears in the first layers of the slab but extends very deep under the surface. This behaviour was previously observed in an experimental study of a liquid slab of In [10]. The transverse current in the z direction in principle should contain contributions from capillary waves and from shear waves. The latter are only supported starting from a non-zero value of q, whereas the former should disperse as commented above in the small q region. The results shown in figure 8 follow this behaviour for $q \to 0$, while for larger wavevectors it is not directly possible to disentangle the contribution of each type of waves. However, comparison with the bulk transverse DR, which has lower energy, suggests that we are observing the dispersion of the capillary waves moving from the hydrodynamic into the kinetic regime.

4 Conclusions.

We have been able to construct a LPS to study through OF-AIMD simulations the free liquid surface of Sn, using the force-matching method developed recently. Using this LPS, a liquid slab of 2000 Sn atoms has been studied at 600 K and 1000 K, obtaining very good results for the reflectivity as compared to experimental data, including the presence of the characteristic shoulder at $q \simeq 0.8$ Å^{-1}. Due to the large number of particles, dynamic properties have also been studied in the liquid slab. We have obtained an increase in sound velocity of around 30% when going from the central region (behaving like a bulk) to the surface, suggesting a denser and less compressible surface as compared to the bulk. When studying the dispersion relation of longitudinal and transverse waves we obtain a second contribution, related to the presence of capillary waves, extending even to the central region of the slab.

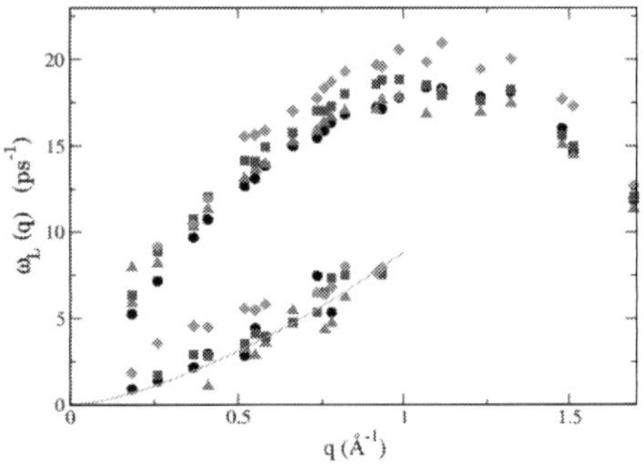

Figure 7. DRs of $C_L(q,\omega)$ at different depths: the symbols have the same meaning as in figure 6. The higher frequencies correspond to sound propagation while the lower ones are related to capillary waves. Full curve: theoretical DR of capillary waves.

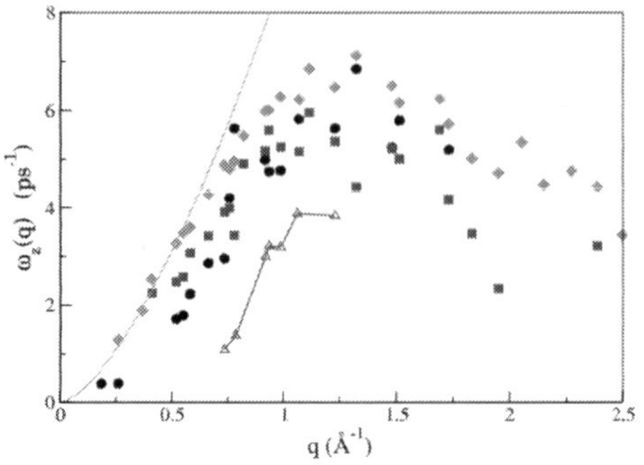

Figure 8. DR of $C_{Tz}(q,\omega)$ at different depths: the symbols have the same meaning as in figure 6. Full curve: theoretical DR of capillary waves. Full curve with open triangles: DR from bulk l-Sn.

We acknowledge the support of the MEC (FIS2014-59279-P) and JCyL (VA104A11-2). BGR acknowledges the financial support of Universidad de Valladolid.

References

[1] W. Kohn and L. J. Sham, Phys. Rev. **140**, A1133 (1965).

[2] P. Hohenberg and W. Kohn, Phys. Rev. **136**, B864 (1964).

[3] D. J. González *et al*, Phys. Rev. B **65**, 184201 (2002).

[4] D. J. González *et al*, J. Chem. Phys. **115**, 2373 (2001); L. E. González, D. J. González and J. M. López, J. Phys.: Condens. Matter **13**, 7801 (2001); A. Delisle, D. J. González and M. J. Stott, Phys. Rev. B **73**, 064202 (2006); L. E. González and D. J. González, Phys. Rev. E **77**, 064202 (2008).

[5] D. J. González, L. E. González and M J Stott, Phys. Rev. Lett. **92**, 085501 (2004); **94**, 077801 (2005); Phys. Rev. B **74**, 014207 (2006).

[6] M. P. D'Evelyn and S. A. Rice, Phys. Rev. Lett. **47**, 1844 (1981).

[7] O. M. Magnussen *et al*, Phys. Rev. Lett. **74**, 4444 (1995); E. DiMasi *et al*, Phys. Rev. B **58**, R13419 (1998)

[8] P. S. Pershan and M. L. Schlossman, *Liquid Surfaces and Interfaces. Synchrotron X-ray Methods* (New York: Cambridge University Press, 2012).

[9] P. S. Pershan, Eur. Phys. J. - Special Topics, **196**, 109 (2011).

[10] B. Wehinger, M. Krisch and H. Reichert, New J. Phys. **13**, 023021 (2011).

[11] J. P. Perdew and A. Zunger, Phys. Rev. B **23**, 5048 (1981).

[12] G. M. Bhuiyan, L. E. Gonzalez and D. J. Gonzalez, Eur. Phys. J. - Web of Conferences **15**, 01011 (2011).

[13] B. G. del Rio and L. E. González, J. Phys.: Condens. Matter **26**, 465102 (2014).

[14] L. Calderín, L. E. González and D. J. González, Phys. Rev. B **80**, 115403 (2009).

[15] D. J. González, L. E. González and M. J. Stott, Phys. Rev. Lett. **94**, 077801 (2005).
[16] L. E. González, D. J. González and M. J. Stott, J. Chem. Phys. **123**, 201101 (2005).
[17] E. Chacon and P. Tarazona, Phys. Rev. Lett. **91**, 166103 (2003).
[18] O. G. Shpyrko *et al*, Phys. Rev. B **70**, 224206 (2004).
[19] S. Blairs, Int. Mater. Rev. **52**, 321 (2007).

EPJ Web of Conferences **151**, 03004 (2017) DOI: 10.1051/epjconf/201715103004
LAM-16

Structure and dynamics of liquid Zn: an analysis of ab-initio simulations.

B. G. del Rio[1,*] and *L. E.* González[1]

[1]Departamento de Física Teórica, Facultad de Ciencias, Universidad de Valladolid, Valladolid, Spain

Abstract. The static and dynamic properties of liquid Zn have been studied using an ab initio molecular dynamics method. Results are reported for the thermodynamic states at 723K near which inelastic neutron and x-ray scattering data are available. The calculated static structure shows very good agreement with experimental measurements, including an asymmetric main peak. The dynamic structure reveals the existence of propagating density fluctuations, and the associated dispersion relation has also been calculated. The possible coupling between longitudinal and transverse excitation modes has been investigated by looking at specific signatures in two wavevector regions: the first one is located around the position of the main peak of the structure factor, q_p, as suggested by the recently reported appearance of high frequency transverse waves in several liquid metals under high pressures; the second region is around $q_p/2$, as suggested by inelastic scattering experiments for liquid Zn and other metals.

1 Introduction.

In the periodic table, zinc is at the end of the transition metals series, presenting a full d shell that contributes in an essential way to its properties. The study of Zn in different states, ranging from small clusters of atoms to the metallic liquid, has shown it to be quite an anomalous system. In small clusters [1] the structure is rather peculiar with a dense core and a separated surface at an expanded distance with respect to the in-core and in-surface interatomic distances. Moreover for some sizes a single atom appears strongly bound to another particular atom via a covalent bond, which coexists with the metallic bonding in the rest of the cluster. When coming to the solid, the ground state is a hexagonal close packed structure, but the ratio between the c and a axis is extremely large, 1.856, around 15 % higher than the ideal value corresponding to hard spheres. Moreover, this value increases with temperature. Such a behaviour is only shared with Cd among single-component solids. The reason lies in an extremely flat energy surface, related to electronic correlation [2]. The liquid structure is also unusual among elemental systems, showing near melting a structure factor with a highly asymmetric main peak [3], a feature only present additionally again in liquid Cd. This asymmetry decreases upon increasing temperature. Finally,

*e-mail: beatriz@metodos.fam.cie.uva.es

concerning its dynamic properties, recent measurements performed using either inelastic x-ray scattering (IXS) at 713 K or inelastic neutron scattering (INS) at 773 K, have concluded that the data is compatible with the existence of two types of propagating excitations in the density fluctuations for wavevectors around $q_p/2$, where q_p denotes the position of the main peak of $S(q)$ [4, 5].

These complexities pose therefore the requirement of a highly accurate description of the interactions present in the system in order to carry out a theoretical study of the properties of liquid Zn (l-Zn). Such requirement is certainly met by ab initio molecular dynamics (AIMD) methods based on density functional theory [6]. Curiously, the only study of l-Zn within AIMD simulations [7] considered the orbital-free version of DFT, which, in spite of allowing large simulation samples and long runs, shows some shortcomings, particularly in treating d-band electrons. Here we present the results of a full AIMD study of the properties of l-Zn, in particular its complex dynamics.

2 Computational method.

The simulated sample consists in a set of N bare ions with valence Z, enclosed in a volume V and interacting with $N_e = NZ$ valence electrons. The total potential energy of the system, with ionic positions $\{\vec{R}_l\}$, is written, within the Born-Oppenheimer approximation, as the sum of the direct ion-ion coulombic interaction energy, $E_{i-i}(\{\vec{R}_l\})$, plus the ground state energy of the electronic system subject to the external potential created by the ions, $V_{\rm ext}(\vec{r}, \{\vec{R}_l\})$,

$$E(\{\vec{R}_l\}) = E_{i-i}(\{\vec{R}_l\}) + E_g[n_g(\vec{r}), V_{\rm ext}(\vec{r}, \{\vec{R}_l\})] \tag{1}$$

where $n_g(\vec{r})$ is the ground state electronic density. According to DFT, the ground state electronic density minimizes the energy functional

$$E[n(\vec{r})] = T_s[n] + E_{\rm ext}[n] + E_H[n] + E_{xc}[n] \tag{2}$$

where the terms represent, respectively, the electronic kinetic energy, $T_s[n]$, of a non-interacting system of density $n(\vec{r})$, the classical electrostatic energy (Hartree term), the exchange-correlation energy, $E_{xc}[n]$, for which we have used the generalized gradient approximation [8] and finally the electron-ion interaction energy, $E_{\rm ext}[n]$.

The ground state density and energy and the forces acting on the atoms are obtained for each ionic configuration, and therefrom the atoms are moved according to Newton's equations of motion, as implemented in the VASP software package [9].

The electron-ion potential has been modeled through an ultrasoft pseudopotential that considers 12 valence electrons, as incorporated in the dataset included in the VASP utility.

3 Results and discussion.

After the 90 Zn atoms, enclosed in a cubic simulation box of side $L = 11.35$ Å, have equilibrated at the desired temperature of 723 K, the system is allowed to evolve for 36000 additional configurations with a timestep of 4 fs, from which the properties of interest are sampled and analyzed.

3.1 Static structure.

The atomic positions are obtained directly from the AIMD run, and the simulation box fixes the wavevectors compatible with the periodic boundary conditions used. Consequently, both the pair distribution function, $g(r)$, and the static structure factor, $S(q)$, can be straighforwardly computed. The results obtained, depicted in figure 1, agree well with all the availabe experimental measurements, obtained either through neutron diffraction [10], or from x-ray diffraction [3, 11], which are remarkably close to each other. In particular, the asymmetric shape of the main peak of $S(q)$ at $q_p \approx 2.92$ Å^{-1} is well reproduced in the simulations.

3.2 Dynamic properties

The collective dynamics of density fluctuations in the liquid is described by the intermediate scattering function, $F(q,t)$, which can be computed, for the q vectors allowed by the simulation box, in terms of the time evolution of the particle coordinates.

The dynamic structure factor, $S(q,\omega)$ is the Fourier Transform (FT) into the frequency domain of $F(q,t)$. The scattered intensity obtained in an IXS experiment is directly related to $S(q,\omega)$, although a proper comparison between the AIMD result and IXS data requires taking into account the difference between the simulations, where atoms are classical particles, and the photon scattering which is a quantum process. Therefore the $S(q,\omega)$ is corrected so as to fulfill the detailed balance condition. Moreover the experimental setup is characterized by an energy resolution, which is taken into account through the convolution of the detail-balanced $S(q,\omega)$ with the experimental resolution function. Additionally, IXS intensity is usually given in arbitrary units, and therefore a proper rescaling is needed in the final AIMD vs IXS comparison. Figure 2 shows the results of such comparison for an experimental $q = 1.06$ Å^{-1}, which lies between two values allowed in the simulations. The central peak obtained from AIMD is lower than the experimental data, but the positions of the side-peaks and the general form of the curve are well reproduced. Particularly, in the region between the quasielastic and inelastic peaks both the IXS data and our results show a behavior somewhat different from the expectations based on a simple sum of a diffusive term to account for the central peak and a propagating mode to account for the side peak. This prompted the experimentalists to suggest the existence of a second propagating mode of energy lying within this region, and showed that a fit based on two propagating modes produces better results than a fit based on just one of them. Specifically, they used a Lorentzian function to account for the central peak and a damped harmonic oscilator (DHO) for each of the propagating modes [4]. The variation with q of the frequencies of the modes obtained from the fits leads to a dispersion relation (DR) with two branches, a high frequency one that displays a positive dispersion with respect to the hydrodynamic sound velocity, and that is similar to the usual DR in liquid systems, plus a lower energy branch, which has been interpreted as a fingerprint of tranverse waves in the dynamics of density fluctuations. The existence of these two modes was first suggested for liquid Ga and later also for other liquid metals (Sn, Fe, Cu, Na) [4, 13], based on the similarity between the frequencies observed and those of transverse excitations either in the corresponding solid or in the liquid. However, at present no theoretical approach has been able to explain their appearance. The DR observed for l-Zn is shown in the inset of figure 2. INS experiments (although using a different model for fitting) also led to similar conclusions and the corresponding DR is shown in figure 2.

 DOI: 10.1051/epjconf/201715103004

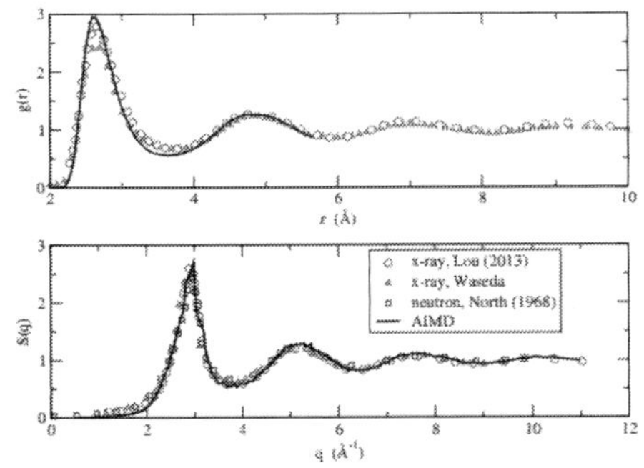

Figure 1. Pair distribution function (up) and static structure factor (down) of l-Zn obtained from AIMD at 723 K (line) compared with several experimental measurements. Circles: x-ray data from Lou *et al* at 710 K [11]; triangles: x-ray data from Waseda at 723 K [3]; squares: neutron data from North *et al* at 723 K [10].

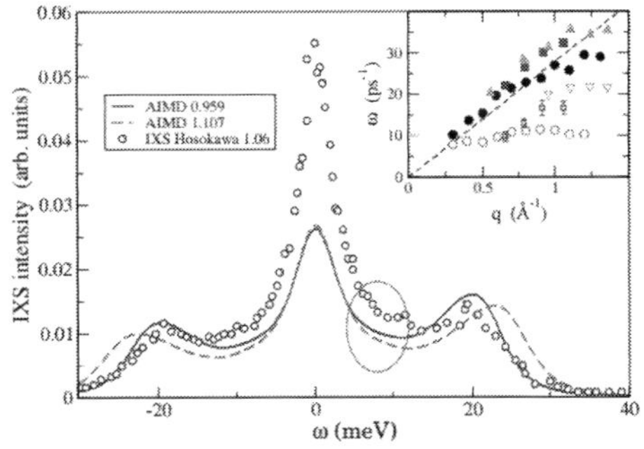

Figure 2. Comparison between AIMD and IXS data for wavevectors (in Å^{-1}) shown in the legend. Circles: experimental results [4]. Lines: AIMD data (modified for proper comparison with experiment, see text). The encircled region corresponds to the anomalous shape associated to the low energy excitation. The inset shows the dispersion relations for $q < q_p/2$ as obtained from INS (circles) [5], IXS (squares) [4], and AIMD (triangles). The dashed line corresponds to the hydrodynamic sound velocity (2780 m/s) [12].

We have analyzed the $F(q,t)$ obtained in our simulations with models similar to those used by experimentalists, slightly modified to cope with the correct time reversal symmetry of the classical $F(q,t)$. The analytical form used for the diffusive term is given by

$$f^{\mathrm{dif}}(t) = \frac{\alpha}{2}\mathrm{sech}(\beta t)\,. \tag{3}$$

For long times it behaves as a decaying exponential, which is the expected asymptotic behavior for a diffusive term, but for $t = 0$ has null odd derivatives, which is the correct behaviour (contrary to how exponentials behave at $t = 0$).

The DHO model corresponds to a sum of two complex-conjugate exponential functions, subject to the condition of null first derivative at $t = 0$. It reads

$$f^{\mathrm{DHO}}(t) = A\,\exp[-\gamma t]\left(\cos(\omega_o t) + \frac{\gamma}{\omega_o}\sin(\omega_o t)\right)\,. \tag{4}$$

This is exactly the form in which a damped harmonic oscillator oscillates for initial position given by A and null initial velocity. It is however important to remember that the oscillation frequency ω_o is not the natural frequency of the oscillator, since it is affected by the damping coefficient γ. The natural frequency is in fact $\Omega_+ = \sqrt{\omega_o^2 + \gamma^2}$. This is therefore the characteristic frequency, or what we call the mode frequency. It is also interesting to consider the quantity $\Omega_- = \sqrt{\omega_o^2 - \gamma^2}$. The reason is that the FT of $f^{\mathrm{DHO}}(t)$, namely $\tilde{f}^{\mathrm{DHO}}(\omega)$, shows a peak at Ω_- whenever this magnitude is real, i.e. if $\gamma < \omega_o$. Additionally it should be noted that $\omega^2\tilde{f}^{\mathrm{DHO}}(\omega)$ always has a peak, located at Ω_+.

Summarizing, by fitting $F(q,t)$ to a sum of a diffusive term and two DHOs we are obtaining the damping coefficients (γ) and the oscillation frequencies corresponding to the two propagating modes (ω_o), and from them we can construct the frequencies where in absence of other terms the dynamic structure factor would show a peak (Ω_-) and the mode frequencies (Ω_+) which coincide with the frequencies where the longitudinal current spectrum (see below) would show a peak in absence of other terms. In fact one should remember that the presence of other terms (the diffusive one and the other propagating mode) will in general modify the positions of the peaks, or may even make them disappear turning them into shoulders or masking them completely if the amplitude is small.

For completeness we mention that in experiments the fit is performed for $S(q,\omega)$ and the parameter obtained from the fit is directly Ω_+.

Coming to the results, we again obtain a better fit to the AIMD calculated $F(q,t)$ when 2 DHOs are used, and their corresponding frequencies are displayed in figure 2. Our data are more in line with the IXS results, but in any case support the existence of two propagating modes in this wavevector region.

We have extended the analysis to higher wavevectors, where $F(q,t)$ starts being dominated by strong structural relaxations that ultimately lead to the de Gennes slowing down near q_p, by looking at the current correlation functions. We want to study the possible existence of two propagating modes in the longitudinal current correlation functions, whose FT, $C_L(q,\omega)$, is directly related to the dynamic structure factor, since $C_L(q,\omega) = \omega^2 S(q,\omega)/q^2$ [note that $C_L(q,\omega=0)$ must therefore be zero]. This relation is a consequence of the continuity equation relating density and current, which in the time domain means that $C_L(q,t) = -(d^2F(q,t)/dt^2)/q^2$. Also the spectra of the transverse current correlation functions, $C_T(q,\omega)$, were analyzed in order to compare with the characteristic frequencies that appear in the longitudinal currents. The transverse current does not fulfill a continuity equation, so $C_T(q,\omega=0)$ can have any non-zero value.

The longitudinal spectra always shows one maximum, whose frequency, $\omega_L(q)$, determines the DR of the longitudinal peaks, shown in figure 3. Even so, the evolution of $\omega_L(q)$ with q displays a quite abrupt change around $q \approx 2$ Å^{-1}, and also around $q \approx 3.4$ Å^{-1}. This suggests the existence of two modes of varying amplitude rather than the rapid variation of the frequency of a single mode. This possibility is further analyzed below. The transverse spectra, on the other hand, behaves in a peculiar way, since in the q-region around q_p two maxima appear. The low energy peak has the usual behaviour for liquids: it first appears at a non-zero value of q, where shear waves start being supported, and then shows a quite small dispersion until eventually the peak returns to zero frequency in the single particle regime of large q. The appearance of the second high frequency peak in $C_T(q,\omega)$ had indeed been observed before, but only for liquids under high pressure [14], so it is rather surprising to find them at ambient pressure as we do here. However, we note that this behaviour is also observed in other metals at ambient pressure, such as liquid Ni [15].

The double peak structure of $C_T(q,\omega)$ near q_p as well as the rapid variation of $\omega_L(q)$ in the two regions mentioned above suggest the existence of two propagating modes both in the longitudinal and in the transverse current correlation functions for $q > q_p/2$. In order to analyze this possibility we have performed fits based on the DHO model for both functions.

The models used for the fits of the longitudinal and transverse currents must be different for the reasons commented above. In order to be consistent with the fit of $F(q,t)$, the model used in the case of $C_L(q,t)$ includes the second derivative of

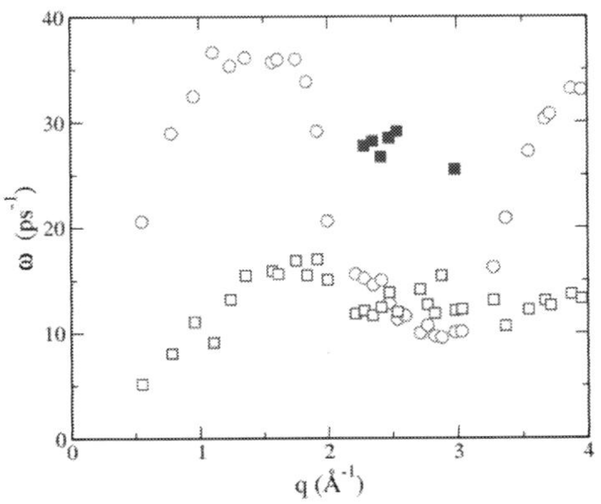

Figure 3. DR of the peaks in the longitudinal (circles) and transverse (squares) spectra.

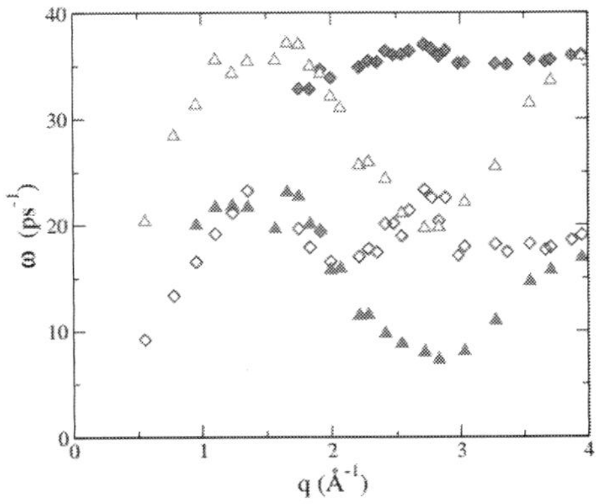

Figure 4. DR of the DHO modes found from the fits to the longitudinal (triangles) and transverse (diamonds) current correlation functions.

the diffusive model used for $F(q,t)$ and two functions that correspond to the second derivative of the DHO model. In the case of $C_T(q,\omega)$ we have found that two DHOs provide a good fit to the calculated functions above approximately $q_p/2$, but for lower q one DHO plus one diffusive mode represents better the data.

The DR of the DHO mode frequencies is shown in figure 4. We want to remark three points regarding the comparison between figures 3 and 4. First, the observed "jumps" in $\omega_L(q)$ in fact correspond to transitions from one of the longitudinal modes to the other, with the amplitude of one mode decreasing and the amplitude of the other mode increasing. Second, there is a large difference between the position of the peaks in $C_T(q,\omega)$, shown in figure 3, and the mode frequencies, shown in figure 4. This is due to the large damping coefficients of the transverse modes, that makes Ω_- (which is near the position of the peak) rather smaller than Ω_+. Finally, we observe that the low frequency longitudinal mode has a numerical value, for $q < 2$ Å^{-1}, very close to the frequency of the transverse mode when only one is present, or to the low frequency transverse mode when there are two. This fact suggests, once more, that the weak low frequency excitations observed in the dynamic structure factor or in the longitudinal current spectra are somehow related to the transverse waves that propagate in the liquid in this range of q.

4 Conclusions.

We have performed the first full AIMD study of the complex dynamics of liquid Zn, recovering the main characteristics of the experimental data, and analyzed the results obtained through the use of models similar to those used experimentally. The analysis unveils the existence of two propagating modes for the longitudinal and the transverse waves, leading near the position of the main peak to two peaks in the transverse current spectra, a feature previously observed only for high pressures. Finally the numerical similarity between the frequencies obtained in the region $q < 2$ Å^{-1} suggests a transverse-like character of the low energy excitations observed in the dynamic structure factor. Nevertheless there is a clear call for a theoretical analysis that clarifies which mechanisms render this possible.

We acknowledge the support of MECD (FIS2014-59279-P) and JCyL (VA104A11-2). BGR acknowledges the financial support of Universidad de Valladolid.

References

[1] A. Aguado *et al*, Angew. Chem. **54**, 2111 (2015).
[2] N. Gaston *et al*, PCCP **12**, 537 (2010).
[3] Y. Waseda, http://res.tagen.tohoku.ac.jp/~waseda/scm/LIQ/all/zn.html
[4] S. Hosokawa *et al*, J. Phys.: Condens. Matter **27**, 194104 (2015).
[5] M. Zanatta *et al*, Phys. Rev. Lett. **114**, 187801 (2015).
[6] P. Hohenberg and W. Kohn, Phys. Rev. **136**, B864 (1964); W. Kohn and L.J. Sham, Phys. Rev. **140**, A1133 (1965).
[7] M. R. Molla *et al*, J. Non-Cryst. Sol. **406**, 45 (2014).
[8] J. P. Perdew, K. Burke, and M. Ernzerhof, Phys. Rev. Lett. **77**, 3865 (1996).
[9] G. Kresse and J. Hafner, Phys. Rev. B **47**, 558(R) (1993); **49**, 14251 (1994); G. Kresse and J. Furthmuller, Comput. Mater. Sci. **6**, 15 (1995); Phys. Rev. B **54**, 11169 (1996).
[10] D. M. North, J. E. Enderby and P. A. Egelstaff, J. Phys. C: Solid State Phys. **1**, 1075 (1968).
[11] H. Lou *et al*, PNAS **110**, 10069 (2013).
[12] T. Iida and R. I. L. Guthrie, *The Thermophysical Properties of Metallic Liquids* (Oxford Univ. Press, 2015)
[13] S. Hosokawa *et al*, Phys. Rev. Lett. **102**, 105502 (2009); V. M. Giordano and G. Monaco, PNAS **107**, 21985 (2010); S. Hosokawa *et al*, J. Phys.: Condens. Matter **25**, 112101 (2013).
[14] T. Bryk *et al*, J. Chem. Phys. **143**, 104502 (2015); M. Marqués, L. E. González and D. J. González, Phys. Rev. B **94**, 024204 (2016); M. Marqués, L. E. González and D. J. González, J. Phys.: Condens. Matter **28**, 075101 (2016).
[15] B. G. del Rio, L. E. González and D. J. González, *these proceedings*.

Molecular dynamics study on glass and molten state of AgI-$AgPO_3$

Shigeki Matsunaga*

National Institute of Technology, Nagaoka College, 940-8532 Nagaoka, Japan

Abstract. Molecular dynamics (MD) simulation on molten and glass state of AgI-$AgPO_3$ have been performed to investigate the structural features and transport properties. In MD, the screened Born-Mayer type potentials including the effect of polarizability of ions have been used. The structure, conductivity, shear viscosity, and Voronoi polyhedron are discussed in relation with the temperature change.

1 Introduction

The superionic conducting glasses have attracted much attention for many years as a promising candidate for the application to solid state batteries or indication devices [1,2]. Above all, a lot of examinations have been performed on the silver iodide – silver phosphate series, $(AgI)_x(AgPO_3)_{1-x}$, because of some advantages. For example, the low glass transition temperature varies from 450K to 350K as x increases [3,4]. The ionic conductivity at room temperature is about 10^{-2} S cm^{-1}, which is comparable to that of liquid electrolytes [3,5].

Neutron scattering experiment and Reverse Monte-Carlo (RMC) analysis have revealed that the large structure change has occurred around the glass transition temperature to form the tetrahedral structure PO_4 on the glass state [6-10]. In the glass state, the infinite random chain of PO_4 is formed, with Ag^+ ions embedded in the PO_4 network. Ag^+ ions hop one Ag-O bond to another. Part of AgI added to $AgPO_3$ form small clusters in the network structure of PO_4. Additional AgI clusters are expected to makes space for Ag^+ ions to move easily.

The molecular dynamic (MD) simulation studies have been performed to simulate the glass transition of AgI based conducting glass. By means of MD, not only the structure of glass state, but also the transport properties of ions can be discussed. For $(AgI)_x(AgPO_3)_{1-x}$ glass, however, no MD simulation is available in literature, to the best of our knowledge. In this study, we wish to show the MD simulation results for $(AgI)_x(AgPO_3)_{1-x}$, x=0.5 in its molten and glass state. The structure and transport properties will be discussed.

2 Simulation Procedure

The basic simulation procedure of MD of molten salt is similar to our previous simulation studies [11-14]. The background of a series of study will be briefly described as follows. The Born-Mayer-Huggins (BMH) type inter-ionic potentials are often used in MD study for AgI

* Corresponding author: matsu@nagaoka-ct.ac.jp

based molten salts, superionic conductor, and glasses [11-17], as well as Rahman, Vashishta and Parrenello (RVP) type potentials [18,19]. However, the former potentials seem more profitable in describing the interaction between AgI and alkali halide, chalcogenide, or pnictogenide. In the ion conducting glasses including AgI, the conduction cation is Ag, on the other hand the other cation and anions form the glass network structure. Under these circumstances, we have proposed the polarised and screened inter-ionic potential model based on BMH type potentials [11,12,13]. The screening parameters have been introduced to take into account the environmental effect by the surrounding ions, and to avoid the dispersion of potentials, *i.e.* the polarisation catastrophe [11,20].

The used screened BMH type potentials in MD are expressed as,

$$\Phi_{ij}(r) = \Phi_{ij}^{0}(r) - \{1 - f_{ij}(r)\}^{2} \frac{P_{ij}}{r^{4}}, \tag{1}$$

$$\Phi_{ij}^{0}(r) = f^{2} z_i z_j \frac{e^2}{r} + A_{ij}\exp(-B_{ij} r) - \{1 - f_{ij}(r)\}^{2} \frac{C_{ij}}{r^{6}}, \tag{2}$$

and

$$P_{ij} = \frac{1}{2}\left(\alpha_i z_j^2 e^2 + \alpha_j z_i^2 e^2\right). \tag{3}$$

In Eq.2, f is a reduction parameter of Coulomb interaction, which represents the screening effect by the surrounding ions. f is assumed to be 0.4. The second term at the right hand side of Eq.2 is a repulsive term attributed to the overlap of electronic charges. P_{ij} is the charge dipole interaction by the polarizability of ions α_i. $f_{ij}(r)$ in Eq.1 and 2 is the short range dumping factor introduced to avoid the dispersion of the potential [11,20]. $f_{ij}(r)$ is expressed as,

$$f_{ij}(r) = \frac{2}{\exp(+k_{ij} r) + \exp(-k_{ij} r)}, \tag{4}$$

where the parameter k_{ij} is given a value of 0.77, then $f_{+-}(r)$ is about 0.2 when the inter ionic distance is around 3.0Å. In addition, the three body potentials are used for O-P-O and P-O-P interactions, which is expressed as,

$$V_{ijk} = K_{ijk}(\theta_{ijk} - \theta_0)^2, \tag{5}$$

where θ_{ijk} is the bond angle for O-P-O and P-O-P [21,22]. The used potential parameters are listed in Table 1 and 2. The MD cell contains 11200 atoms (1600 AgI-$AgPO_3$). The Ewald method is used for the calculation of Coulomb interaction. The numerical integration has been done using Gear algorithm for every 1 fs time steps. The system is at first kept equilibrium at 6000K in NTV, *i.e.* the number of atoms, the temperature, and the volume of the cell are constant. Next, the system is cooled to 2000K at the rate 1K/fs. Further cooling was performed at the rate of 0.1K/fs to 1000K-400K, then to the temperature below the glass transition temperature Tg, 362K [23].

3 Results and Discussion

The partial pair distribution function $g_{ij}(r)$ and the integrated coordination number $n_{ij}(r)$ for the ion pairs *i* and *j* are calculated by MD to analyze the short range configuration of ions.

Table 1. Used pair potential parameters.

ion pair	z_i	z_j	A_{ij} (10^{-19} J)	B_{ij} (Å^{-1})	P_{ij} (10^{-19}J A^4)	C_{ij} (10^{-19}J A^6)
Ag - Ag	1	1	1.120E+03	3.4483	0.000E+00	0.000E+00
Ag - I	1	-1	5.633E+03	3.4483	7.146E+01	0.000E+00
Ag - P	1	2.6	8.731E+02	3.4483	3.107E+01	0.000E+00
Ag - O	1	-1.2	2.236E+03	3.4483	2.306E+01	0.000E+00
I - I	-1	-1	2.728E+04	3.4483	1.429E+02	4.825E+02
I - P	-1	2.6	4.844E+03	3.4483	5.141E+02	2.108E+02
I - O	-1	-1.2	1.076E+04	3.4483	1.260E+02	1.761E+02
P - P	2.6	2.6	4.909E+02	3.4483	4.200E+02	9.208E+01
P - O	2.6	-1.2	1.953E+03	3.4483	2.000E+02	7.596E+01
O - O	-1.2	-1.2	4.237E+03	3.4483	6.640E+01	6.542E+01

Table 2. Used three body potential parameters.

ion trio	K_{ijk}(10^{-19} J)	θ_0(deg)
O-P-O	2.80382	109.47
P-O-P	2.40327	135.50

All $g_{ij}(r)$'s are shown in Fig.1 as the examples of molten state at 1000K and glass state at 280K, respectively. Large first peaks are seen in $g_{IAg}(r)$, $g_{OAg}(r)$ and $g_{PO}(r)$, which show significant correlations between I and Ag, O and Ag, and P and O ions. The first peaks of $g_{ij}(r)$'s at 280K are enhanced. Small shoulders are also seen in the larger r side of the first and second peaks at 280K, which is attributed to the fine structure in glass state.

In order to examine the detailed correlation between Ag and anions, the temperature dependence of $g_{IAg}(r)$ and $n_{IAg}(r)$ (left), and $g_{OAg}(r)$ and $n_{OAg}(r)$ (right) are shown in Fig.2, respectively. In both figures, the differences between molten state 1000K-400K and glass state 325K and 280K are observed especially in the first and second peak hights.

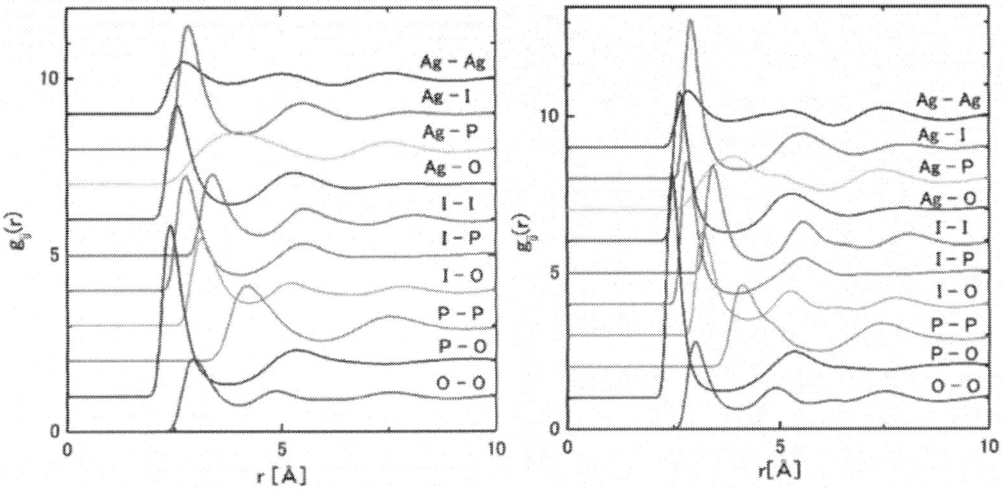

Fig.1. $g_{ij}(r)$'s of AgI-$AgPO_3$ melt at 1000K (left) ,and glass at 280K (right).

 DOI: 10.1051/epjconf/201715103005

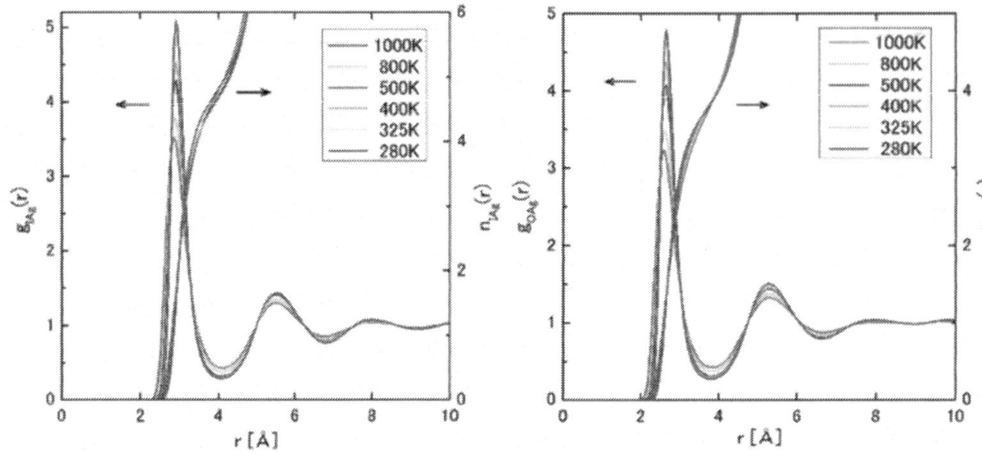

Fig.2. $g_{IAg}(r)$ and $n_{IAg}(r)$ (left), and $g_{OAg}(r)$ and $n_{OAg}(r)$ (right) for molten state 1000K-400K, and glass state 325K and 280K, respectively.

The temperature dependence in molten phase 1000K-400K is quite large, however, the structure change in glass state is rather small. The slight increase of $n_{IAg}(r)$ at the distance of the first minimum is obseved, which may be corresponding to the decreasing tendency of the conductivity as temperature decrease. We will discuss this point later in this section.

The Voronoi polyhedron is used to analyse the structure change by glass formation. When many ions are dispersed in space, you can split the line between adjacent ions by the vertical bisection to two equal parts, then you can sepalate the whole space into Voronoi polyhedron. Therefore, the Voronoi polyhedron is a sort of an extension of the Wigner-Seitz cell for a regular crystal to the amorphous state [25]. Using the present MD results, the destribution of number of vertices of polygons, and faces surrounding Voronoi polyhedra are obtained, which are shown in Fig.3. The most superiour number of vertices is 5 in both molten and glass state, which is one of the features of the disordered system. The most superior number of faces is 13 in melt at 1000K, then the distribution sifts to larger number of faces in glass at 280K. These structure changes also suggests the glass transition of the system.

Next, we will discuss the transport properties. The conductivity σ is connected with the diffusion coefficient D_i by the Nernst-Einstein (N-E) relation [24], as,

$$\sigma = \sum_i \frac{\rho_i z_i^2 e^2 D_i}{k_B T} \tag{6}$$

where ρ_i is the number density of the i-th ion species. The diffusion coefficient of ions are obtainable from the slope of the mean square displacement (MSD), as,

$$D_\alpha = \lim_{t\to\infty} \frac{r_\alpha^2(t)}{6t} = \lim_{t\to\infty} \frac{1}{6tN_\alpha} \sum_{i=1}^{N_\alpha} \langle |\boldsymbol{r}_i(t) - \boldsymbol{r}_i(0)|^2 \rangle_\alpha \tag{7}$$

where i stands for the ith α-type ion, and N_α represents its number, respectively. The angular brackets mean the average over all atomic position $\boldsymbol{r}_i(t)$ and the time average, *i.e.* the ensemble average.

 DOI: 10.1051/epjconf/201715103005

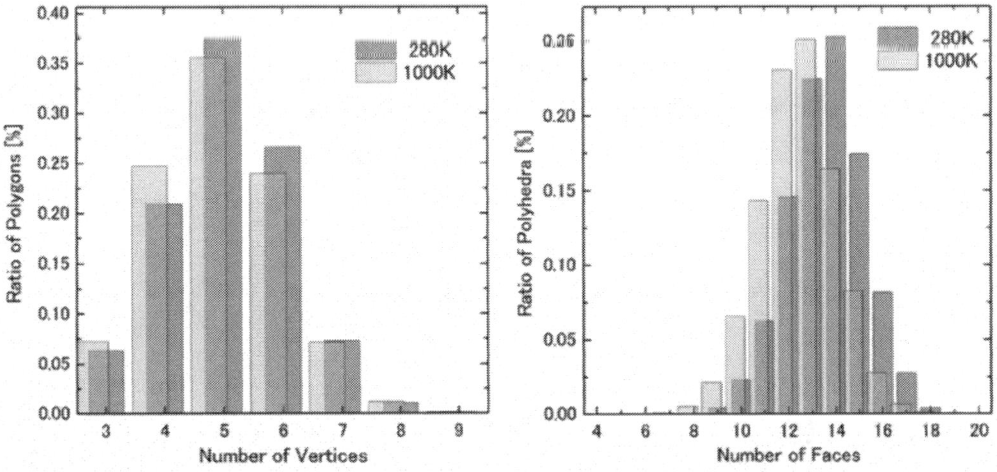

Fig. 3. Number of vertices of polygons (left) and number of faces of polyhedra (right).

In order to compare the MD results with the experimental conductivity data, the total conductivity σ in molten phase is estimated by the diffusion coefficients of all constituent ions. In glass phase, σ is obtained from the diffusion coefficient of Ag. The obtained σ's are shown in Fig.4 in comparison with the experimental data. Although there are some discrepancies between MD results and the experimental data in Fig.4 (left), the temperature dependence of σ is reproduced in both molten and glass phases to a certain extent. A large gap of σ is seen around the glass transition temperature. Although screening and reduction parameters are used in the pair potentials, rather smaller value of σ comparing to the experimental data may be attributed to some overestimation of interaction between ions.

According to the Green-Kubo (G-K) formula, the shear viscosity η is represented by the integration of the autocorrelation function of an off-diagonal element of the stress tensor in the long wave length limit, as [24],

$$\eta = \frac{1}{k_B T V} \int_0^{\infty} \lim_{k \to 0} \langle \sigma_{\boldsymbol{k}}^{xz}(t) \sigma_{-\boldsymbol{k}}^{xz}(0) \rangle \, dt \, . \tag{8}$$

The MD results of η at various temperatures in molten phase are shown in Fig.4 (right) alongside the experimental data. Although the MD results are somewhat higher than the experimental data, the temperature dependence of η by MD agrees well with the experimental data. This fact also suggests the rather larger interaction between ions.

The frequency dependent diffusion coefficient, $D_i(\omega)$, can be calculated from the velocity auto-correlation function (VAF), $\langle \boldsymbol{v}_i(t) \cdot \boldsymbol{v}_i(0) \rangle$, as,

$$D_i(\omega) = \frac{1}{3} \int_0^{\infty} \langle \boldsymbol{v}_i(t) \cdot \boldsymbol{v}_i(0) \rangle \, cos\omega t \; dt. \tag{9}$$

$D_{Ag}(\omega)$'s for Ag ion obtained by MD at various temperatures are show in Fig.5. The peak heights are lowered as the temperature decreases. Although the distributions of $D_{Ag}(\omega)$'s spread over wide frequency range, the peak positions distribute around 1100 1/cm at higher temperatures, then slightly shift toward 1200 1/cm as the temperature decreases. According

DOI: 10.1051/epjconf/201715103005

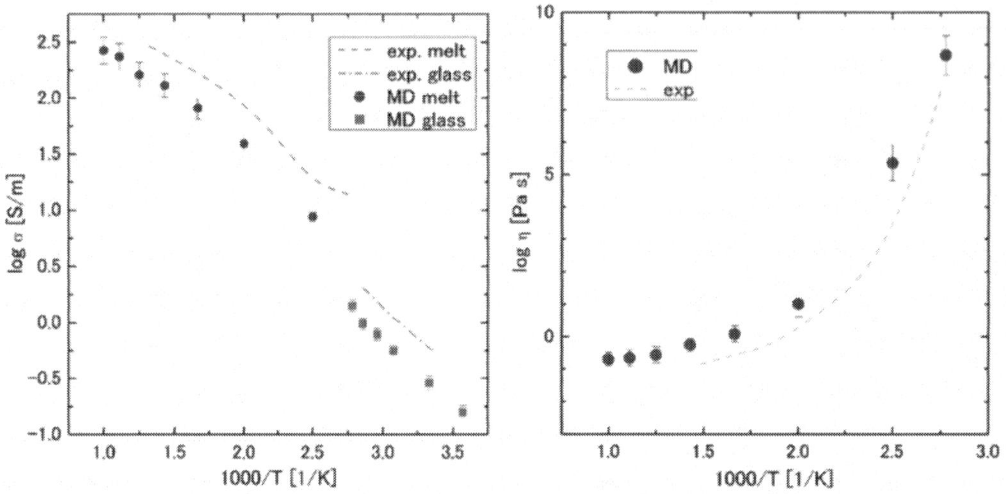

Fig. 4. log σ (left) and log η (right) obtained by MD alongside the experimental data.

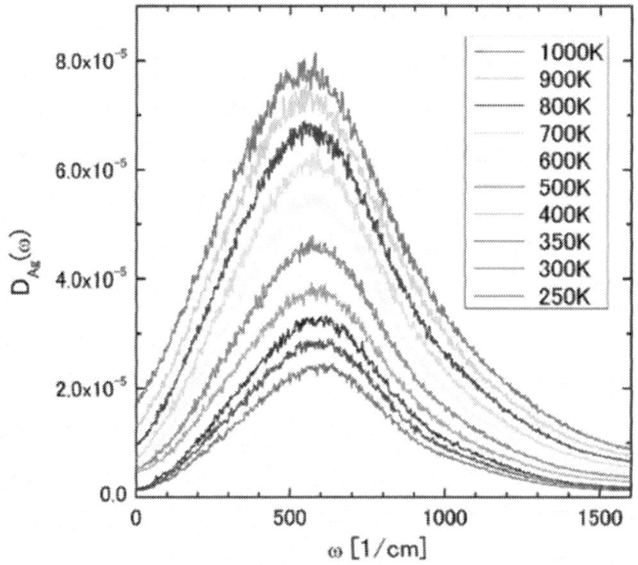

Fig. 5. $D_{Ag}(\omega)$ for molten state 1000K - 400K, and glass state 350K – 250K.

to the Raman spectra experiment [26,27], the peak positions are very close to the P-O symmetric vibration mode of a long $(P\text{-}O\text{-}P)_n$ chain 1140 1/cm, which is obviously different from the modes of Ag oscillations inside of O cage (85 1/cm) nor in Ag-I bonding (130 - 120 1/cm). This fact suggests the following scenario stands to a certain extent. As AgI is doped to $AgPO_3$, then part of Ag is trapped in P-O bond to form a temporal P-O-Ag bond. The diffusion of Ag occurs by hopping from one P-O-Ag bonding to another. In other words, most of Ag is vibrating in the mode of P-O-P long chain. These vibrations may affect Ag ions trapped in AgI clusters in the polymeric chains of PO_4 tetrahedral units.

4 Conclusion

MD simulations have been performed for AgI-$AgPO_3$ in molten and glass state. The structural difference between molten and glass stated have been examined by $g_{ij}(r)$ and Voronoi polyhedron analysis. The conductivity and shear viscosity have been obtained by N-E relation and G-K formula, respectively. The temperature dependence of MD results agree well with the experimental data. The distribution of $D_{Ag}(\omega)$ suggests that Ag ions form the P-O-Ag bond.

Acknowledgements

The author expresses his thanks to Professor S. Tamaki for his fruitful comments and encouragement on this study. The computation was mainly carried out using the super computer facilities at Research Institute for Information Technology, Kyushu University. This work was supported by JSPS KAKENHI Grant Number JP15K05136.

References

1. M.Z.A. Munshi (ed.), *Handbook of Solid State Batteries and Capacitors* (World Scientific, 1995)
2. K. J. Rao, *Structural Chemistry of Glasses* (Elsevier, Amsterdam, 2002)
3. J.P. Malugani, A. Wasniewski, M. Doreau, G. Robert, A. Al Rikabi, Mat. Res. Bull. **13**, 427 (1978)
4. C. Tomasi, P. Mustarelli, A. Magistris, Ma. P.I. Garcia, J. Non-Cryst. Sol. **293-295**, 785 (2001)
5. K. Pathmanathan, R. Mlcak, P.G. Johari, Phys. Chem. Glass. **30**, 180 (1989)
6. M. Tachez, R. Mercier, J.P. Malugani, P. Chieux, Solid State Ionics **25**, 263 (1987)
7. J.D. Wicks, L. Börjesson, G. Bushnell-Wye, W. S. Howells, and R. L. McGreevy, Phys. Rev. Lett. **74**, 726 (1995)
8. J. Swenson, S. Adams, Phys. Rev. B **64**, 024204 (2001)
9. A. Thazin, Y. Fujishima, M. Arai, T. Sakuma, H. Takahashi, Solid State Ionics, **175**, 675 (2004)
10. R. Magaraggia, S. Kennedy, T. Tedesco, E. Kartini, M. Collins, K. Itoh, Physica B: Condens. Matter, **385**, 263 (2006)
11. S. Matsunaga, S. Tamaki, EPI Web of Conferences **15**, 02010 (2011)
12. S. Matsunaga, Molecular Simulation **39**, 119 (2013)
13. S. Matsunaga, Physics and Chemistry of Liquids **51**, 414 (2013)
14. S. Matsunaga, Ionics **21**, 161 (2015)
15. M.C. Abramo, G. Pizzimenti, A. Consolo, Phil. Mag. B **64**, 495 (1991)
16. M. C. Abramo, C. Caccamo, G. Pizzimenti, J. Phys.: Condens. Matter **5** , 397 (1993)
17. A. Karthikeyan and K.J. Rao, J. Phys. Chem. B **101**, 3105 (1997)
18. P. Vashishta and A. Rahman, Phys. Rev. Lett. **40**, 1337, (1978)
19. M. Parrinello, A. Rahman, and P. Vashishta, Phys. Rev. Lett. **50**, 1073 (1983)
20. V. Bitrián, J. Trullàs, J. Phys. Chem. B, 110, 7490, (2006)
21. E.B. Clark, R.N. Mead, G. Mountjoy, J. Phys.: Conden. Matter, **18**, 6815 (2006)
22. J.-J. Liang, R.T. Cygan, T.M. Alam, J. Non-Cryst. Sol. 263–264, 167, (2000)
23. H. Kobayashi, Y. Hiki, H. Takahashi, J. Appl. Phys. **80**, 122 (1996)
24. J.P. Hansen, I.R. McDonald, *Theory of Simple Liquids* (Academic Press, London, UK, 1986)

25. N.F. Cusack, *The Physics of Structurally Disordered Matter* (Adam Hilger, Bristol, 1987)
26. E.I. Kamitsos, J.A. Kapoutsis, G.D. Chryssikos, J.M. Hutchinson, A.J. Pappin, M.D. Ingram, J.A. Duffy, Phys. Chem. Glasses **36**, 141 (1995)
27. D.I. Novita, P. Boolchand, M. Malki, M. Micoulaut, J. Phys.: Condens. Matter **21**, 205106 (2009)

Structure and Ionic Diffusion in Molten NaI, RbI, and NaI-RbI mixture

Shuta Tahara[1,2,*], *Koji* Ohara[2], *Hironori* Shimakura[3], and *Takanori* Fukami[1]

[1]Department of Physics and Earth Sciences, Faculty of Science, University of the Ryukyus, Okinawa 903-0213, Japan
[2]Research and Utilization Division, Japan Synchrotron Radiation Research Institute (JASRI, SPring-8), Hyogo 679-5198, Japan
[3]Niigata University of Pharmacy and Applied Life Sciences, Niigata 956-8603, Japan

Abstract. Static structure factors of molten NaI, RbI, and their mixture of $(RbI)_{0.3}(NaI)_{0.7}$ are measured up to high-Q region by using the high-energy X-ray diffraction technique. Moreover, molecular dynamics (MD) simulations are carried out, and the simulation results well reproduce the diffraction data. The partial structure factors, partial pair distribution functions, and ionic diffusion coefficients calculated by the MD simulations are reported in detail. The mixing effects of cations on the structure and ionic diffusion are also discussed.

1 Introduction

Both experimental and theoretical approaches are important to investigate the physical properties of the liquid materials including molten salts. For example, the structure of molten NaCl has been studied by using the neutron diffraction measurements with isotopic substitution technique by Edwards et al [1], and their partial structure factors and partial pair distribution functions have been well reproduced by the molecular dynamics (MD) simulation by Dixon and Sangster [2], and Ciccotti et al. [3]. The ionic transport properties have also been discussed on the basis of the simulation results.

NaI is one of the simple salts, however the experimental structure-data for the molten phase is rare. Specifically, McGreevy et al. have measured the dynamic structure factor $S(Q,\omega)$ by using the inelastic neutron scattering technique in 1984. They have reported the broadening of the quasi-elastic peak associated with diffusion together with an extended inelastic spectrum associated with corrective motion [4]. They have also estimated the total static structure factor $S(Q)$ for molten NaI up to approximately $Q = 2$ Å^{-1} from the analysis of $S(Q,\omega)$, and they pointed out that the simulation results in the low-Q region by Dixon and Sangster show the good agreement with the experimental $S(Q)$ [4]. Recently, Alcaraz et al. have also carried out the MD simulations with new models, and their simulation have well reproduced the experimental $S(Q)$ obtained by McGreevy et al. in the low-Q region [5,6]. Although McGreevy's experimental $S(Q)$ for molten NaI is very important and pioneering, an experimental $S(Q)$ up to relative high-Q is needed to investigate deeply the physical properties, and to confirm the validity of the theoretical model in a wide Q-range.

* Corresponding author: tahara@sci.u-ryukyu.ac.jp

In this paper, we report *S(Q)* for molten NaI, RbI measured by high energy X-ray diffraction (HEXRD) technique up to Q = 15 $Å^{-1}$. Moreover, we carry out the MD simulation to compare with the experimental data with the simple model and parameters suggested by Ciccotti et al. [3], and the information on the partial structures and ionic transport properties are obtained. The *S(Q)* for the mixture melt of $(RbI)_{0.3}(NaI)_{0.7}$ is also obtained by HEXRD experiment and MD simulation to investigate the mixture effects of different cation on the structure and ionic transport.

2 Methods

2.1 High-energy X-ray diffraction measurement

Each powder sample of NaI, RbI and their mixture of $(RbI)_{0.3}(NaI)_{0.7}$ sealed in a quartz tube under vacuum was once melted at 1023K, and then cooled to ambient temperature to form a bulk. The bulk samples were sealed in a fused silica capillary with 2.4 mm inner diameter and 0.25 mm wall thickness. The HEXRD experiments for molten states of these samples were carried out at 973K using a two-axis diffractometer installed at the BL04B2 beamline [7] at SPring-8. The incident X-ray beam from a Si(111) monochromator with the third harmonic reflection was 113.4 keV in energy and 0.1093 Å in wavelength. The measured X-ray diffraction data were corrected for polarization, absorption, and background contribution from the capillary and the instrument, and then the contribution of Compton scattering was also subtracted. The corrected data sets were normalized to give the Faber-Ziman [8] structure factor *S(Q)*.

2.2 Molecular dynamics simulation

MD simulations were carried out for molten NaI, RbI and $(RbI)_{0.3}(NaI)_{0.7}$ at the same temperature (973K) as the experiments. In this study, we used the Born-Mayer potential which is often used for simple alkali halide melts expressed as

$$\phi_{ij} = z_i z_j e^2/r + B_{ij}\exp(-\gamma_{ij}r) - C_{ij}/r^6 - D_{ij}/r^8. \qquad (1)$$

The first term on the right-hand side is Coulomb interaction between ionic charges where z_{Na}, z_{Rb} and z_I were set as +1, +1 and -1, respectively. The second term is a repulsion between the ions arising from the overlap of the outer shell of electrons. The B_{ij} in the second term is expressed as $B_{ij} = \beta_{ij} b\exp[\gamma_{ij}(\sigma_i+\sigma_j)]$, where σ denotes ionic radius. The third and forth terms are van der Waals interactions. The potential parameters for pure NaI and RbI were taken from the earlier work [3]. For molten $(RbI)_{0.3}(NaI)_{0.7}$, the parameters were estimated from those for pure melts as listed in Table 1.

	B_{ij}	C_{ij}	D_{ij}
Na-Na	7.862	0.073	0.035
Na-Rb	37.578	0.433	0.351
Na-I	49.549	1.128	1.303
Rb-Rb	225.488	2.575	3.554
Rb-I	282.196	6.705	13.094
I-I	317.685	17.459	48.980

The potential parameters for the molten mixture were estimated as follows. For the repulsion interaction term, the Pauling factors were $\beta_{++} = 1.25$, $\beta_{+-} = 1$ and $\beta_{--} = 0.75$, and the strength parameter b was the same value of 0.211 eV for all interactions as written in Ref. 3. The γ_{ij} for the mixture was estimated from $\gamma_{ij} = 1 / (\varrho_i + \varrho_j)$, which is the same manner in Ref. 9. The C_{II} and D_{II} for the mixture were linearly interpolated between pure NaI and RbI. The C_{ij} and D_{ij} for unlike-ion pairs were estimated by geometric mean of those for like-ion interactions, which is similar manner to the Lorentz-Berthelot combining rules for the Lennard-Jones potential as written in Refs. 10 and 11. Since any experimental number density of the NaI-RbI mixture has not been reported, we attempted to estimate the number density of the mixture melt with several methods; (a) linear interpolation of the number densities of the pure melts (0.0200 [1/Å^3]) where the number densities of pure molten NaI and RbI are 0.0217 Å^{-3} and 0.0161 Å^{-3}, respectively [12], (b) linear interpolation of the molar volume of the pure melts (0.0196 [1/Å^3]), (c) the inversion from experimental $S(Q)$ to $g(r)$ (0.0200±0.0005[1/Å^3]). In this case, the number density estimated by the method (c) is almost the same as that estimated by the method (a). Therefore, we used 0.0200 Å^{-3} in this simulation. The simulation step was set as 5 fs and 1000 ionic particles (N_{Na} : 350, N_{Rb} : 150, N_I : 500) were placed in the simulation cell. The quantities of interest were obtained by taking the average over 20000 time steps which corresponds to 100 ps.

3 Results and discussion

3.1 Static structure

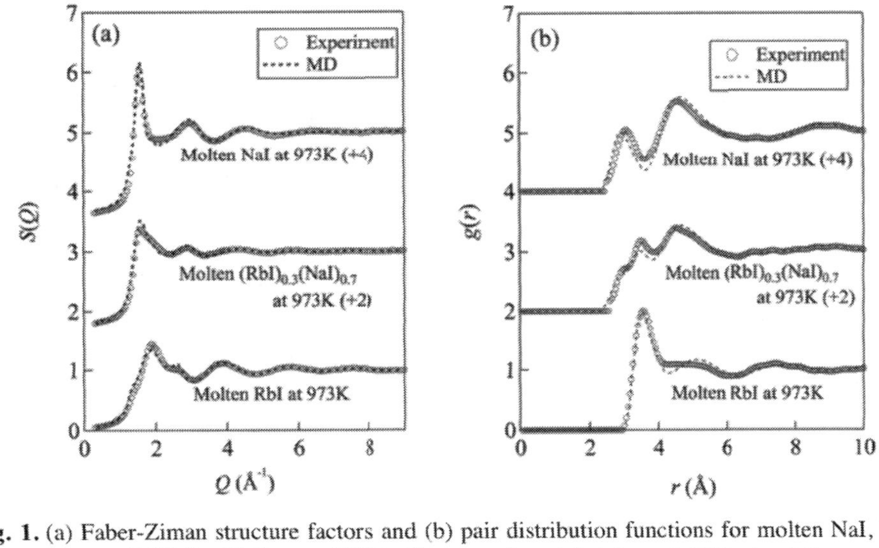

Fig. 1. (a) Faber-Ziman structure factors and (b) pair distribution functions for molten NaI, RbI and their mixture of $(RbI)_{0.3}(NaI)_{0.7}$ at 973K. The experimental data and MD simulation results are denoted by open circles and dashed curves, respectively.

The Faber-Ziman $S(Q)$ for molten NaI, RbI and $(RbI)_{0.3}(NaI)_{0.7}$ obtained by HEXRD are exhibited in Fig. 1(a). Although the $S(Q)$ were measured up to 15 Å^{-1}, the $S(Q)$ are basically unity for $Q > 9$ Å^{-1}. The first, second and third peaks for molten NaI are observed at around $Q = 1.5$, 2.9 and 4.5 Å^{-1}, respectively. The first peak of molten NaI was observed as a small pre-peak in $S(Q)$ estimated by the inelastic neutron scattering [4]. The differences in peak heights between HEXRD and neutron scattering are due to the difference in weighting factor of partial correlations between X-ray and neutron scatterings. The first peak in $S(Q)$ for molten RbI is observed at around Q = 1.9 Å^{-1} with a hump (Q = 1.4 Å^{-1}) and a shoulder (Q = 2.6 Å^{-1}). The second and third peaks are observed at around Q = 3.9 and 5.8 Å^{-1}, respectively. While the peak positions in $S(Q)$ for $(RbI)_{0.3}(NaI)_{0.7}$ are similar to those for pure NaI because of the high concentration of NaI, the peak widths of the first and third peaks are broader than those for pure molten NaI. These peak broadenings would be caused by the differences in peak positions between pure NaI and RbI. For example, the shoulder-like feature of the first peak for $(RbI)_{0.3}(NaI)_{0.7}$ at around Q = 2 Å^{-1} corresponds to the first peak position of pure RbI. The MD results well reproduce the experimental data as shown in Fig. 1(a), which suggests that the potentials and parameters used in this study are reasonable.

Figure 1(b) shows pair distribution functions $g(r)$ for molten NaI, RbI, and $(RbI)_{0.3}(NaI)_{0.7}$ at 973K together with the results of MD simulations. These $g(r)$ were obtained by Fourier transform to the $S(Q)$. The first and second peaks in $g(r)$ for molten NaI are observed at around r = 3.1 and 4.6 Å, respectively. For RbI, the intense first peak and the vague second peak are observed at around r = 3.6 and 5 Å, respectively. We refer to the positions of the first and second peaks in $g(r)$ as r_1 and r_2 in the following discussions, respectively. Since r_1 corresponds to the first neighboring distance of unlike-ion pairs [3], the difference between r_1^{NaI} and r_1^{RbI} would reflect the difference in ionic radii between Na and Rb. In molten $(RbI)_{0.3}(NaI)_{0.7}$, two peaks are observed at around r = 3.0 and 3.5 Å which correspond to the first neighboring distances of Na-I (r_1^{NaI}) and Rb-I (r_1^{RbI}) correlations, respectively. The peak for molten $(RbI)_{0.3}(NaI)_{0.7}$ at around r = 4.5 Å corresponds to r_2^{NaI} which is mainly contributed by the first neighboring distance of Na-Na and I-I pairs [3]. These results suggest that the effects of mixing of two pure melts on the first neighboring distance of each correlation are small. The MD results are also shown in Fig. 1(b) and well reproduce the experimental $g(r)$.

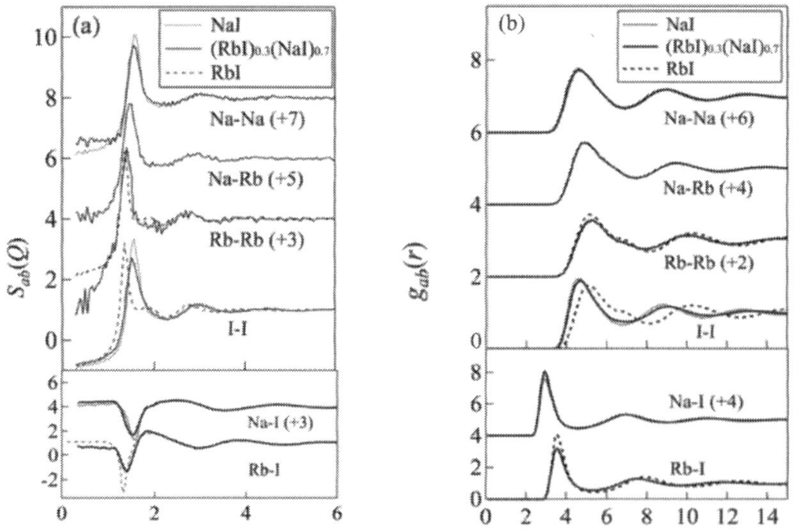

Fig. 2. (a) Faber-Ziman partial structure factors and (b) partial pair distribution functions for molten NaI (dotted curve), RbI (dashed curve) and their mixture of $(RbI)_{0.3}(NaI)_{0.7}$ (solid line) at 973K.

The Faber-Ziman partial structure factors $S_{ab}(Q)$ obtained by the MD simulations for molten NaI, RbI and $(RbI)_{0.3}(NaI)_{0.7}$ are shown in Fig. 2(a). Overall features of $S_{NaRb}(Q)$ is similar to $S_{NaNa}(Q)$ and $S_{RbRb}(Q)$, because Na and Rb ions are largely affected by the repulsion force between their positive charges. The first peak position of $S_{II}(Q)$ for the mixture is observed at between those for the pure melts, suggesting that the dense random packing of I ions are easily affected by the variety of density.

Figure 2(b) shows the partial pair distribution functions $g_{ab}(r)$ obtained by MD simulations for molten NaI, RbI and $(RbI)_{0.3}(NaI)_{0.7}$. The $g_{NaNa}(r)$ for pure NaI and the mixture coincide with each other, demonstrating that Na-Na correlation is not affected by Rb ions. On the other hand, Rb-Rb correlations show a slight difference between pure RbI and the mixture. For example, the shoulder observed at around r = 6.8 Å in $g_{RbRb}(r)$ for pure RbI becomes vague in the mixture. A slight shift of the second peak in $g_{RbRb}(r)$ from r = 10.1 Å for pure melt to 9.8 Å for the mixture is also observed, which would be affected by difference in the density between the pure RbI and the mixture. Each peak position for Na-Rb correlation for the mixture corresponds to the average of the peak positions for $g_{NaNa}(r)$ and $g_{RbRb}(r)$, for example the first peaks in $g_{NaNa}(r)$, $g_{NaRb}(r)$ and $g_{RbRb}(r)$ are observed at r = 4.5, 4.8 and 5.1 Å, respectively. The $g_{II}(r)$ is the most affected by the difference in densities among these melts. The first neighboring distances of unlike-chage pairs Na-I and Rb-I are 3 and 3.5 Å, respectively, which are also affected by the difference in ionic radii of Na and Rb.

3.2 Ionic transport

The diffusion coefficients D_α (10^{-5} cm^2/s) (α = Na, Rb, I) and ionic conductivities σ calculated by the MD simulations for molten NaI, RbI and their mixture at 973 K are summarized in Table 2 with experimental literature values for pure NaI [12] and RbI [13]. Ionic conductivities of the MD simulations are estimated by using Nernst-Einstein equation expressed as

$$\sigma = e^2 \left(\Sigma \rho_\alpha D_\alpha\right) / k_B T$$

where e, k_B, and ρ_α denote elemental charge, Boltzmann constant, and partial number density, respectively. The experimental ionic conductivities and diffusion coefficients are reasonably reproduced by MD simulations. Since ionic conductivities and diffusion coefficients for the mixture have not been measured, the MD results for the mixture cannot be compared to experimental data at this time. Although we used the rigid ion model in this study, if we consider a polarizable ion model, the cation diffusion would become large. However, polarization effects on NaI, RbI, and their mixture would be weak compared to silver (or cupper) halides [13-20] and zinc halides [21,22].

Ciccotti et al. have reported that D_{Na} and D_I for pure NaI at 1081K are 9.4 × 10^{-5} and 6.8 × 10^{-5} cm^2/s, respectively [3]. Therefore, the D_{Na} and D_I for pure NaI at 973K in this study are decreased by approximately 26% and 22% compared to the results at 1081K because of the decrease of the temperature. The D_{Rb} and D_I for pure RbI at 973K are also decreased by approximately 44% and 31% in comparison with the D_{Rb} (4.3 × 10^{-5} cm^2/s) and D_I (3.5 × 10^{-5} cm^2/s) at 1081K, respectively [3]. The D_{Rb} for the mixture is 2.25 times to that for pure RbI, while D_{Na} for the mixture is decreased by approximately 19% compared to that for pure NaI at the same temperature of 973K.

From the comparisons between the pure melts and the mixture melt, Na ions in the mixture diffuse slower than Na ions in pure NaI, while Rb ions in the mixture diffuse faster than Rb ions in pure RbI. These results suggest that the diffusion of Na ions would be disturbed by slow Rb ions, and the diffusion of Rb ions would be promoted by fast Na ions because of the repulsive interaction between Na and Rb ions, and resultant D_{Na} and D_I for the mixture would show similar values each other. Since the D_I for the mixture is twice as large as that for pure RbI, diffusion of I ions might also be affected by the attractive interaction from fast Na ions. Although the mixture effects of cations on Na-Na and Rb-Rb partial structures are very small, the effects on the ionic diffusion are large.

Table 2. Diffusion coefficients D_α (10^{-5}cm^2/s), for Na, Rb and I ions in molten NaI, RbI and $(RbI)_{0.3}(NaI)_{0.7}$ at 973K. Parentheses denote experimental literature values.

	NaI	RbI	$(RbI)_{0.3}(NaI)_{0.7}$
D_{Na} (10^{-5}cm^2/s)	7.0 (7.35 [12])		5.7
D_{Rb} (10^{-5}cm^2/s)		2.4	5.4
D_I (10^{-5}cm^2/s)	5.3 (4.05 [12])	2.4	4.5
σ (S/cm)	2.55 (2.38 [12])	0.74 (0.84 [13])	1.93

4 Summary

We have measured the $S(Q)$ for molten NaI, RbI, and their mixture of $(RbI)_{0.3}(NaI)_{0.7}$ up to Q = 15 Å^{-1}. These $S(Q)$ have been well reproduced by the MD simulations with the simple Born-Mayer potentials and parameters reported by the literature [3]. The $S_{ab}(Q)$ and $g_{ab}(r)$ have been obtained by the MD simulations. The structural feature for Na-Rb correlation in the mixture is similar to those for Na-Na and Rb-Rb correlations. The $S_{II}(Q)$ and $g_{II}(r)$ are largely depend on the materials, suggesting that the dense random packing of I ions are easily affected by the variety of density, which is the mixing effect of the cations on the structure. The D_α for the pure and mixture melts have also been calculated by the MD simulations. The D_α for the pure melts at 973 K in this study shows the decrease compared to that at 1081 K reported by the earlier work [3]. The D_α of all the ionic species in the mixture is different from that in the pure melts, which is the mixing effect of the cations on the ionic diffusions. While the mixture effects of cations on Na-Na and Rb-Rb partial structures are very small, the effects on the ionic diffusion are large.

The synchrotron radiation experiments were performed at the BL04B2 at SPring-8 with the approval of the Japan Synchrotron Radiation Research Institute (JASRI, proposal 2011B1531).

References

1. F. G. Edwards, J. E. Enderby, R. A. Howe, D. I. Page, J. Phys. C: Solid State Phys. **8**, 3483 (1975)
2. M. Dixon, M. J. L. Sangster, J. Phys. C: Solid State Phys. **9**, 909 (1976)
3. G. Ciccotti, G. Jacucci, I. R. McDonald, Phys. Rev. A **13**, 426 (1976)
4. R. L. McGreevy, E. W. J. Mitchell, F. M. A. Margaca, J. Phys. C: Solid State Phys. **17**, 775 (1984)
5. O. Alcaraz, J. Trullàs, J. Mol. Liq. **136**, 227 (2007)
6. O. Alcaraz, V. Bitrián, J. Trullàs, J. Chem. Phys. **127,** 154508 (2007)
7. S. Kohara, M. Itou, K. Suzuya, Y. Inamura, Y. Sakurai, Y. Ohishi, M. Takata, J. Phys.: Condens. Matter **19**, 506101 (2007)
8. T. E. Faber, J. M. Ziman, Phil. Mag. **11**, 153 (1965)

9. R. Takagi, F. Hutchinson, P. Madden, A. K. Adya, M. Gaune-Escard, J. Phys.: Condens. Matter **11**, 645 (1999)
10. H. A. Lorentz, Ann. Phys. **12**, 127 (1881)
11. D. C. R. Berthelot, Hebd. Seanc. Acad. Sci. **126**, 1703 (1898)
12. G. J. Janz, *Molten salts handbook* (Academic Press, New York, 1967)
13. S. Tahara, S. Ohno, T. Okada, Y. Kawakita, S. Takeda, EPJ Web Conf. **15**, 02005 (2011)
14. J. Trullàs, O. Alcaraz, L. E. González, M. Silbert, J. Phys. Chem. B **107**, 282 (2003)
15. V. Bitrián, J. Trullàs, J. Phys. Chem. B **110**, 7490 (2006)
16. V. Bitrián, J. Trullàs, M. Silbert, T. Enosaki, Y. Kawakita, S. Takeda, J. Chem. Phys. **125**, 184510 (2006)
17. V. Bitrián, J. Trullàs, J. Phys. Chem. B **112**, 1718 (2008)
18. V. Bitrián, O. Alcaraz, J. Trullàs, J. Chem. Phys. **134**, 044501 (2011)
19. S. Tahara, T. Fukami, Journal of the Physical Society of Japan **84**, 024602 (2015)
20. S. Tahara, Y. Kawakita, H. Shimakura, K. Ohara, T. Fukami, S. Takeda, J. Chem. Phys. **143**, 044509 (2015)
21. M. Wilson and P. A. Madden, Phys. Rev. Lett. **72**, 3033 (1994)
22. M. Wilson and P. A. Madden, Phys. Rev. Lett. **80**, 532 (1998)

DOI: 10.1051/epjconf/201715104001

Shear Viscosity of Liquid Potassium and Cesium: a Simulation Study

Nadège Meyer[1], *Hong* Xu[1], and *Jean-François* Wax[1*].

[1]Laboratoire de Chimie et de Physique - A2MC, Université de Lorraine - Metz, 1, boulevard Arago 57078 Metz Cedex 3, France

Abstract. The density and temperature dependences of the shear viscosity of liquid potassium and cesium are studied. The stress autocorrelation function is calculated from equilibrium molecular dynamics simulations. Using the Green-Kubo formula, the shear viscosity is obtained. Interionic interactions are calculated by Fiolhais potential and are validated by comparison between simulation and experimental data along the liquid-gas coexistence curve for K and Cs. For both metals, three isochors and one isotherm are investigated. The recently proposed relation in [*Phys. Rev. B* **93**, 214203 (2016)] is tested in the cases of K and Cs and it appears that this function reproduces qualitatively and quantitatively well the behavior of each element.

1 Introduction

In our recent article [1], the density and temperature dependences of the shear viscosity of liquid sodium were studied. Along isochoric lines, the viscosity presents a minimum, while it monotonically increases along isotherms. An expression was proposed for the viscosity as a function of temperature and density which reproduced our data for liquid sodium at any density in the range [1000; 2000 kg.m^{-3}] and any temperature in the range [700; 7000 K].

Sodium is one of the alkali metals and it was reported that several properties such as the diffusion coefficient have an universal behavior among these metals. In this spirit, potassium and cesium are now studied. These metals have a high chemical reactivity especially at high temperatures. As a consequence, there are little or no experimental measurements of transport properties in the literature, even near the triple point. Adding to this difficulty, an increase of the density makes measurements very impractical to do. In this context, prediction of these properties using numerical simulation appears to be an interesting alternative to the experimental determination.

The purpose of this work is threefold: (1) investigate the behavior of the density and temperature dependences of the viscosity for K and Cs, (2) compare them with the results of sodium and (3) test the relation proposed for sodium on its counterparts.

*e-mail: jean-francois.wax@univ-lorraine.fr

2 Formalism

2.1 Simulation method

We performed equilibrium molecular dynamics simulations, carried out in the NVE ensemble with N=2048 particles. The velocity Verlet algorithm was used to calculate positions and velocities of each particle. Fiolhais potential [2] has been chosen for our simulations, given that it has proven its capability to describe the alkali metals in the liquid state [3]. In addition, the previous study on sodium showed its validity even in the high density region. The effective interionic potential in the metal can be expressed, in atomic units, by

$$u(r) = \frac{Z^2}{r} - \frac{2Z^2}{\pi}\int_0^\infty F_N(q)\frac{\sin(qr)}{qr}dq. \tag{1}$$

where Z is the ionic valency. This potential is density dependent, but does not depend on the temperature. It displays a minimum, followed by Friedel oscillations due to the electron gas screening. The first term in the right side of Eq. 1 corresponds to the direct Coulomb repulsive interaction between two ions while the second one describes electron-ion and electron-electron interactions. The total expression behaves asymptotically as r^{-3}, so that there is no need to take into account long range interactions and that the potential can be truncated (at the node just below half the box size in our case). Complete analytical expressions, as well as references to earlier studies, can be found in Ref. [1].

Viscosity values are obtained by integration of the stress autocorrelation function (SACF) thanks to the Green-Kubo relation:

$$\eta = \frac{V}{k_B T}\int_0^\infty \langle \delta\sigma_{\alpha\beta}(0)\delta\sigma_{\alpha\beta}(t)\rangle dt, \tag{2}$$

where V is the volume of the system, T, its temperature and k_B, the Boltzmann constant. $\delta\sigma_{\alpha\beta}(t)$ denotes the difference between the instantaneous value of the stress tensor $\alpha\beta$ component, $\sigma_{\alpha\beta}(t)$, and its statistical average value $<\sigma_{\alpha\beta}>$ (with $\alpha, \beta = x, y, z$) [1]. The thermalisation stage lasted at least 60 000 steps during which the velocities were rescaled to the desired temperature every 50 steps. In order to reach the required accuracy when computing the viscosity, production steps are about 6 000 000 for K and Cs, that is a duration of 1 and 2 ns, respectively. During the production stage, we observed that the drift of the temperature was less than 1%. As a consequence, it was not necessary to perform simulation within NVT ensemble in order to control the temperature. The SACF fluctuates strongly and three techniques were used to improve its accuracy. (1) We considered many time origins (at least 590 000 in Eq. 2). (2) An average was done over three off-diagonal elements of the stress tensor σ_{xy}, σ_{yz}, σ_{xz}, and three directions obtained by 45 degrees rotations of the axes, namely $\frac{1}{2}(\sigma_{xx} - \sigma_{yy})$, $\frac{1}{2}(\sigma_{yy} - \sigma_{zz})$, and $\frac{1}{2}(\sigma_{xx} - \sigma_{zz})$. (3) 8 independent microscopic states corresponding to the same macroscopic state point were considered. Thus, we reach an error of less than 5%. For further details about this point, see Ref. [1] where the method for viscosity calculation is exactly the same and more detailed.

2.2 Phase diagram

At ambient conditions, alkali metals are considered as simple because they only have one valence electron and crystallize in the body-centered cubic structure. Under pressure, they transform first into face-centered cubic. On further compression, a wide variety of more or less complex phases has been observed and reviewed in recent years. Degtyareva [4] summarized all the sequences of structural transformations on pressure increase for each

Table 1. Experimental critical (T_c, ρ_c) and melting (T_f, ρ_f) points values [7] of K and Cs.

		T (K)	ρ (kg.m^{-3})
K	Critical point	2239	192
	Melting point	337	829
Cs	Critical point	1924	380
	Melting point	301	1835

alkali element. In the work of Falconi *et al.* [5], the phase diagram of cesium is presented in the (T,P) plane. The Cs melting curve shows drastic changes in slope with pressure. Starting from ambient pressure, a positive slope is observed then two maxima at (2.25 GPa; 470 K) and (3.05 GPa; 471 K). Further increase of the pressure induces a strongly negative slope of the melting curve until (4.2 GPa; 361 K) before again becoming strongly positive. This study ends at the highest pressure investigated by experimental measurements, equivalent to 9.8 GPa. The phase diagram of K is drawn up to 22 GPa by Narygina *et al.* [6]. From the melting point, the positive gradient of the melting temperature plateaus at (5.8 GPa; 550 K). The melting curve decreases up to the bcc-fcc-liquid triple point found at (13.6 GPa; 466 K). For pressures higher than this point, the melting temperature remains constant up to 15.6 GPa. Further pressure increase induces a negative slope up to a clear minimum (19 GPa; 390 K). The experimental critical and melting points of K and Cs are summarized in the Table 1. For all alkali metals, simulations are prevented in the low density region by the metal - non metal transition at about $2\rho_c$. In order to remain in the metallic region, densities lower than $2.5\rho_c$ are not considered in this study.

In the framework of molecular dynamics simulation, it is more convenient to use temperature T and density ρ. However, published experimental and ab-initio studies do not specify the density values as a function of T and P. Knowledge of the density values is needed because above a given pressure, the physical properties of liquid potassium and cesium are expected to change, namely above 19 GPa [6] and 3.9 GPa [5], respectively. The potential we used to describe the atomic interactions is a priori not able to account for these changes. To our knowledge, the only study which indicates approximately the density at high pressures was made by molecular dynamics simulation on liquid cesium using embedded atom model [8]. At 493 K and 3330 kg.m^{-3}, the authors found a pressure corresponding to 2.98 GPa. This justifies our study of density dependence up to 3600 kg.m^{-3}. Concerning potassium, density specifications do not exist in the literature to the best of our knowledge.

3 Results

3.1 Comparison with literature

In Figure 1, we compare the calculated shear viscosity with the experimental data of K (panel a) and Cs (panel d) along the liquid-gas coexistence. We can see that there exists a slight shift of the same order between both sets of curves. As outlined in our study of Na, the difference can be explained by the possible inaccuracies in the description of the interactions. However, it is important to note that measurements are done indirectly and that the uncertainties are very large. Moreover, other parameters such as the presence of impurities in the material can influence the viscosity value. Measures date back to the sixties or earlier and to our knowledge, there are no updated data available in the literature. So, these experimental data should be confirmed. Nevertheless, experimental and numerical studies show similar results

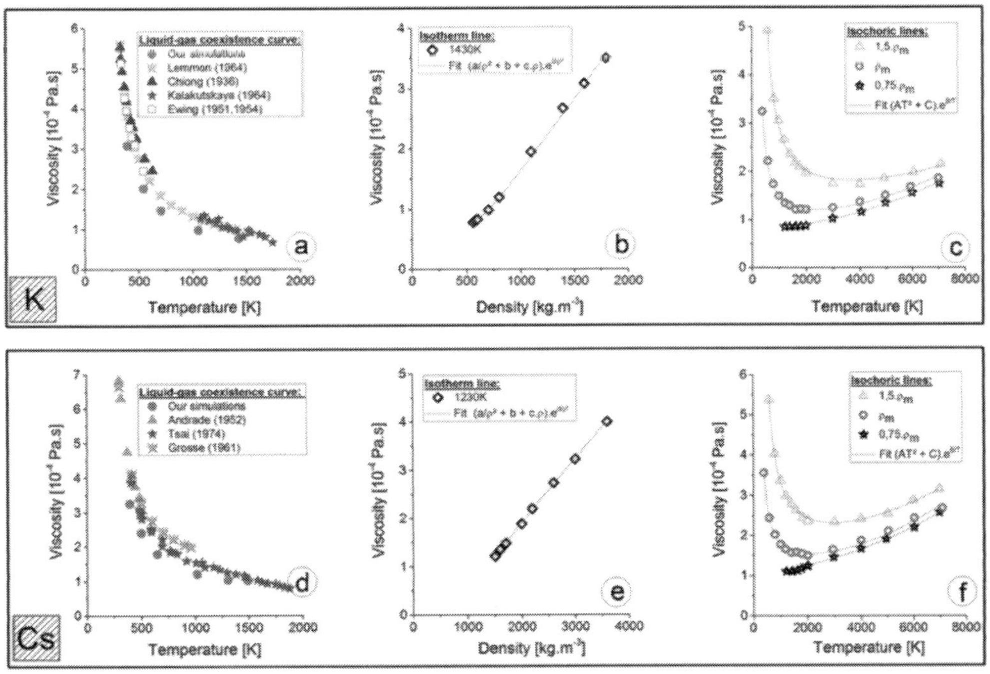

Figure 1. Shear viscosity along the liquid-gas coexistence line (panel a for K, panel d for Cs); along an isotherm line (panel b for K, panel e for Cs); along isochoric lines (panel c for K, panel f for Cs). Experimental data correspond to Refs [9] - [16].

and the same trend is observed along this line. This constitutes the required condition for discussing the density and temperature dependences of the viscosity in the following.

3.2 Temperature dependence

Panels c and f of Figure 1 show the temperature dependence of the viscosity of K and Cs respectively. For both, the viscosity is studied along three isochors: at the density of the melting point ρ_m, at a lower density corresponding to $0.75\rho_m$ and at a higher density equal to $1.5\rho_m$. For densities equal or higher than ρ_m, we obtain the same qualitative behavior of the viscosity. As temperature increases, viscosity decreases quickly at low temperatures, reaches a minimum, and increases slowly in the high-temperature region. At low densities, the viscosity behavior of these metals can be modeled by a simple linear law. For both density ranges, the same typical behaviors were observed in the case of Na. Each isochoric line was fitted by the empiric relation proposed in the study of Na [1]:

$$\eta(T) = (AT^2 + C).e^{B/T}. \tag{3}$$

As we can see, a very good agreement between the data and the fit over the whole temperature interval is found. Thus, in addition to the results of Na, this expression is also able to reproduce qualitatively and quantitatively the temperature dependence of the viscosity for K and Cs over the whole temperature range studied.

Table 2. Parameters values of Eq. 3 for K and Cs.

	K			Cs		
	A (K^{-2})	B (K)	C (Pa.s)	A (K^{-2})	B (K)	C (Pa.s)
$0.75\rho_m$	$1.84.10^{-12}$	-144.6	$0.90.10^{-5}$	$2.99.10^{-12}$	-204.3	$1.24.10^{-5}$
ρ_m	$1.84.10^{-12}$	527.8	$0.87.10^{-5}$	$2.99.10^{-12}$	466.5	$1.10.10^{-5}$
$1.5\rho_m$	$1.37.10^{-12}$	850.0	$1.22.10^{-5}$	$2.59.10^{-12}$	728.5	$1.59.10^{-5}$

3.3 Density dependence

In panels b and e of Figure 1, shear viscosity is represented as a function of density for K and Cs. The density dependence is studied along an isotherm corresponding approximately to $0.64T_c$ (at 1430 K for potassium and 1230 K for cesium). For both elements, an increase of the density implies a higher viscosity. At high density, the behavior of the viscosity is linear while in the low density region, viscosity values are not accessible due to the metal-non metal transition. Nevertheless, at intermediate densities, the behavior does not follow a linear law anymore. According to Chapman-Enskog theory [17], the viscosity must remain positive as ρ tends to zero; this would not be recovered by extrapolating our high density results. In our study of Na, we obtained a relation from Eq. 3 which accounts for the density dependence as

$$\eta(\rho) = \left(\frac{a}{\rho^2} + b + c.\rho\right).e^{d/\rho^2}, \quad (4)$$

where a, b, c, and d were obtained from the density dependence of A, B, and C [1]. In Figure 1, the isotherms are directly fitted using Eq. 4. At high density, the above mentioned linear behavior is recovered. For K and Cs, a good agreement between our simulations and Eq. 4 is observed. As in the case of Na, we are able to model the density dependence of the viscosity. Consequently, we believe that this relation can be applied to the different alkali metals.

4 Conclusion

We have presented a systematic investigation of shear viscosity for K and Cs, obtained by equilibrium MD simulations. We confirmed that the relation proposed previously (Eq. 3) for Na is able to reproduce quantitatively the behavior of the viscosity for K and Cs over the wide temperature and density ranges studied. For both elements, the behavior of the temperature and density dependences is in qualitative agreement with those of Na. Investigations about the universality of viscosity behavior among the alkali metals are underway.

Acknowledgments The PMMS (Pôle Messin de Modélisation et de Simulation) is greatly acknowledged for providing us with computer time.

References

[1] N. Meyer, H. Xu, J.-F. Wax *Phys. Rev. B* **93**, 214203 (2016).

[2] C. Fiolhais, J.P Perdew, S.Q. Armster, J.M. MacLaren and M. Brajczewska *Phys. Rev. B* **51**, 14001 (1995), *Phys. Rev. B* **53**, 13193 (E) (1996).

[3] J.-F. Wax, R. Albaki, J.-L. Bretonnet *Phys. Rev. B* **65**, 014301 (2001).

[4] V. F. Degtyareva *Solid State Sciences* **36**, 62 (2014).

[5] S. Falconi, L. F. Lundegaard, C. Hejny, M. I. McMahon *Phys. Rev. Lett.* **94**, 125507 (2005).

[6] O. Narygina, E. E. McBride, G. W. Stinton, M. I. McMahon *Phys. Rev. B* **84**, 054111 (2011).

[7] R. G. Chapman, N. H. March *Phys. Chem. Liq.* **16**, 77 (1986).

[8] D. K. Belashchenko, N. Yu Nikitin *Journal of Physics: Conference Series* **98**, 042020 (2008).

[9] A. W. Lemmon, H. W. Deem, E. H. Hall, J. F. Walling *ORNL Report* **3605**, (1964).

[10] Y. S. Chiong *Proc. Roy. Soc.* **A156**, 264 (1936).

[11] N. A. Kalakutskaya *Teplofizika vysokikh temperatur 6* **3**, 455 (1968).

[12] C. T. Ewing, J. A. Grand, R. R. Miller, *J. Amer. Chem. Soc.* **73**, 1168 (1951).

[13] C. T. Ewing, J. A. Grand, R. R. Miller, *J. Phys. Chem.* **58**, 1086 (1954).

[14] E. N. Andrade, E. R. Dobbs *Proc. Roy. Soc.* **A211**, 12 (1952).

[15] H. C. Tsai, D. R. Olander, *High Temp. Sci. 6* **2**, 142 (1974).

[16] A. V. Grosse *Science* **147**, 1438 (1965).

[17] E. N. Andrade *Phil. Mag.* **17**, 497 (1934).

Magnetic Properties of Liquid RE-Co (RE = Dy, Er) Alloys

Satoru Ohno[1,*], *Hironori* Shimakura[1] , *Shuta* Tahara[2], and *Tatsuya* Okada[3]

[1]Niigata University of Pharmacy and Applied Life Sciences, Niigata 956-8603, Japan
[2]Department of Physics and Earth Sciences, Faculty of Science, University of the Ryukyus, Okinawa 903-0213, Japan
[3]Niigata College of Technology, Niigata 950-2076, Japan

The magnetic susceptibilities χ of liquid Dy-Co and Er-Co alloys have a large and negative temperature coefficient, which suggests that the Dy, Er, and Co ions in their liquid alloys are in the magnetic state. However, the temperature dependence of χ in both systems becomes weak near the content of 70 at% Co. It is interesting that the compositional dependence of χ for liquid Dy-Co and Er-Co alloys has a minimum at content of 80 at% Co, respectively. On the Co-rich side, the magnetic susceptibility of liquid Dy-Co and Er-Co alloys obeyed the Curie-Weiss law with regard to their temperature dependence of χ. On the rare earth-rich side, the magnetic susceptibilities of liquid Dy-Co and Er-Co alloys also exhibited Curie behavior with a reasonable value for the effective number of Bohr magnetons. The compositional dependence of χ_{4f} for liquid Dy-Co and Er-Co alloys was extracted by subtracting the corresponding data for liquid La-Co alloys from χ for the liquid Dy-Co and Er-Co alloys, respectively.

1 Introduction

Busch and colleagues [1] reported the electronic properties of liquid transition metal (TM), rare earth metal (RE), and their alloys. Schlapbach [2] reported more detailed dependence of magnetic susceptibility on composition with a minimum for liquid Ce-TM (TM = Fe, Co, Ni) alloys and found a continuous transition of Ce ions into the nonmagnetic state. Moreover, Singer et al. [3] extended and explained the compositional dependence of χ for liquid Ce-TM, Pr-TM, La-TM (TM = Fe, Co, Ni), and Sm-Fe alloys. In their paper [3], the compositional dependence of χ for liquid La-TM alloys plays an important role in understanding the magnetic susceptibility of liquid RE-TM (RE = Ce, Pr, Sm) alloys. Actually, the compositional dependence of χ_{4f} can be obtained by subtracting the corresponding data for liquid La-TM alloys from χ for liquid RE-TM (RE = Ce, Pr, Sm) alloys.

We attempt to obtain the magnetic susceptibilities as the functions of temperature and composition for liquid Dy-Co and Er-Co alloys. The magnetic susceptibility of liquid Dy-Co and Er-Co alloys consists of the two large terms due to the 3d- and 4f-electron states.

* Corresponding author: ohno@nupals.ac.jp

The contribution of these two electron states provides a convenient sectionalization in the each region on the Co-rich and RE rich (RE = Dy, Er) sides. Therefore, these two terms can be separated to obtain the temperature dependence of χ_{3d} and χ_{4f} for liquid RE-Co (RE = Dy, Er) alloys. The P_{eff} values for the Dy, Er, and Co ions in liquid Dy-Co and Er-Co alloys can thus be estimated from the temperature dependence of χ_{3d} and χ_{4f}. We investigate the trend of P_{eff} for Dy and Er ions on the basis of the s-f and indirect d-f interactions and compare the trend of Peff for Ce, Pr, Gd, Dy, and Er ions in liquid RE-Co alloys (RE = Ce, Pr, Gd, Dy, Er), respectively.

2 Experimental

Magnetic susceptibility measurements were made using the Faraday method with a torsion balance [4, 5]. The force applied to the sample in the cell was quantified by an automatic method of feedback current control. Mohr's salt (χ = 1.26 × 10^{-2} cm^3/mol at room temperature) was employed as a standard sample. The field strength H of the electromagnet was 10 kOe with a 6.0 cm gap between pole pieces, and $H{\cdot}dH/dx$ was 6.25 ± 0.12 $(kOe)^2$/cm.

The measurement system can be evacuated in a vacuum of 1.2 × 10^{-2} Pa and filled with a purified Ar gas at a pressure of 7 kPa. The RE-Co alloy sample was placed in a magnesia (MgO) crucible packed loosely with a MgO rod, and was completely melted and homogenized in a silicon carbide furnace under the above condition. The measurements of liquid Co and Er were carried out up to about 1570 °C. The measurements on the Co-rich side were carried out from the temperature between 1500 °C and 1550 °C to the melting point. The chemical composition of each sample was determined by accurately measuring the mass of each component before mixing. After this measurement, the decrease in weight of a 0.2 g sample was less than 4 mg. The temperature of the liquid alloy was determined using Pt-Pt13%Rh thermocouples placed close to the sample. The elemental purity was 99.9% for Dy, Er, and Co.

3 Results

As shown in Fig. 1(a), liquid Co has a large and negative temperature coefficient of magnetic susceptibility. On the Co-rich side, the magnetic susceptibilities χ of liquid Dy-Co alloys decrease with increasing Dy content up to 20 at% Dy and these temperature coefficients of χ become weak up to 30 at% Dy. With further increasing Dy content, the magnetic susceptibilities of liquid Dy-Co alloys increase with a large and negative temperature coefficient of χ. This suggests that the Co and Dy ions in the liquid Dy-Co alloys are in magnetic states. The values of χ for liquid Co and Dy are in good agreement with those obtained in the previous works [5, 6].

As shown in Fig. 1(b), the magnetic susceptibilities χ of liquid Er-Co alloys also decrease with increasing Er content up to 20 at% Er and these temperature coefficients of χ become weak up to 30 at% Er. With further increasing Er content, similarly, the magnetic susceptibilities of liquid Er-Co alloys increase with a large and negative temperature coefficient of χ. This suggests that the Co and Er ions in the liquid Er-Co alloys are in magnetic states. The limit of utility of MgO crucible may be up to the temperature of 1570 °C for liquid Er.

Figure 2 shows the compositional dependence of χ for liquid Dy-Co and Er-Co alloys at 1400 and 1500 °C. We added the compositional dependence of χ of liquid Gd-Co and La-Co alloys at 1350 and 1500 °C to do some more investigating. The compositional dependences of χ for liquid Dy-Co and Er-Co alloys decrease with decreasing Co content and exhibit a minimum at the compositions of 20 at% Dy and Er, respectively. The magnetic susceptibilities of liquid Dy, Er, and Gd are approximately 901, 693, and 663 × 10^{-5} cm^3/mol at the melting point, respectively.

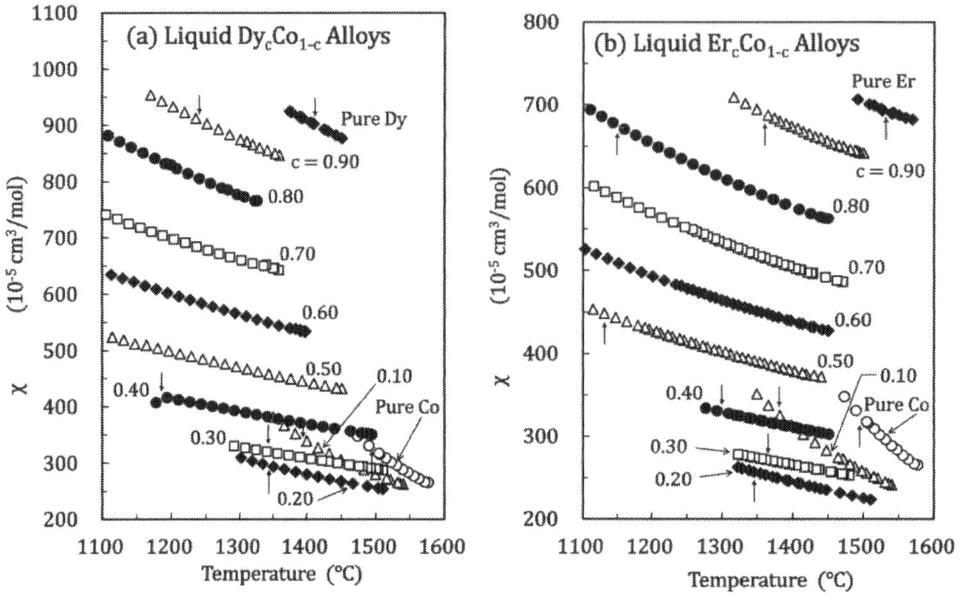

Fig. 1. Magnetic susceptibility of liquid RE$_c$Co$_{1-c}$ alloys as a function of temperature. (a) liquid Dy$_c$Co$_{1-c}$ alloys; (b) liquid Er$_c$Co$_{1-c}$ alloys. The arrows indicate the melting point of the alloys. Experimental error is about 2.5%.

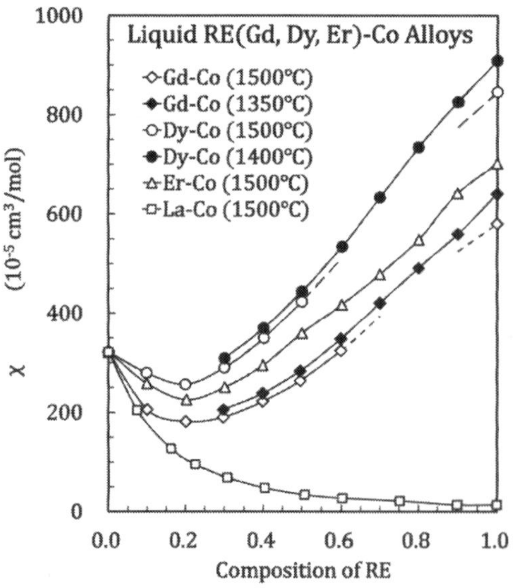

Fig. 2. Dependence of χ on composition for RE$_c$Co$_{1-c}$ (ER = Dy, Er, Gd) alloys. The compositional dependence of χ for liquid La$_c$Co$_{1-c}$ alloys was reported by Singer et al. [3]. The solid lines and dashed lines are drawn as guides for the eye.

4 Discussion

The magnetic susceptibility of a liquid RE-Co alloy is given by [5, 7]

$$\chi_l = [(1-c)\chi_{dia}(Co^+) + c\chi_{dia}(RE^{3+})] + \chi_{el} + \chi_{3d} + \chi_{4f} + \chi_{5d}, \quad (1)$$

where $\chi_{dia}(Co^+)$ and $\chi_{dia}(RE^{3+})$ are the diamagnetic susceptibilities due to the Co^+ and RE^{3+} (RE = Dy, Er, Gd) ions, respectively. The diamagnetic susceptibilities of Co^+, Dy^{3+}, Er^{3+}, and Gd^{3+} ions have been estimated by Bain and Berry [8], and Mendelsohn et al. [9], respectively. The diamagnetic susceptibility of Co ions in metals has been obtained by Banhart et al. [10]. The values of χ_{dia} for Co^+, Dy^{3+}, Er^{3+}, and Gd^{3+} ions employed in this work are -1.82, -1.90, -1.80, and -1.80 × 10^{-5} cm^3/mol, respectively. We assume that the total χ_{dia} for liquid Dy-Co, Er-Co, and Gd-Co alloys varies linearly with composition.

χ_{el} is the paramagnetic susceptibility due to conduction electrons, expressed as [7, 11]

$$\chi_{el} = \alpha(m^*, r_s)\mu_B^2 N(E_F), \quad (2)$$

where $N(E_F)$ is the density of states at E_F, which can be estimated from the nearly free electron model as follows: [7, 11]

$$N(E_F) = (V_a/2\pi^2)(2m/\hbar^2)^{3/2}E_F^{1/2}. \quad (3)$$

Here, V_a is the atomic volume and E_F is given by [7, 11]

$$E_F = (\hbar^2/2m)(3\pi^2 N_e/V_a)^{2/3}, \quad (4)$$

where N_e is the number of conduction electrons.

We assume that the E_F and $N(E_F)$ values are estimated by the linear interpolation of N_e and V_a from the RE side to the Co side. The 4s electrons of Co ions and the $6s^2$ electrons of RE ions are simply treated as conduction electrons. It is assumed that the 5d electrons of RE ions form independently the density of 5d-electron states.

We assume that the m^*/m ratio is 1.0 in liquid Co, RE, and their alloys. In this assumption, the enhancement factor for nearly free electrons is given by [12, 13]

$$\alpha(m, r_s) = \alpha(r_s) - 1/3, \quad (5)$$

where $\alpha(r_s)$ is the enhancement factor of free-electron theory. The dimensionless quantity r_s is defined as (3.64 × 10^8)/k_F, where k_F is the Fermi wave number. The $\alpha(r_s)$ estimated by the random phase approximation increases slightly with increase r_s, and were determined to be 1.28, 1.28, 1.28, and 1.27 for liquid Dy, Er, Gd, and Co, respectively.

The compositional dependences of χ_{el} estimated using this approach deviate above slightly from the straight line connected on the Co side and RE (RE = Dy, Er, Gd) side, respectively. Assuming that the value of χ_{dia} for liquid RE-Co varies linearly with composition, we estimated the compositional dependence of ($\chi_{dia} + \chi_{el}$), as shown in Fig. 3. As a result, the compositional dependence of ($\chi_{dia} + \chi_{el}$) obtained in this procedure approximates a straight line on the scale shown in Fig. 3. As a result, the compositional dependence of ($\chi_{dia} + \chi_{el}$) obtained in this procedure approximates a straight line on the scale shown in Fig. 3. The values of V_a, χ_{dia}, and χ_{el} are shown in Table 1.

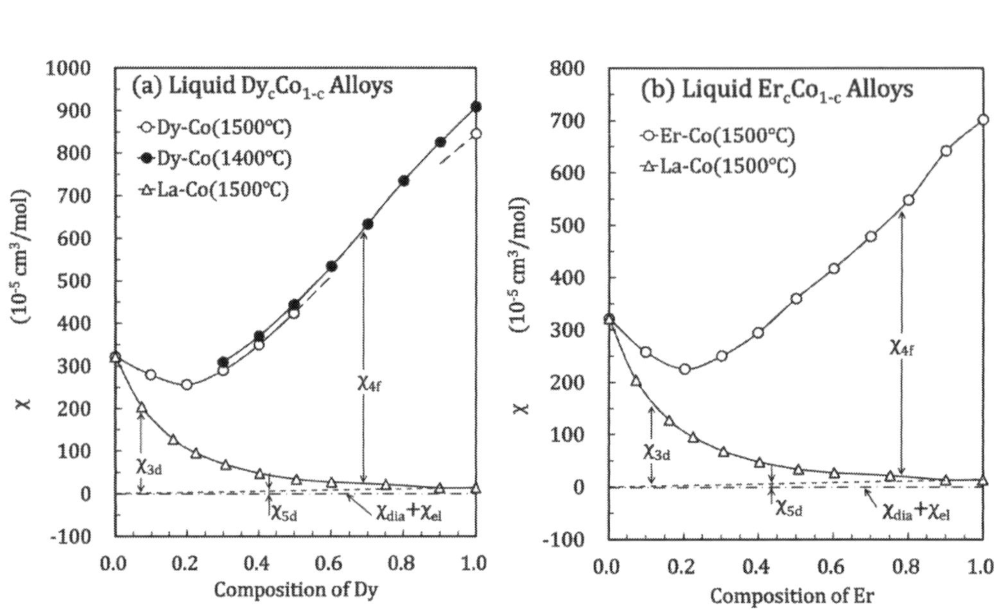

Fig. 3. Dependence of χ on composition for liquid RE$_c$Co$_{1-c}$ alloys. (a) liquid Dy$_c$Co$_{1-c}$ alloys; (b) liquid Er$_c$Co$_{1-c}$ alloys. The compositional dependence of χ for liquid La$_c$Co$_{1-c}$ alloys was reported by Singer et al. [3]. The base line is the compositional dependence of (χ_{dia} + χ_{el}). The solid lines and dashed lines are drawn as guides for eye.

Table 1. V_a, χ_{dia}, and χ_{el} for Co^+, Dy^{3+}, Er^{3+}, La^{3+}, Gd^{3+}, Pr^{3+}, and Ce^{3+} ions.

	Co^+	Dy^{3+}	Er^{3+}	La^{3+}	Gd^{3+}	Pr^{3+}	Ce^{3+}
V_a (10^{-23} cm^3/mol)	1.26	3.32	3.32	3.94	3.80	3.63	3.56
χ_{EXP} (10^{-5} cm^3/mol)	321	846	702	14	579	147	85
χ_{dia} (10^{-5} cm^3/mol)	-1.82	-1.90	-1.80	-2.00	-2.00	-2.00	-2.00
χ_{el} (10^{-5} cm^3/mol)	0.68	1.65	1.65	1.86	1.81	1.75	1.87
$\chi_{dia}+\chi_{el}$ (10^{-5} cm^3/mol)	-1.14	-0.25	-0.15	-0.14	-0.19	-0.25	-0.13

As mentioned above, the 5d-electrons of RE and La ions seems to form independently the density of 5d-electron states. Therefore, the magnetic susceptibility of χ_{5d} due to 5d-electron is given by [14]

$$\chi_{5d} = N_A\mu_B^2\rho_{5d}(E_F)/[1-U\rho_{5d}(E_F)/2], \tag{6}$$

where U is the intra-atomic Coulomb interaction. The experimental χ value of liquid La corresponds to (χ_{dia} + χ_{el} + χ_{5d}) value. We can obtain the $\rho_{5d}(E_F)$ value of liquid La under the assumption that U = 2.0 eV. The $\rho_{5d}(E_F)$ obtained is 0.81 eV^{-1}·atom^{-1}, which is larger than the $\rho_{6s}(E_F)$ value of 0.60 eV^{-1}·atom^{-1}. The $\rho_{5d}(E_F)$ value may be really admixed by $\rho_{6s}(E_F)$, which forms the width of 5d-state. The magnetic susceptibilities of liquid TM-La alloys consists of (χ_{dia} + χ_{el}) value and χ_{5d} of eq. (7). The dependence of (χ_{dia} + χ_{el} + χ_{5d}) on composition obtained in this procedure approximates a straight line on the scale shown in Fig. 3.

According to their suggestion of March and Sayers [15], the magnetic susceptibility of liquid Dy-Co, Er-Co, and Gd-Co alloys consists of the two large terms due to the 3d-electron and 4f-electron states on the TM-rich side and RE-rich side. We compare the magnetic susceptibility of liquid RE-Co and La-Co alloys. The magnetic susceptibility of La-Co alloy is given by [5, 7]

$$\chi_2 = [(1-c)\chi_{dia}(Co^{+}) + c\chi_{dia}(La^{3+})] + \chi_{el} + \chi_{3d} + \chi_{5d}, \quad (7)$$

where $\chi_{dia}(La^{3+})$ are the diamagnetic susceptibilities due to La^{3+} ions. Roughly speaking, we can obtain the χ_{4f} value by subtracting eq. (7) from eq. (1) as follows;

$$\chi_{4f} \approx \chi_1 - \chi_2. \quad (8)$$

Assuming that ($\chi_{dia} + \chi_{el}$) is considerably smaller than χ_{3d} and χ_{4f}, we can obtain the χ_{4f} values of eq. (8) with a high degree of accuracy.

Figure 3(a) shows the dependence of χ on composition for liquid Dy-Co alloys. Assuming that the χ_{3d} of Co ions in liquid Dy-Co alloys may be little different from that in liquid La-Co alloys, we also plotted the dependence of χ on composition for liquid La-Co reported by Singer et al. [3]. This difference between the lines of liquid Dy-Co and La-Co alloys corresponds to the dependence of χ_{4f} on composition due to 4f-electrons. The base line exhibits the dependence of ($\chi_{dia} + \chi_{el}$) on composition due to the Dy and Co ions, and conduction electrons, which is a very small value compared with χ_{4f} and χ_{3d}. The dependence of χ_{3d} on composition decreases rapidly with decreasing Co content. The variation of χ_{4f}, χ_{5d}, and χ_{3d} are reasonable for the composition of Dy and Co.

Figure 3(b) shows the dependence of χ on composition for liquid Er-Co alloys. The dependence of χ on composition for liquid La-Co alloys exhibits the data reported by Singer et al. [3]. As shown in Fig. 3(b), the significant point of comparison is that the difference between the lines for liquid Er-Co and La-Co increases with increasing Er (or La) content. This difference corresponds to the dependence of χ_{4f} on composition due to 4f-electrons. The base line exhibits the dependence of ($\chi_{dia} + \chi_{el}$) on composition due to the Er (or La) and Co ions, and conduction electrons, which is a very small value compared with χ_{4f} and χ_{3d}. The dependence of χ_{3d} on composition decreases rapidly with decreasing Co content. The variations of χ_{4f} and χ_{3d} obtained in Fig. 3(b) are also reasonable for the composition of Er and Co.

Figure 4 shows the dependence of χ_{4f} on composition for liquid RE-Co (RE = Dy, Er, Gd) alloys. For liquid Gd-Co alloys, the dependence of χ_{4f} on composition was obtained by similar manner. Liquid RE-Co alloys have similar dependence of χ_{4f} on composition, which deviates below from straight line. The χ_{4f} values of liquid RE-Co alloys decrease with increasing temperature, which reflects the temperature dependence of the experimental χ value. It suggests that the RE ions in liquid RE-Co alloys are in the magnetic state over the wide composition range. It is known that the Curie-Weiss law describes the magnetic susceptibility in the magnetic state fairly well. For liquid RE-Co in the magnetic state, the third and fourth terms on the right-hand side of eq. (1) is given by [5, 7]

$$\chi_{4f,\ or\ 3d} = [N_A P_{eff}^2 \mu_B^2 / 3k_B(T-\theta)] + \alpha, \quad (9)$$

where N_A is the number of RE or Co atoms per mole. P_{eff} is the effective number of Bohr magnetons. The symbols θ and α are the paramagnetic Curie temperature and a residual susceptibility at infinite temperature, respectively [9].

In this work, the P_{eff} values of RE ions in liquid RE-Co alloys were obtained from the equation of $P_{eff}(RE^{3+}) = (3k_B T\chi_{4f}/N_A)^{1/2}\mu_B$ under the assumptions that $\theta = 0$ and $\alpha = 0$. On the Co side, the P_{eff} and θ for the Co ions in liquid RE-Co (RE = Dy, Er, Gd) alloys were obtained from the plots of $(1/\chi_{3d})$ vs T. The θ values obtained from this approach are 1450, 1340, and 920 K for liquid Co, $Dy_{0.1}Co_{0.9}$, and $Dy_{0.2}Co_{0.8}$ alloys, respectively. The θ values obtained for liquid $Er_{0.1}Co_{0.9}$, $Er_{0.2}Co_{0.8}$, $Gd_{0.1}Co_{0.9}$, and $Gd_{0.2}Co_{0.8}$ alloys are 1340, 790, 1140, and 730 K, respectively. The rapid decrease in χ of liquid RE-Co (RE = Dy, Er, Gd) alloys on the Co-rich side corresponds to the rapid decrease in θ which results from the reduction of exchange d-d interaction between Co ions.

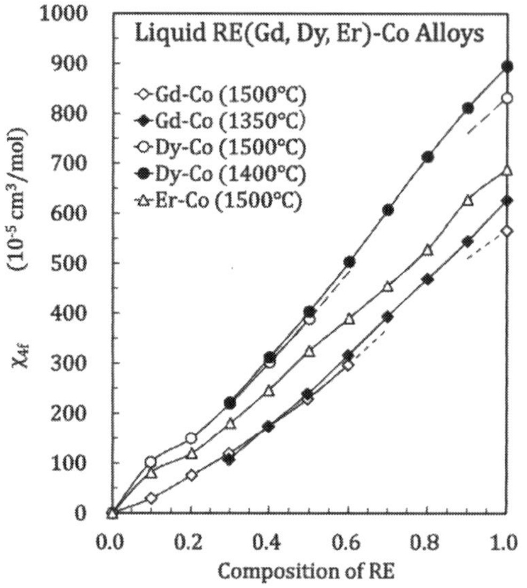

Fig. 4. Dependence of χ 4f on composition for liquid RE_cCo_{1-c} (RE = Dy, Er, Gd) alloys. The solid lines and dashed lines are drawn as guides for eye.

As shown in Fig. 5, we integrate the trends of P_{eff} (RE ions) in liquid RE-Co (RE = Dy, Er, Gd, Pr, Ce) alloys. The P_{eff} values of Ce and Pr ions were estimated from their data [2, 3] by the same manner. The P_{eff} values of Dy and Er ions are approximately 10.9 and 9.9, which are somewhat larger than the theoretical prediction of 10.6 and 9.6 estimated for Dy^{3+} and Er^{3+} ions [16], respectively. The P_{eff} value of Gd^{3+} ions is approximately 8.9, which is larger than the theoretical prediction of 7.9 estimated for Gd^{3+} ions [16]. A swell in P_{eff} for Gd^{3+} ions on the Gd-rich side may be attributable to the assumption of $\theta = 0$. If we can obtain an accurate θ value, we apply the general equation of $P_{eff} = [3k_B(T - \theta)\chi_{4f}/N_A]^{1/2}$ for the analysis of P_{eff}.

It is supposed that metallic RE alloys can be represented by an assembly of RE^{3+} ions embedded in a sea of conduction electrons [17]. In the case of heavy RE-TM alloys [17], the spin of TM ion lying near Gd^{3+} ion take the orientation contrary to the total spin of Gd^{3+} ion. Therefore, a gradual decrease in P_{eff} for liquid RE-Co (RE = Dy, Er, Gd) alloys may be attributed to the alignment of antiparallel moment due to the indirect d-f interaction by way of the conduction electrons [17]. It is well known that liquid Ce-Co alloys give rise to the continuous transition of Ce^{n+} ions with increasing Co content [2, 3]. The trend of P_{eff} for Ce^{n+} ions in liquid Ce-Co alloys is closely related to the continuous transition of Ce ions into the nonmagnetic state. We confirmed the transition of Ce^{3+} to Ce^{4+} ions by this treatment. On the other hand, the Pr ions are stable in liquid Pr-Co alloys. On the Co-rich side, the P_{eff} values correspond to the intermediate state between Co^{+} and Co^{2+} ions with the total spin $S = 1$ and 3/2, respectively. This indicates that the s-d interaction dominates the magnetic susceptibilities of liquid RE-Co alloys on the Co-rich side.

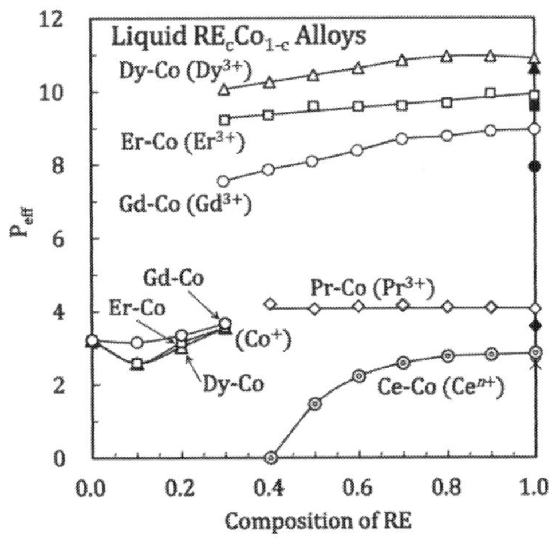

Fig. 5. Dependence of P_{eff} on composition for RE (RE = Dy, Er, Gd, Pr, Ce) and Co ions obtained in liquid RE-Co alloys. The closed traiangle, closed square and so on plotted on the RE side are the calculated values due to Van Vleck [16]. The solid lines are drawn as guides for the eye.

5 Conclusions

Based on the compositional dependence of χ for liquid La-Co alloys, those for liquid Dy-Co and Er-Co alloys can be divided into the two main terms due to the 4f- and 3d-lectrons. Using the Curie and the Curie-Weiss law, P_{eff} for Dy, Er, and Co ions can be estimated from the temperature dependence of χ divided partially. By the addition of Co, the gradual decrease in P_{eff} of Dy and Er ions in these liquid alloys may be attributed to the indirect d-f interaction by way of the conduction electrons. In the liquid RE-Co systems, the 4f-electron states involved with the s-f interaction predominates steadily over the wide composition rage.

References

1. G. Busch, H. -J. Güntherodt, H. U. Künzi, H. A. Meier, L. Schlapbach, J. Phys. Colloque **35**, C4-329 (1974)
2. L. Schlapbach, Phys. Cond. Matter **18**, 189 (1974)
3. V. V. Singer, A. E. Vafin, L. M. Sandratskii, L. P. Mokhracheva, S. V. Konovalov, I. Z. Radovskii, P. V. Geld, Phys. Status. Solidi A **117**, 577 (1990)
4. S. Tahara, S. Ohno, and T. Okada: J. Phys. Soc. Jpn. **78**, 074702 (2009)
5. S. Ohno, H. Shimakura, S. Tahara, T. Okada, J. Phys. Soc. Jpn. **84**, 014706 (2015)
6. M. Müller, E. Huber, H. -J. Güntherodt, J. Phys. Colloque. **C5**, 260 (1979)
7. C. Kittel, *Introduction to Solid State Physics*, 5th ed. (Wiley, New York, 1953) Chs. 14 and 15
8. G. A. Bain, J. F. Berry, J. Chem. Educ. **85**, 532 (2008)
9. L. B. Mendelsohn, F. Biggs, J. B. Mann, Phys. Rev. **A2**, 1130 (1970)
10. J. Banhart, H. Ebert, J. Voitländer, H. Winter, J. Magn. Magn. Mater. **61**, 221 (1986)

11. T. E. Faber, *An Introduction to the Theory of liquid Metals* (Cambridge University Press, London, 1972) Chs. 1 and 4
12. S. D. Silverstain, Phys. Rev. **130**, 1703 (1963)
13. D. Pines, *Elementary Excitations in Solids* (W. A. Benjamin, New York, 1964) Chap. 5
14. P. W. Anderson, Phys. Rev. **124**, 41 (1961)
15. N. H. March, C. M. Sayers, Adv. Phys. **28**, 1 (1979)
16. J. H. Van Vleck, *The Theory of Electric and Magnetic Susceptibilities* (Oxford University Press, London, 1932) p. 226
17. K. N. R. Taylor, Adv. Phys. **20**, 551 (1971)

Density of liquid Ti-6Al-4V

Alexander Schmon[1], *Kirmani* Aziz[1], and *Gernot* Pottlacher[1,*]

[1]Institute of Experimental Physics, Graz University of Technology, NAWI Graz, Austria

Abstract. Ti-6Al-4V is due to its high strength-density ratio a commonly used alloy in aerospace industry applications. But liquid phase data are scarce as preventing contaminations of the reactive high temperature melt during the investigation process poses a challenge. The thermophysical quantity density is of special interest since it is necessary input parameter in modern numerical casting and solidification simulations. Liquid phase density of Ti-6Al-4V as function of temperature was determined employing a fast resistive pulse-heating technique based on the approach to avoid contaminations of the specimen by extremely reducing the experimental duration of the investigation process. Temperature dependent density of liquid Ti-6Al-4V was determined in a temperature range between 2050 K and 2590 K and is presented.

1 Introduction

Titanium Ti-6Al-4V is also known as Grade 5, TA6V or Ti64. This α-β titanium-based alloy is the most commonly of all titanium alloys. Phase stabilizers are aluminum and vanadium. Al reduces density, stabilizes and strengthens α while vanadium provides a greater amount of the more ductile β phase for hot-working. Ti-6Al-4V exhibits strength-toughness combinations between those of steel and aluminum alloys, has the same stiffness than commercially pure titanium but is significantly stronger. [1]. It is widely used in industrial applications such as aeronautic and aerospace due to its good mechanical properties at high temperatures. Within an earlier publication [2] experiments on two different resistive pulse-heating devices (CEA Valduc and TU Graz) have been carried out in order to study thermophysical properties such as electrical resistivity, volume expansion, heat of fusion, heat capacity, normal spectral emissivity, thermal diffusivity, and thermal conductivity of both solid and liquid Ti-6Al-4V. At this time the fast CCD-camera system at TU Graz was not operable, therefore volume expansion was only measured at CEA Valduc. The graph density versus temperature could not be given within publication [2]. As already mentioned in the abstract there is a need of density data of Ti-6Al-4V in the liquid phase. Therefore, within this short note we want to add recently at TU Graz new measured density data for the solid and the liquid phase of Ti-6Al-4V.

* Corresponding author: pottlacher@tugraz.at

2 Experimental

The specimens - Ti-6Al-4V wire samples of about 0.5 mm diameter and 55 mm length were purchased from Fa. GoodFellow, No: TI015260/1, typical analysis in ppm: Al 6%, V 4%, Fe 300, C 220, H 100, N 100, O 650, rest Ti. The initial diameter of the specimen is measured by a high accuracy laser micrometer KEYENCE LS - 7001 before the experiment. The wire-samples were resistively self-heated as part of an electrical discharge circuit by a current of about 5000 A. Heating rates of about 10^8 K/s reduced the experimental duration to below 50 µs. Within this time the specimen material was heated from room temperature up to the end of the liquid phase and concurrently monitored by an adapted CCD camera system working with a frame rate of 200.000 images per second.

The recorded images describing the thermal expansion of the specimen were evaluated to calculate the density of the specimen material. Temperature was determined with a pyrometer operating at a wavelength of 1500 nm by measuring the surface radiance emitted from the sample surface. For temperature determination as reference point to calibrate the pyrometer signals a liquidus temperature of T_{liq} = 1923 K was adopted from [2]. An argon atmosphere with 1.5 bar above ambient pressure was maintained for all measurements. More information regarding the pulse-heating system please find in [3].

2.1. Thermal radial expanding

Thermal expanding of the specimen during pulse-heating is monitored by shadowgraph imaging. The cylindrical formed liquid specimen column is backlit by a photoflash and the silhouette is pictured and recorded. To monitor the radial thermal expansion of the specimen an adapted CCD camera setting, capable to record several images during the short experimental duration of less than 50 µs, is necessary. This requires a fast successive image recording within time-intervals of just a few µs, which is accomplished by modifying the CCD array controlling. For more information on the operation of the fast CCD array see [4, 5].

Prior each pulse-heating experiment an image sequence at room temperature with the before measured initial diameter is recorded to be able to scale the sequence recorded during heating. Since this setting performs a fast monitoring of a segment of the specimen wire, only the radial expansion of the wire is determined. Experimental series in the past [5] indicate that in the setup at TU Graz no longitudinal expansion of the vertically clamped wire occurs, provided that the heating rate used for the experiment is high enough.

Thus density as function of temperature D(T) can be calculated from density at room temperature only by considering thermal radial expansion of the wire sample.

3 Results and Discussion

3.1. Density as a function of temperature

The material's pronounced mechanical strength eased the handling regarding specimen preparation and assembling of the specimen in the pulse-heating setting and led to a straightforward operating when performing the pulse-heating experiments. The stability of the vertical aligned liquid metal column could be facilitated without difficulties. Image series were recorded with relatively long exposure times (633 ns) leading to images of pronounced contrast. The recorded image series were depicted as intensity profiles and evaluated according the procedure described in [4]. Thermal radial expansion was obtained by relating the image sequence recorded during the pulse-heating to an image sequence recorded prior starting the heating process. Density was calculated and depicted as function of temperature, presented in Figure 1.

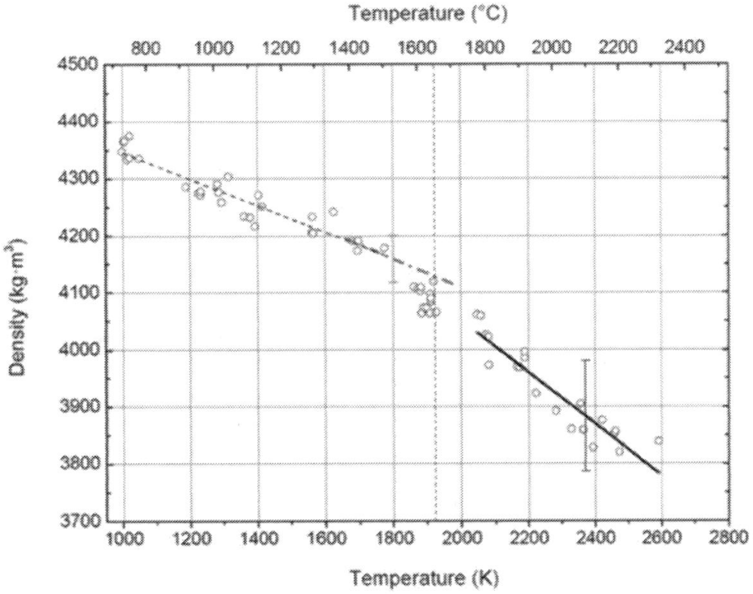

Fig. 1. Density of Ti-6Al-4V as function of temperature. Open circles with dedicated line in black: Results for density with linear least square fit obtained by pulse-heating. Pink dash-dotted line literature data for comparison [6]. Vertical dashed line: liquidus temperature of $T_{liq} = 1923$ K.

3.2 Least squares fit for the liquid phase

The linear least square fit for density *D* versus temperature *T* was obtained to be:

$$(D) = 4955 - 0.452 \cdot T\,(2046\ \ K < T < 2592\ K) \tag{1}$$

For comparison literature data of Li et al. [6] are plotted as dash-dotted line in the liquid phase. There is a good agreement of the pulse-heating results with the data of Li. We could not find any additional density data of liquid Ti-6Al-4V for comparison.

3.3 Numerical density values in the liquid phase

Density was calculated from the ratio of the determined specimen diameters:

$$D(T) = D_{\mathrm{RT}}\left(\frac{d(T)}{d_{\mathrm{RT}}}\right)^{-2} \tag{2}$$

With $d_{\mathrm{RT}}; d(T)$ ...diameter of the specimen at room temperature; diameter as function of temperature and $\frac{d}{d_{RT}}$ the relative radial expansion of the wire; $D_{\mathrm{RT}}; D(T)$ ... density at room temperature; density as function of temperature (for data evaluation a density at room temperature of 4420 kg·m^{-3} was adopted [2]).

Table 1. Tabulated density results of liquid Ti-6Al-4V determined with the pulse-heating setting

Temperature (K)	$\left(\frac{d(T)}{d_{RT}}\right)^2$	Temperature (K)	D ($kg \cdot m^{-3}$)
2046	1,088	2046	4030
2057	1,089	2050	4028
2072	1,098	2100	4006
2079	1,099	2150	3983
2081	1,113	2200	3961
2165	1,113	2250	3938
2173	1,114	2300	3915
2188	1,109	2350	3893
2189	1,106	2400	3870
2222	1,127	2450	3848
2282	1,136	2500	3825
2328	1,145	2550	2802
2393	1,155	2592	3783
2356	1,132		
2364	1,145		
2420	1,141		
2457	1,147		
2461	1,146		
2472	1,157		
2592	1,152		

4 Uncertainties

The uncertainty of density is dominated by the uncertainty of the diameter determination and results to an expanded uncertainty of 3%. Within Figure 1 an uncertainty bar is indicated for the liquid phase. According to GUM [7], this uncertainty is an expanded relative uncertainty with a coverage factor of k = 2. Literature data of Li et al. [6] were reported with an uncertainty of 1%.

Work partially funded by the Austrian Science Fund (FWF) Project P 23838-N20 and by the „Forschungsförderunggesellschaft mbH, Sensengasse 1, 1090 Wien, Austria", Project 855678.

References

1. http://www.farinia.com/additive-manufacturing/3d-materials/how-can-aerospace-benefit-from-3d-printed-titanium-Ti6Al4V (site visited at 07.09.2016)
2. M. Boivineau, C. Cagran, D. Doytier, V. Eyraud, M.-H. Nadal, B. Wilthan, and G. Pottlacher, Int. Journal of Thermophysics, **27, 2,** 507–529 (2006)
3. B. Wilthan, W. Schützenhöfer, and G. Int. J. Thermophysics, **36, 8,** 2259–2272 (2015)
4. A. Schmon, *Dissertation* at TU Graz, (2016)
5. G. Pottlacher and T. Huepf. *Thermal Conductivity 30: Thermal Expansion 18:* Joint Conferences: August 29-September 2, 2009, Pittsburgh, Pennsylvania, USA, volume **30**, page 192. DEStech Publications, Inc, (2010)
6. J. J. Z. Li, W. L. Johnson, and W.-K. Rhim. Applied Physics Letters, **89,** 111913 (2006)
7. Bureau International des Poids et Mesures, Commission électrotechnique internationale, and Organisation internationale de normalisation. Guide to the Expression of Uncertainty in Measurement. International Organization for Standardization, (1995)

EPJ Web of Conferences **151**, 04004 (2017) DOI: 10.1051/epjconf/201715104004
LAM-16

Magnetic Susceptibility of liquid Gd-NM (NM = Cu, Ga, Ge) alloys

Hironori Shimakura[1*], *Shuta* Tahara[2], *Tatsuya* Okada[3], and *Satoru* Ohno[1]

[1]Niigata University of Pharmacy and Applied Life Sciences, Niigata 956-8603, Japan

[2]Department of Physics and Earth Sciences, Faculty of Science, University of the Ryukyus, Okinawa 903-0213, Japan

[3]Niigata College of Technology, Niigata 950-2076, Japan

Abstract. For rare earth alloys, the indirect interaction of RKKY is at work between rare-earth atoms. Therefore, the magnetism of them depends on the number of conduction electrons and the distance between rare-earth metals. In this work, to reveal the relationship between the number of conduction electrons and magnetic property of rare earth metal alloys, magnetic susceptibility measurements for liquid Gd-NM (NM = Cu, Ga, Ge) was performed by Faraday method. As the results, it was observed that the sign of paramagnetic Curie temperature of Cu-Gd alloys are positive at all composition, while Ga-Gd and Ge-Gd alloys show negative paramagnetic Curie temperature at certain composition. Moreover, it was indicated when the alloy at certain composition shows highest melting temperature, it has the lowest paramagnetic Curie temperature.

1 Introduction

Rare earth alloys are used for making various materials and they are known as useful materials. Among the rare-earth alloys, Al-REM (rare earth metal) alloys have been investigated for a long time, and it is widely known that they have interesting physical properties. Specifically, the rare earth-Al phase diagrams [1] contain a number of intermetallic compounds ranging from AlR_3 to $Al_{11}R_3$.

For rare-earth alloys, it is common sense that rare-earth atoms exist in R^{3+} or R^{2+} states at high temperatures in Al-based alloys and their magnetic moment per R-atom in the alloys shows the same value of pure rare-earth metals. However, former experimental results by Sidorov et al. indicated that the magnetic moment per R-atom of dilute Al-REM alloys is rather lower than for ions R^{3+} [2-4]. Given that alloys make intermetallic compounds, they claim this fact can be explained if REM ions exist not in form R^{3+}. And they claim that the existence of Al_2R quasimoleculars is highly probable in these alloys and their destruction starts only above melting point of Al_2R compound. Moreover, Uporova et al. reported the property of magnetic susceptibility of Al_2R alloys in a wide range of temperature [5] and

* Corresponding author: shimakura@nupals.ac.jp

they discovered the magnetic susceptibility of Al_2R alloys increase over the melting temperature of Al_2R.

The Al_2Gd alloy shows really high melting temperature and it is naturally considered that the Al_2Gd is the most stable in the Al-Gd alloys. In addition, Ga_2Gd alloy also show the highest melting temperature. On the other hand, in Ge-Gd systems, Ge_2Gd_5 shows the highest melting temperature in Ge-Gd alloys. In addition, in the case of Cu-Gd systems, they have lower melting temperature than the melting points of pure Gd and Cu. Cu, Ga and Ge belong to fourth period in periodic table and the valence are 1, 3 and 4, respectively. So we think the difference of height of melting temperature between Cu-Gd, Ga-Gd and Ge-Gd would depend on the number of conduction electrons of their alloys. In this study, we performed the magnetic susceptibility measurements for liquid Cu-Gd, Ga-Gd and Ge-Gd. The aim of this work is to reveal the relationship between the number of conduction electrons and magnetic susceptibility.

2 Experimental procedure

Magnetic susceptibility measurements were carried out using the Faraday method with a torsion balance [6]. The force applied to the sample in the cell was quantified by an automatic method of feedback current control. A Mohr's salt was used as a standard. The field strength, H, of the electromagnet was approximately 10 kOe with a 6.0 cm gap between pole pieces, and $H \cdot dH/dx$ was approximately 6.34 ± 0.04 $(k \cdot Oe)^2/cm$.

A Gd-NM (NM = Cu, Ga and Ge) alloy sample was placed in an magnesia cell with a magnesia stick, and was completely melted and homogenized under purified helium gas at a pressure of 20 kPa in a silicon carbide furnace. All measurements were carried out in the liquid state from about 1450°C down to the melting point. The chemical composition of each sample was determined by accurately measuring the mass of each component before mixing. After these measurements, we measured the sample mass and we re-calculate the accurate composition. The temperature of the liquid alloy was determined using Pt-Pt13%Rh thermocouples placed close to the sample.

3 Results

3.1 Experimental results

Figure 1 shows that the temperature dependence of magnetic susceptibility of liquid Gd_{1-c}-NM_c (Cu, Ga, Ge) alloys. The arrows show the melting temperature at the compositions. Since all of them decrease with increasing temperature, it is naturally considered that all of them follow Curie-Weiss law.

Figure 2 shows that the compositional dependence of magnetic susceptibility from 4f electrons, χ_{4f}, of liquid Gd_{1-c}-NM_c (Cu, Ga, Ge) alloys. In the case of Cu-Gd and Ga-Gd alloys, the magnetic susceptibility decrease linearly to some extent with increasing the amount of Cu and Ga composition. On the other hand, the magnetic susceptibility of $Ge_{0.2}Gd_{0.8}$ shifts downward from linear change. However, the magnetic susceptibility of Ge-Gd alloys roughly follows linear change.

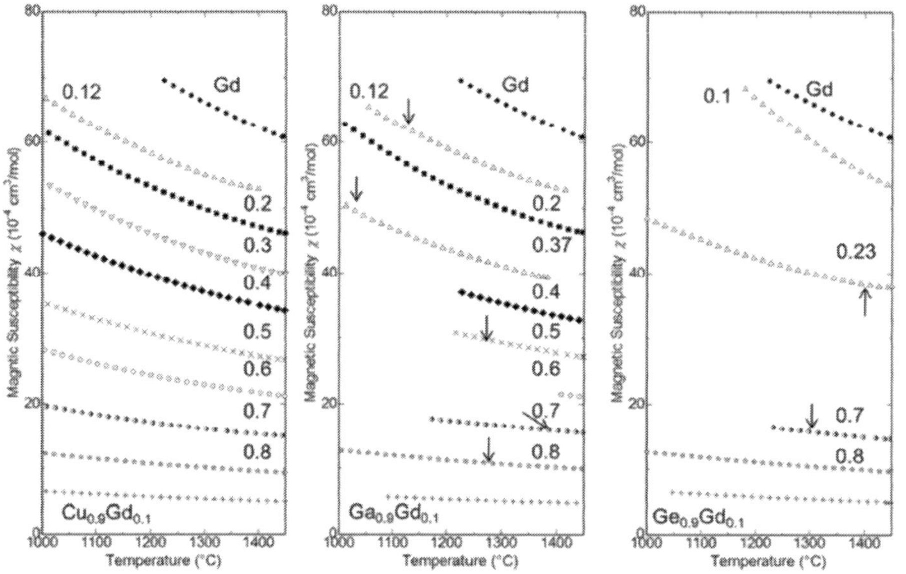

Fig. 1. Temperature dependence of magnetic susceptibility of $Gd_{1-c}NM_c$ (NM = Cu, Ga, Ge) alloys.

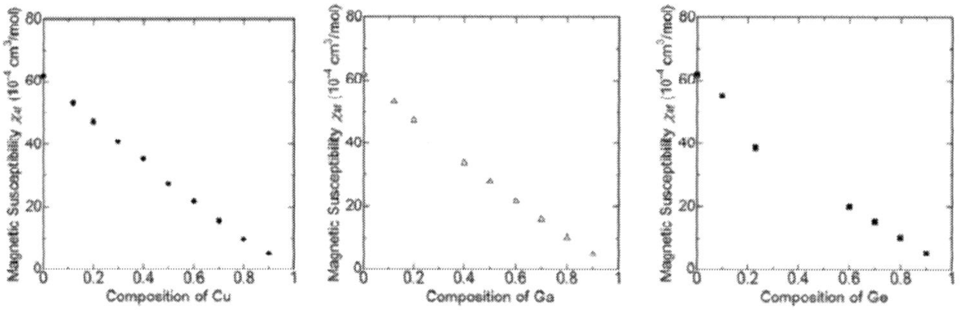

Fig. 2. Compositional dependence of magnetic susceptibility of $Gd_{1-c}NM_c$ (NM = Cu, Ga, Ge) alloys.

3.2 Analysis

In our experiment, the magnetic susceptibility of a liquid Gd-NM alloy is given by

$$\chi_{exp} = \chi_{para} + \chi_{dia} + \chi_{4f}, \quad (1)$$

where χ_{para} is the paramagnetic susceptibility due to conduction electrons, χ_{dia} is the diamagnetic susceptibilities due to Gd and NM ions and χ_{4f} denotes the magnetic susceptibility due to the localized 4f electrons. The diamagnetic susceptibility, χ_{dia}, was estimated by following formula.

$$\chi_{dia} = (1\text{-}c)\,\chi_{dia}(Gd^{3+}) + c\chi_{dia}\,(\text{NM ion}) \quad (2)$$

We utilized the values of $\chi_{dia}(Gd^{3+})$ and χ_{dia} (NM ion) adopted in previous papers. We assumed that the total χ_{dia} for liquid Gd-NM alloys varies linearly with composition dependence. And the paramagnetic susceptibility, χ_{para}, was estimated by nearly free electron approximation [7, 8].

For all the compounds, since the magnetic susceptibility decreases with increasing temperature, they were approximated by the generalized Curie-Weiss law:

$$\chi_{4f} = \frac{C}{T-\theta} + \alpha \tag{3}$$

where C – Curie constant, θ – paramagnetic Curie temperature, α – the residual susceptibility at infinite temperature, respectively. And it is widely known that the Curie constant, C, can be denoted by

$$C = \frac{N_A p^2 \mu_B^2}{3k_B}, \tag{4}$$

where N_A is the number of Gadrinium atoms per mole, p is the effective number of Bohr magnetons, μ_B is Bohr magneton and k_B is Boltzmann constant.

Curie constants and paramagnetic Curie temperatures were obtained by fitted with Curie-Weiss law. In our fitting, we assumed the residual susceptibility for pure Gd as 0.00075 cm^3/mol. It was decided as the Curie constant of liquid Gd become appropriate value, which compared with the Curie constant in solid state. And we assumed that the residual susceptibility of other alloys decrease linearly with increasing the amount of NM composition.

3.3 Curie constant and the effective number of Bohr magnetons

Figure 3 shows the compositional dependence of Curie constant. The black circles, red triangles and blue squares show the Curie constant of Cu-Gd, Ga-Gd and Ge-Gd, respectively. In the case of Cu-Gd system, the curie constant decrease linearly with increasing the amount of Cu composition. However, in the case of Ga-Gd, the Curie constant at $Ga_{0.2}Gd_{0.8}$ shifts down and that at $Ga_{0.6}Gd_{0.4}$ shifts up from linear change. Moreover the Curie constant at $Ge_{0.1}Gd_{0.9}$ also shifts down with linear change.

Figure 4 shows the compositional dependence of the effective number of Bohr magnetons. In the case of Cu-Gd alloys, almost all the P values are around 8.0, which is equal to the pure Gd one. However the P of $Ga_{0.6}Gd_{0.4}$ is 9.0 and the P of $Ge_{0.1}Gd_{0.9}$ become 6.9. It can be interpreted that the surrounding environment around Gd atoms is different with pure liquid Gd atoms as shown in former studies [2-5].

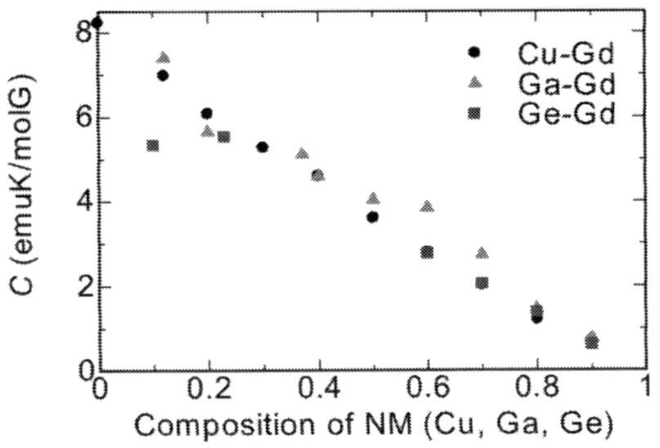

Fig. 3. The compositional dependence of Curie constant

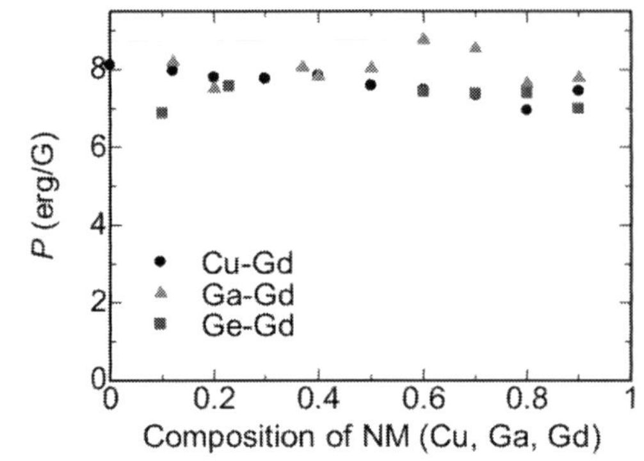

Fig. 4. The compositional dependence of the effective number of Bohr magnetons

3.4 Paramagnetic Curie temperature

The compositional dependence of paramagnetic Curie temperature is shown in Figure 5. In the case of Cu-Gd alloys, the paramagnetic Curie temperatures are positive at all composition. However, in the case of Ga-Gd alloys, they are negative at around $c = 0.6$ compositions. And in Ge-Gd alloys, it is easily considered that there may be negative paramagnetic Curie temperature at around $c = 0.4$.

For rare earth metals the indirect interaction of RKKY(Ruderman and Kittel [9], Kasuya [10], Yoshida [11]) type is at work between the rare-earth atoms. RKKY exchange integrals have oscillating and weak decay on distance:

$$H_{\mathrm{RKKY}} = -9\pi \frac{J^2}{\varepsilon_{\mathrm{F}}} \left(\frac{N_e}{N} \right)^2 f(2k_F R)\mathbf{S}_1 \cdot \mathbf{S}_2 \tag{5}$$

$$f(x) = \frac{-x\cos x + \sin x}{x^4}, \tag{6}$$

where $f(x)$ determines the space dependence of RKKY interaction.

The composition dependence of calculated $f(x)$ of Ga-Gd was shown in Figure 6. To estimate the $f(x)$, we assumed that the distance between Gd ions become longer linearly with decreasing the amount of Gd composition. And the both valence of Gd and Ga was approximated as 3. At a glance, the compositional dependence of paramagnetic Curie temperature looks like the composition dependence of $f(x)$. Specifically, both of them has local minimum at $c = 0.6$.

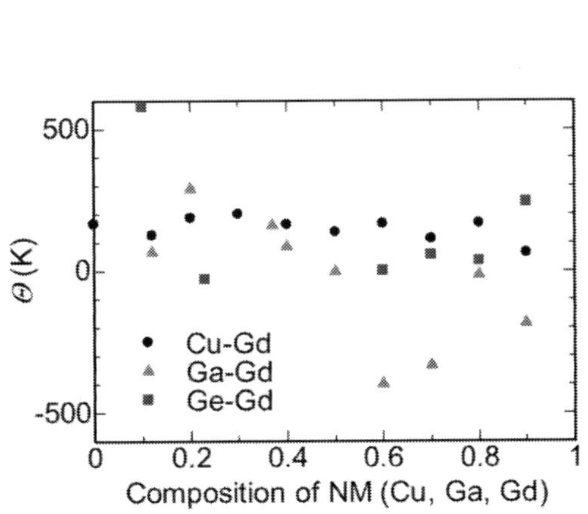

Fig. 5. The compositional dependence of paramagnetic Curie temperature

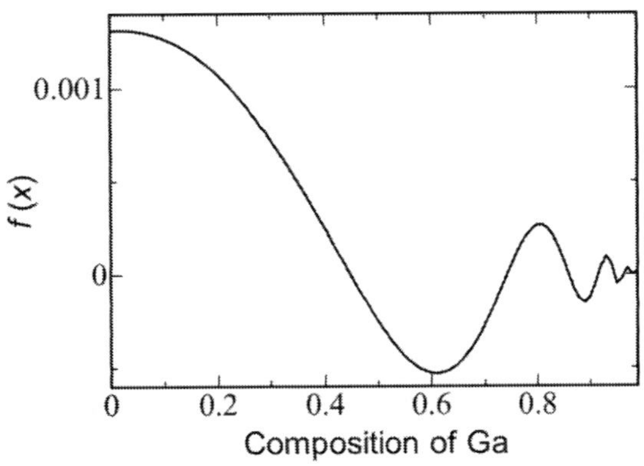

Fig. 6. The compositional dependence of $f(x)$

4 Discussion

For rare-earth alloys, the magnetism includes the effect of RKKY interaction. In brief, when the sign of $f(x)$ is positive, the magnetism become ferromagnetism and *vice versa*. In liquid state, although all of them show paramagnetism, there are differences whether it shows ferromagnetism or anti-ferromagnetism below Curie temperature or Néel temperature.

In our results, it is revealed that the paramagnetic Curie temperatures of Cu-Gd systems are positive in all composition. On the other hand, the sign of paramagnetic Curie temperature of Ga-Gd and Ge-Gd alloys show negative at specific compositions, which are c = 0.6-0.7, 0.9 in Ga_c-Gd_{1-c} systems and may be c = 0.2-0.5 in Ge_c-Gd_{1-c} systems. And, it is revealed that the compositional dependence of paramagnetic Curie temperature consists with the compositional dependence of $f(x)$.

While the melting temperature of Cu-Gd systems show lower value than the melting temperature of pure Cu and Gd, the melting temperature of $Ga_{0.66}Gd_{0.34}$ and $Ge_{0.38}Gd_{0.62}$ show the highest melting temperature in their system. In Ga-Gd and Ge-Gd systems, the sign of paramagnetic Curie temperature around $Ga_{0.66}Gd_{0.34}$ and $Ge_{0.38}Gd_{0.62}$ show negative and the sign of paramagnetic Curie temperature of eutectic composition show positive. The latter tendency is also shown in Cu-Gd systems. From these, we think that the sign of paramagnetic Curie temperature relates to the height of melting temperature.

4 Summary

The magnetic susceptibility measurements of Gd-NM (NM = Cu, Ga and Ge) alloys were carried out by using Faraday method. In all alloys, it is revealed that the magnetic susceptibility from 4f electrons follows Curie-Weiss law. Although the Curie constant of Cu-Gd systems decrease linearly with increasing the amount of Cu to some extent, the Curie constant become larger or smaller than linear change for Ga-Gd and Ge-Gd system. It can be interpreted that the surrounding environment of Gd shows difference with pure Gd systems. Moreover, while the sign of paramagnetic Curie temperature of Cu-Gd system is positive at all composition, for Ga-Gd and Ge-Gd liquid alloys paramagnetic Curie temperature become positive or negative depending on alloy composition. In addition it is indicated that the composition dependence of melting temperature follow the sign of paramagnetic Curie temperature and we think that the sign of paramagnetic Curie temperature relates to the height of melting temperature.

Acknowledgement

This work was supported by a grant for the Uchida Energy Science Promotion Foundation (28-1-54).

References

1. H. K. J .Buschow and J. H. N. Van Vucht, Philips Res. Rep., **22**, 233 (1967)
2. V. Sidorov, O. Gornov, V. Bykov, L. Son, R. Ryltsev, S. Uporov, V Shevchenko, V. Kononenko, K. Shunyaev, N. Ilynykh, G. Moiseev, T. Kulikova and D. Sordelet, Mat. Sci. Eng. A., **449**, 586 (2007)
3. L. D. Son, R. E. Ryltsev, V. E. Sidorov, J. Non. Crys. Sol. **353**, 3722 (2007)
4. V. E. Sidorov, O. A. Gornov, V. A. Bykov, L. D. Son, V. G. Shevchenko, V. I. Kononenko, K. Yu. Shunyaev, J. Non. Cryst. **353**, 3094 (2007)
5. N. Uprova, S. Uprova, V. Sidorov, K. Shunyaev, R. Zakharov and S. Petrova, EPJ Web of Conf. **15**, 01023 (2011)
6. S. Ohno, H. Shimakura, S. Tahara, and T. Okada, J. Phys. Soc. Jpn. **84**, 014706 (2015).
7. T. E. Faber, An Introduction to the Theory of Liquid Metals (Cambridge University Press, London, 1972). p. 294
8. C. Kittel, Introduction to Solid State Physics, 5th ed. (Wiley, New York, 1953). Chap.15.
9. A. Ruderman and C. Kittel, Phys. Rev., **96**,99 (1954)
10. K. Yoshida, Phys. Rev., **106**, 893 (1957)
11. T. Kasuya, Progr. Theor. Phys. (Kyoto) **14**, 45 (1956)

Ionic Conductivities of Molten CuI and AgI-CuI Mixtures

Shuta Tahara[1,*], *Hironori* Shimakura[2], *Satoru* Ohno[2], and *Takanori* Fukami[1]

[1]Department of Physics and Earth Sciences, Faculty of Science, University of the Ryukyus, Okinawa 903-0213, Japan
[2]Niigata University of Pharmacy and Applied Life Sciences, Niigata 956-8603, Japan

Abstract. Ionic conductivities σ for molten CuI and AgI-CuI mixtures were measured in the temperature ranges of approximately 580-800 and 500-850 °C, respectively. The value of σ for molten CuI in the range is smaller than that for molten CuBr and CuCl. σ for molten AgI-CuI mixtures decreases with increasing CuI-concentration. The activation energies E_a for molten AgI-CuI system were determined from the analysis of temperature dependence of σ by using the by Arrhenius type equation. E_a for molten AgI-CuI gradually increase with increasing CuI-concentration.

1 Introduction

AgI and CuI are well known as superionic conductors. In these materials, Ag and Cu ions can migrate through the interstitial space of the sublattice formed by I ions in high temperature solid phase under the melting point [1-3]. It is also known that AgI-CuI mixtures form solid solutions, and both Ag and Cu ions can migrate through the anion sublattice in the high temperature superionic phase [4,5]. In the superionic phase of AgI-CuI mixtures, I ions form lattice, and the lattice type depends on the cationic concentration (b.c.c. for Ag-rich region and f.c.c. for Cu-rich region) [4,5]. Although the structure and ionic conductivity of AgI-CuI solid solutions have been investigated by several scholars using experimental and simulation techniques [5-7], those in molten phase have not been investigated yet.

In the present study, we report the ionic conductivities of molten CuI and molten $(AgI)_{1-x}(CuI)_x$ mixtures measured by the four-probe method, and discuss the detailed information on ionic conduction with activation energies.

2 Experimental

AgI and CuI powder samples were purchased from Wako Pure Chemical Industries. Samples of $(AgI)_{1-x}(CuI)_x$ (x = 0.2, 0.5, 0.8) mixtures were prepared by mixing powder materials of AgI and CuI. A mixture sample was put into an original quartz cell with 4 mm

* Corresponding author: tahara@sci.u-ryukyu.ac.jp

inner diameter, and four carbon electrodes were inserted in small tapered holes of the cell [8,9]. The four carbon electrodes were fastened by nickel bands to prevent a leakage of the liquid sample at higher temperature. Each nickel band has a lead wire made of tungsten to connect each carbon electrode to the 3532-80 Chemical Impedance Meter (HIOKI E. E. Corporation). The sample was heated and melted by kanthal electric heating wire up to approximately 900°C. Ionic conductivity measurements were performed by four-probes technique under during the cooling process, where the measurement current was supplied to the two outside electrodes, and resulting voltage drop along the sample was measured across the two inside electrodes. Before the measurement, the cell constant of each original sample-cell was determined by using a standard aqueous solution of potassium chloride (76.5829 g-KCl/kg-H_2O). The original cell with the sample was put in a process chamber of stainless steel and alumina, where the conductivity measurements were carried out under an argon atmosphere to prevent oxidation of tungsten electrode wires. Bubbles forming in a liquid sample during the measurements were removed by stirring it with a silica rod.

3 Results and discussion

3.1 Molten CuI

Figure 1 shows the temperature dependence of the ionic conductivity $\sigma(T)$ for CuI together with the literature values of molten CuCl [10] and CuBr [11]. The σ for CuI rapidly increases with the phase transition from superionic to liquid phase around the melting point 605 °C, which is different from the behaviour of σ for AgI and CuBr. In cases of superionic AgI and CuBr, σ decreases when these materials melt, which has been interpreted as that the random distribution of anions in liquid phase disturbs the cationic conduction. Such a difference in the behaviour of σ over the melting point may be associated with the structure-type of anion-sublattice in the superionic phase. Specifically, anions in the superionic phase of AgI and CuBr form a b.c.c. lattice, while I ions in α-CuI form a f.c.c. lattice.

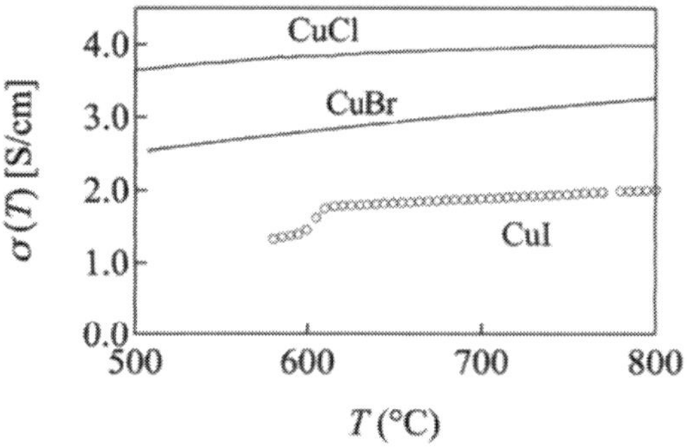

Fig. 1. Temperature dependence of the ionic conductivity σ for CuI (open circles) together with literature values (solid line) for molten CuCl [10] and CuBr [11] (solid lines).

The value of σ for molten CuI is smaller than that for molten CuBr and CuCl in the observed temperature range, which could be affected by the slower diffusion of I ions compared to Br and Cl ions due to the large ionic size and mass. The temperature dependence of σ for molten CuI was fitted by an Arrhenius type equation

$$\sigma T = A\exp(-E_a / k_B T) \qquad (1)$$

to estimate the activation energy E_a, where A and k_B are a pre-exponential factor and the Boltzmann constant, respectively [11]. The estimated E_a for molten CuI is 0.14 eV.

3.2 Molten AgI-CuI mixtures

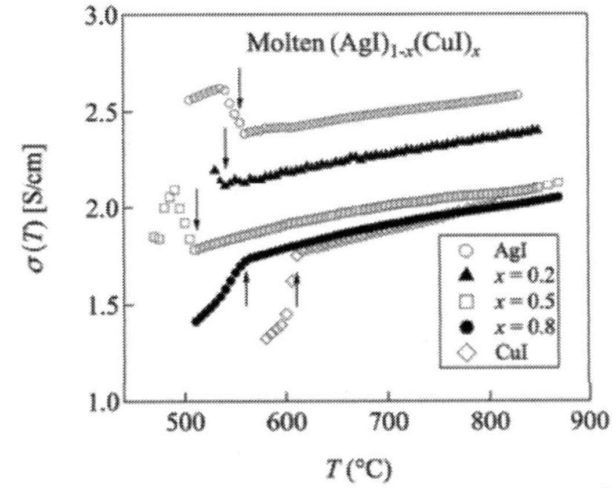

Fig. 2. Temperature dependence of the ionic conductivities for molten $(AgI)_{1-x}(CuI)_x$ mixtures. The arrows denote melting points [4].

The temperature dependence of the ionic conductivities $\sigma(T)$ for molten $(AgI)_{1-x}(CuI)_x$ are shown in Fig. 2. The value of $\sigma(T)$ for pure AgI rapidly decreases at the melting point T_m (= 555 °C) in the heating process as already reported by the literature [1]. The $\sigma(T)$ for the molten phase of AgI gradually increases with increasing temperature, which also shows a good agreement with the literature [1]. The $\sigma(T)$ for the mixtures of $(AgI)_{1-x}(CuI)_x$ in a small x region ($x \leq 0.5$) around the melting point rapidly decreases with increasing temperature, while the $\sigma(T)$ in a large x region ($0.5 < x$) rapidly increases. As written in Sec. 1, I ions in superionic phase of $(AgI)_{1-x}(CuI)_x$ mixture form b.c.c. and f.c.c. lattice in the small and large x regions, respectively. Therefore, the cationic diffusion in the melting process would be affected by the type of anion-sublattice. In all of the CuI-concentrations, the value of σ for the molten phase gradually increases with increasing temperature.

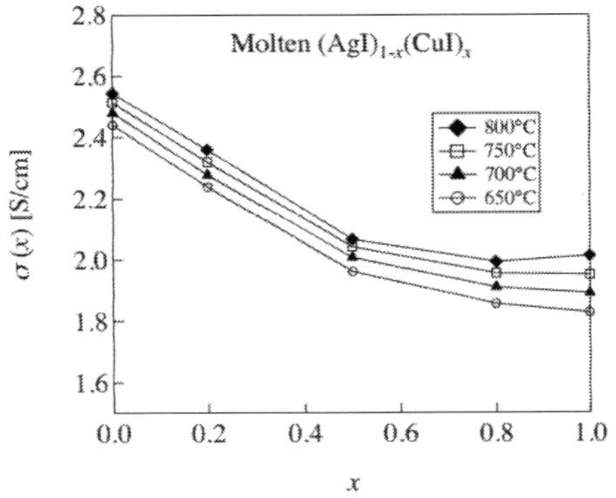

Fig. 3. Composition dependence of the ionic conductivities for molten $(AgI)_{1-x}(CuI)_x$.

The concentration dependence of the ionic conductivities $\sigma(x)$ for molten $(AgI)_{1-x}(CuI)_x$ at several temperatures is shown in Fig. 3. The value of $\sigma(x)$ rapidly decreases with increasing CuI-concentration in a small x region ($x < 0.5$), suggesting that the relative slow conduction of Cu ions in comparison with Ag ions contributes to the decrease of $\sigma(x)$. On the other hand, the concentration dependence of $\sigma(x)$ is small in a large x region ($0.5 \leq x$), suggesting that Ag ions diffuse as slow as Cu conduction in this concentration region.

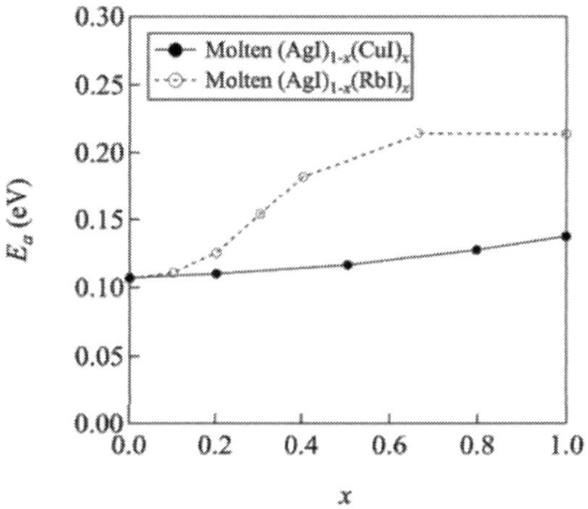

Fig. 4. Concentration dependence of the activation energies E_a for molten molten $(AgI)_{1-x}(CuI)_x$ together with those for molten $(AgI)_{1-x}(RbI)_x$ system [9].

The concentration dependence of E_a for molten $(AgI)_{1-x}(CuI)_x$ mixtures obtained by eq. (1) is shown in Fig. 4 together with that for molten $(AgI)_{1-x}(RbI)_x$ reported in the earlier paper [9]. The E_a for molten $(AgI)_{1-x}(RbI)_x$ rapidly increases in the range of $0.2 < x < 0.4$, suggesting that a large energy is needed to move large and heavy Rb ions, and Rb ions would obstruct the Ag-conduction path in the high RbI-concentration [9]. On the other hand, the E_a for molten $(AgI)_{1-x}(CuI)_x$ mixtures gradually increases with increasing CuI-concentration. It is difficult to explain why the E_a for molten CuI is larger than that for AgI by considering the difference in the ionic size and mass between Ag and Cu, because an Ag ion is larger and heavier than a Cu ion, which seems to be a disadvantage for the ionic diffusion. Furthermore, the valence charges of Ag and Cu ions are almost the same [12,13]. However, the concentration dependence of E_a for $(AgI)_{1-x}(CuI)_x$ mixtures observed in this study would not be the error, because the similar behavior has also been reported for the solid solution of AgI-CuI system [5]. As also discussed in Ref. 5, the activation energy is sometimes not directly related to the ionic size and mass. Therefore, other reasons except for the size, mass, and valence charge of cations have to be considered in this case. For example, a relative strong covalency for Ag-I and Cu-I pairs has been found in molten AgI and CuI by *ab-initio* calculations. Furthermore, weak covalencies of cation-cation pairs have also been reported [14,15], and finally the resulting first correlation lengths of cation-cation and cation-I correlations are different for molten AgI and CuI [16]. Moreover, it has been reported that the electric field generated from the dipole-moment of anions affects the liquid structure and cationic diffusion in molten Ag and Cu halides [12,13,17-19]. These complex properties of Ag and Cu halides compared to typical molten salts might be associated with the activation energy in AgI-CuI system.

4 Conclusion

The electrical conductivities due to the ionic conduction in molten CuI and $(AgI)_{1-x}(CuI)_x$ mixtures were measured in the temperature ranges of approximately 580-800 and 500-

850 °C, respectively. The conductivity of molten CuI in the range was found to be smaller than that of molten CuBr and CuCl previously reported. Moreover, the activation energy of molten CuI was determined as $E_a = 0.14$ eV by the analysis of the temperature dependence of the conductivity. The conductivity and energy of molten $(AgI)_{1-x}(CuI)_x$ mixtures decreases and gradually increases with increasing CuI-concentration, respectively, which is similar behaviour to the AgI-CuI solid solutions. The concentration dependence of the activation energy for molten $(AgI)_{1-x}(CuI)_x$ cannot be explained by considering only the mass, ionic size, and valence charge of Ag and Cu ions. The complex interactions between cation and anion and between cations might affect the activation energies also.

This study was supported by JSPS KAKENHI Grant Number 16K17766.

References

1. C. Tubandt, E. Lorentz, Z. Phys. Chem. **87**, 513 (1914)
2. L. W. Strock, Z. Phys. Chem. B **25**, 441 (1934)
3. J. B. Boyce, B. A. Huberman, Phys. Rep. **51**, 189 (1979)
4. J. Nölting, Ber. Bunsenges, Phys. Chem. **68**, 932 (1964)
5. M. Kusakabe, Y. Shirakawa, S. Tamaki, Y. Ito, J. Phys. Soc. Jpn. **64**, 170 (1995)
6. P. S. Kumar, P. Balaya, P. S. Goyal, C. S. Sunandana, J. Phys. Chem. Solids **64**, 961 (2003)
7. A. K. Ivanov-Schitz, B. J. Mazniker, E. S. Povolotskaya, Solid State Ionics **159**, 63 (2003)
8. S. Ohno, A. C. Barnes, J. E. Enderby, J. Phys.: Condens. Matter **8**, 3785 (1996)
9. S. Tahara, S. Ohno, T. Okada, Y. Kawakita, S. Takeda, EPJ Web. Conf. **15**, 02005 (2011)
10. S. Ohno, S. Tahara, T. Okada, J. Phys. Soc. Jpn. **79**, 114702 (2010)
11. A. Wojakowska, E. Krzyzak, Solid State Ionics **176**, 2711 (2005)
12. V. Bitrián, J. Trullàs, J. Phys. Chem. B, **112**, 1718 (2008)
13. V. Bitrián, O. Alcaraz, J. Trullàs, J. Chem. Phys. **134**, 044501(2011)
14. F. Shimojo, M. Aniya, K. Hoshino, J. Phys. Soc. Jpn. **73**, 2148 (2004)
15. F. Shimojo, T. Inoue, M. Aniya, T. Sugahara, Y. Miyata, J. Phys. Soc. Jpn. **75**, 114602 (2006)
16. Y. Kawakita, S. Tahara, H. Fujii, S. Kohara, S. Takeda, J. Phys. : Condens. Matter. **19**, 335201 (2007)
17. J. Trullàs, O. Alcaraz, L. E. González, M. Silbert, J. Phys. Chem. B, **107**, 282 (2003)
18. S. Tahara, T. Fukami, J. Phys. Soc. Jpn, **84**, 024602 (2015)
19. S. Tahara, Y. Kawakita, H. Shimakura, K. Ohara, T. Fukami, S. Takeda, J. Chem. Phys. **143**, 044509 (2015)

Melt Viscosity of the Soft Magnetic Nanocrystalline $Fe_{72.5}Cu_1Nb_2Mo_{1.5}Si_{14}B_9$ Alloy

Vladmir Tsepelev[1,*], *Yuri* Starodubtsev[1,2], and *Victor* Konashkov[1]

[1]Boris Yeltzin Ural Federal University, Ekaterinburg, Russia
[2]Gammamet Research and Production Enterprise, Ekaterinburg, Russia

Abstract. Temperature dependences of the kinematic viscosity of a multicomponent $Fe_{72.5}Cu_1Nb_2Mo_{1.5}Si_{14}B_9$ melt have been studied. A critical temperature is detected above which the activation energy of the melt' viscous flow changes. Comparison of the temperature dependences of the kinematic viscosity of the melts prepared from the initial ingot and an amorphous ribbon shows that the melt viscosity essentially depends on the initial structural state of the alloy. In amorphous ribbon produced in the mode with overheating, and the melt is above the critical temperature, the enthalpy of crystallization grows; the following heat treatment results in an increase in magnetic permeability.

1 Introduction

Soft magnetic nanocrystalline materials are distinguished by low coercivity and high magnetic permeability magnitudes which are provided by a nanocrystalline, structure [1]. The fabrication of these materials includes three stages. At the first stage, an alloy of a given composition is melted. Then, using ultrarapid quenching, an amorphous metallic ribbon of about 25 μm thickness is produced. At the third stage, in the course of heat treatment, an optimum nanocrystalline structure is formed in the ribbon.

The effect of temperature in preparing the melt before casting, which is aimed at creating an amorphous structure in iron-based alloys, was investigated in a number of works [2–6]. It was shown that with increasing the temperature of the melt before casting the ribbon becomes more plastic, its electrical resistance and crystallization temperature grow, and its density lowers. On the other hand, the increased temperature of the melt deteriorates the quality of the ribbon surface and efficiency of its production. Therefore, it is important to find an optimum mode of preparing the melt before casting. As a characteristic feature of a multicomponent melt, a critical temperature T_k can be specified [6], above which the properties of the melt irreversibly change upon cooling. Irreversibility manifests itself in a hysteresis of kinematic viscosity, surface tension, and other properties of the melt. The hysteresis can arise as a result of breaking interatomic bonds which control the short-range order in the melt.

In the paper, the results of measurements of the $Fe_{72.5}Cu_1Nb_2Mo_{1.5}Si_{14}B_9$ melt viscosity are presented. Using different models of a liquid state, the temperature dependences of

* Corresponding author: v.s.tsepelev@urfu.ru

kinematic viscosity are constructed, which allow us to determine the critical temperature for the nanocrystalline alloy. The parameters of the crystallization process and magnetic properties of ribbons produced after different modes of the melt preparation before casting are investigated.

2 Experimental

Alloys of the $Fe_{72,5}Cu_1Nb_2Mo_{1,5}Si_{14}B_9$ composition were melt in a vacuum induction furnace. Kinematic viscosity of the melts is measured by the torsional vibration method based on measuring the vibration damping factor [7]. The measurement error of kinematic viscosity ν makes up to 3 %. Investigation of the melt properties was performed using samples selected after melting at 1550 °C and cooling in a flat mould, as well as those in the form of a ribbon of an amorphous structure, which was produced from this alloy by rapid quenching from the melt.

A 25 μm thick and 10 mm wide ribbon of an amorphous structure was produced by the planar flow casting process. The ribbon was wound up onto ring-type cores of the 32 mm outer diameter and 20 mm inner diameter. The cores were subjected to annealing at the given temperature of 550°C. The thermomagnetic analysis was performed with the simultaneous recording of the temperature inside the core by a thermocouple and the inductance of the winding wound over the core. The initial permeability μ was calculated from the inductance measurements at 1 kHz. The parameters of the magnetic hysteresis loops were measured by means of point-by-point DC testing. The structural state of specimens was studied using transmission electron microscopy on a JEM 200CX microscope.

3 Results and discussion

It follows from the molecular-kinetic theory of viscosity that the dependence of kinematic viscosity ν on the absolute temperature of the melt T can be expressed through the Arrhenius equation [8]

$$\nu = \nu_0 e^{\frac{-E_a}{RT}}, \tag{1}$$

where ν_0 is the constant factor with the dimensionality of the kinematic viscosity m^2/s, E_a is the energy of activation of the viscous flow J/mol, R is the gas constant J/mol. After taking the logarithm of equation (1) we obtain

$$\ln \nu = \ln \nu_0 - \frac{E_a}{RT}. \tag{2}$$

Thus, the logarithm of the kinematic viscosity, lnν, is a linear function of the inverse absolute temperature, $1/T$.

Fig. 1 shows temperature dependences of the kinematic viscosity in the course of heating the sample of the $Fe_{72.5}Cu_1Nb_2Mo_{1.5}Si_{14}B_9$ alloy subjected to cooling after melting at 1550 °C to cooling in a flat mould. Numerical values of the viscosity are shown on the normal logarithmic scale, whereas those of the absolute temperature are given on the inverse scale. The arrow shows the course of increasing the melt temperature. The curve of heating has two linear portions, which allows a numerical calculation of the energy of viscous flow $E_{a,}$, which is equal for the initial portion of heating to 18 kJ/mol and for a temperature above 1490 °C, 61 kJ/mol.

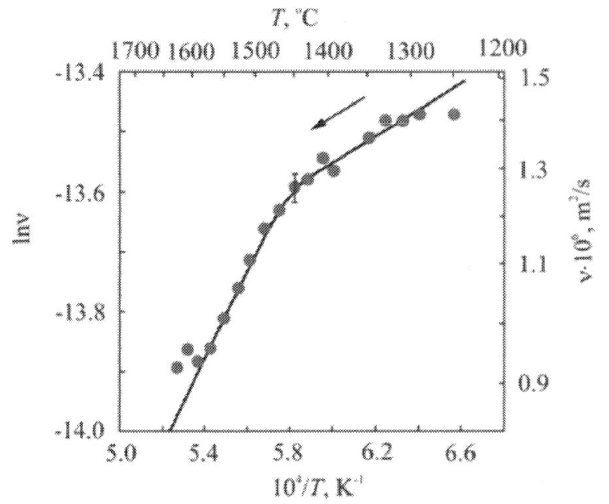

Fig. 1. Dependences of the logarithm of kinematic viscosity, lnν, on the inverse absolute temperature, $1/T$, in the course of heating of the sample of the $Fe_{72.5}Cu_1Nb_2Mo_{1.5}Si_{14}B_9$ alloy subjected to cooling after melting in a flat mould.

Thus, upon heating the $Fe_{72.5}Cu_1Nb_2Mo_{1.5}Si_{14}B_9$ melt above 1490 °C, the viscous flow of the melt changes. These changes can be assigned to breaking interatomic bonds that control the short-range order in the melt [6]. The temperature of 1490 °C can be taken as a critical point T_k for the $Fe_{72.5}Cu_1Nb_2Mo_{1.5}Si_{14}B_9$ alloy above which the viscous flow of the melt changes by the Arrhenius law with another numerical value of the activation energy of the viscous flow E_a.

The results obtained indicate that for the multicomponent $Fe_{72.5}Cu_1Nb_2Mo_{1.5}Si_{14}B_9$ melt, there is the critical temperature T_k = 1490 °C above which the melt can be converted into another structural state with a different magnitude of the activation energy of the viscous flow E_a. To compare two states of the melt, above and below the critical temperature, a ribbon of 25 μm thick was cast from the melt prepared by two modes before the following ultrarapid cooling. In the first mode, the melt of $Fe_{72.5}Cu_1Nb_2Mo_{1.5}Si_{14}B_9$ was heated up to 1480 °C before casting a ribbon. In the second mode, the melt of the same ingot was heated up to 1520 °C, then cooled to 1480 °C, and cast. At each temperature specified the exposure time was 5 min.

In Fig. 2, temperature dependences of kinematic viscosity are presented for the melts produced from the amorphous ribbons that were poured in melt preparation regimes 1and 2. The temperature of the melt was repeatedly changed from the minimal T_{min} = 1200 °C to the maximal T_{max}, which made up 1450 °C in the first cycle and 1600 °C in the second and third ones. At the very beginning of heating melts 1 and 2 from the amorphous ribbon, numerical values of the kinematic viscosity were virtually equal and made up $1.2 \cdot 10^{-6}$ m^2/s. However, on elevating the temperature, the viscosity of the melt produced from the ribbon prepared in the regime of overheating (1520 °C → 1480 °C, melt 1) grows, unlike that without overheating (1480 °C, melt 2). After the second and third "heating-cooling" cycles, the temperature-versus-kinematic viscosity curves become stable; yet, the numerical value of viscosity for the melt undergone overheating is approximately twice as large as that for the regime without overheating. The temperature point corresponding to the change of the curve slope in Fig.2 shifted to 1420 °C compared to 1490 °C for the initial ingot (Fig. 1), i.e. the change in the activation energy of the viscous flow in the melt produced from the amorphous ribbon takes place at a lower temperature. Note that the numerical values of the viscosity of the melt produced from the melt in the regime with overheating are close to the values for the melt prepared from the ingot immediately after casting and cooling the

 DOI: 10.1051/epjconf/201715104006

$Fe_{72.5}Cu_1Nb_2Mo_{1.5}Si_{14}B_9$. The results obtained show that the viscosity of the melt essentially depends on the initial structural state.

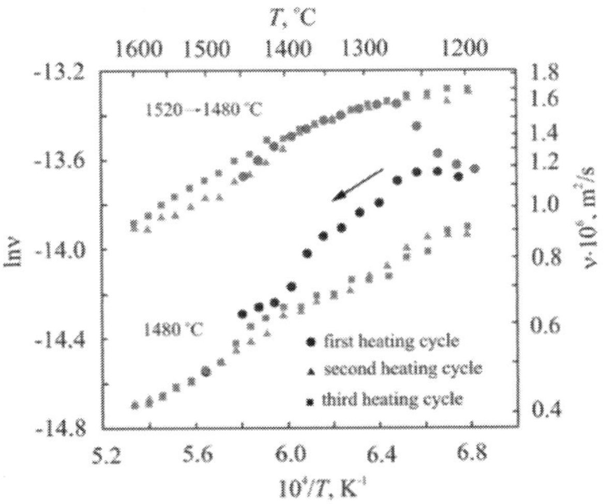

Fig. 2. Dependences of the logarithm of kinematic viscosity, lnv, at on the inverse absolute temperature, $1/T$, in the course of heating the samples of the amorphous ribbon which was produced in preparation regime 1 (heating to 1480 °C before casting a ribbon) and 2 (heating to 1520 °C, then cooling to 1480 °C, and cast).

Investigation in the ribbon's amorphous structure in its initial state and after heat treatment, using transmission electron microscopy, showed that the ribbon's structure differed insignificantly in different regimes of melt preparation (Fig. 3 and 4).

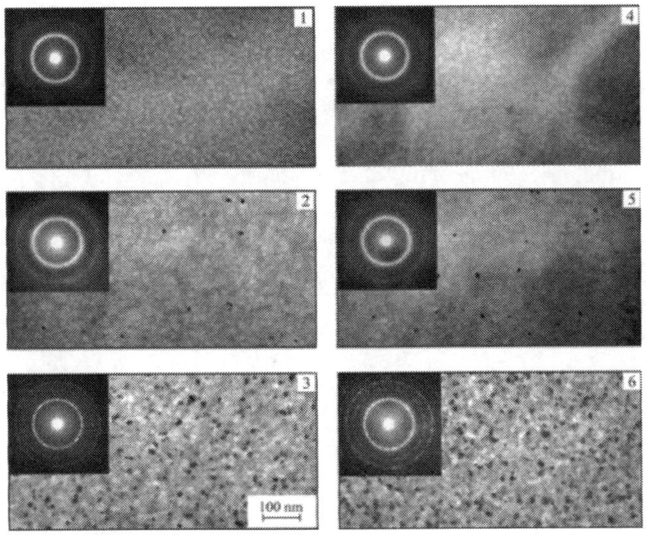

Fig. 3. Bright-field images and micro-diffraction patterns for ribbon prepared by the regime without overheating (1480 °C, 1–3) and with overheating (1520 °C → 1480 °C, 4–6) after heat treatment at 460 (1, 4), 480 (2, 5) and 550 °C, (3, 6).

In accordance with micro-diffraction patterns, the structure of both samples is amorphous right up to the temperature of heat treatment of 460 °C. The first crystallites appear at 480 °C and are of medium size of 3.5 nm. In this case, for the preparation mode of

1520 → 1480 °C (regime 2) the contribution of crystallites of 1-2 nm is greater than for the preparation mode of 1480 °C (regime 1). After heat treatment at 550 °C, the crystallites' medium size was 8 nm; in the process, the contribution of crystallites of 1 – 2 nm size was also greater for the preparation mode of 1520 → 1480 °C.

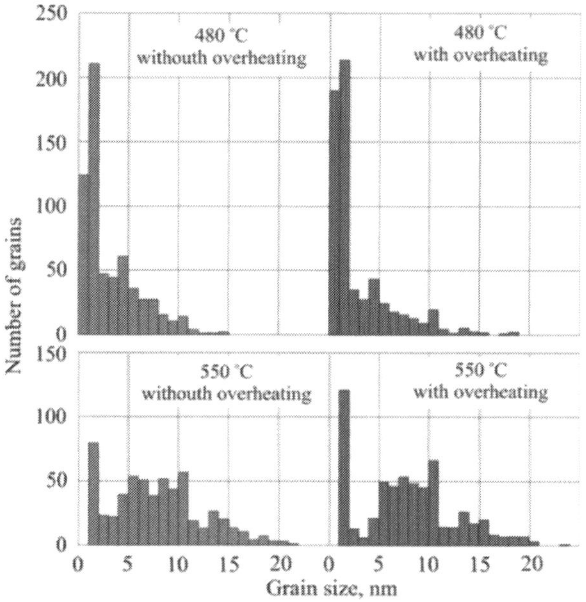

Fig. 4. Histograms of grain-size distributions in ribbon prepared by the regime without overheating (1480 °C, cyan) and with overheating (1520 °C → 1480 °C, red) after heat treatment at 480 and 550 °C.

Figure 5 shows the in-time changes in the temperature and initial permeability in the course of heat treatment at 550 °C for the cores fabricated from the ribbons that were subjected to two modes of preparation of the $Fe_{72.5}Cu_1Nb_2Mo_{1.5}Si_{14}B_9$ melt. From the temperature curve it is possible to determine the initial rate of heating of the core prior to crystallization V_0, the onset temperature of crystallization T_{cr} (the staring point of the nonlinear growth of temperature), the maximal rate of heating the core V_m, and the peak temperature in the course of crystallization T_m. On the curve of magnetic permeability, it is possible to pick out the onset temperature of the fastest growth of permeability T_μ, its peak value μ_m, and stabilized value μ_{st} after completion of the crystallization process. Numerical values of crystallization parameters of the $Fe_{72.5}Cu_1Nb_2Mo_{1.5}Si_{14}B_9$ alloy for two modes of the melt preparation are given in Table 1.

Among the results presented, a special attention is to be paid to the enhanced rate of growing temperature in the course of crystallization V_m = 39 °C/min and peak temperature T_m = 581 °C for the preparation mode 1520 → 1480 °C, which corresponds to a larger enthalpy of crystallization. All other conditions being equal, these data can evidence a higher degree of the structural disorder in the amorphous precursor obtained with overheating above the critical temperature. The more active the process of crystallization is, the higher the peak value of permeability μ_m. What is of great interest is the time difference between peaks of heat release and permeability on the curves depicting the process dynamics, shown in Fig. 3. The permeability peak appears just after the metal temperature is beginning to decrease. The formation of the amorphous-crystalline structure with nanocrystals implanted into the amorphous matrix is over after the peak temperature is reached. α-FeSi-based nanocrystals have a lower content of silicon at the initial state of the crystallization process. This silicon content is gradually increasing up to α-$Fe_{80}Si_{20}$ [9]. Nanocrystals can also contain Mo, the solubility of which is 2.5 % in α-Fe at 700 °C [10, 11]. The amorphous matrix permeability is close to 300 °C. As the core permeability is beginning to grow at the stage of temperature decreasing, then the μ_m peak appearance can be attributed to the transition of nanocrystals from the paramagnetic state into the ferromagnetic one. On the other hand, after reaching the peak of crystallization, residual processes of the structure rearrangement and diffusion of components continue for several more minutes. Therefore, the permeability fall after reaching the peak value can be attributed to the transition of the atomic and magnetic structure into a stabilized state.

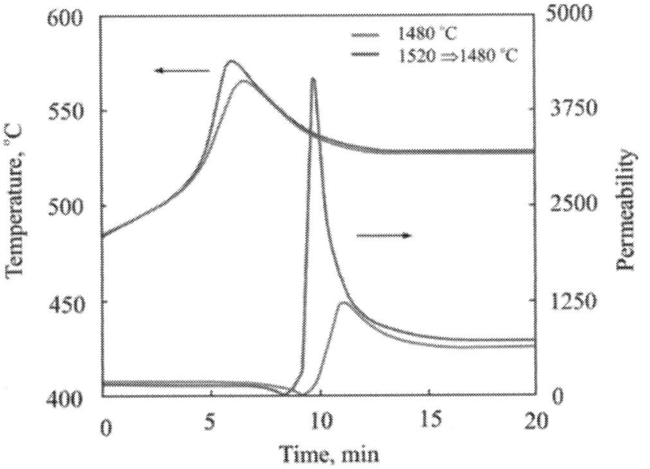

Fig. 5. Temperature and permeability changes in the course of heat treatment of cores made from the ribbons undergone preparation of the $Fe_{72.5}Cu_1Nb_2Mo_{1.5}Si_{14}B_9$ melt by two modes.

Table 1. Parameters of crystallization of the $Fe_{72.5}Cu_1Nb_2Mo_{1.5}Si_{14}B_9$ alloy

Regime of melt preparation	V_0, °C/min	V_m, °C/min	T_{cr}, °C	T_m, °C	T_μ, °C	μ_m	μ_{st}
1480 °C	4.5	22	520	570	548	1270	650
1520 → 1480 °C	4.5	39	520	581	553	4300	760

The cores fabricated from the ribbons obtained by two modes of the melt preparation were subjected together to the heat treatment at the temperature of 550 °C for 1 h after the crystallization peak is reached. In Table 2, demonstrates the data on the permeability $\mu_{0.08}$ at the magnetic field strength of 0.08 A/m, maximal permeability μ_{max}, coercivity H_c of the magnetic hysteresis loop at the maximal magnetic field strength of 800 A/m. It is seen that the mode with overheating of the melt above the critical temperature provides higher magnetic properties.

Table 2. Magnetic properties of cores made from the $Fe_{72.5}Cu_1Nb_2Mo_{1.5}Si_{14}B_9$ alloy

Regime of melt preparation	$\mu_{0.08}$	μ_{max}	H_c, A/m
1480 °C	125000	882000	0.45
1520 → 1480 °C	164000	1024000	0.40

4 Summary

In the work, the temperature dependences of the kinematic viscosity of the multicomponent $Fe_{72.5}Cu_1Nb_2Mo_{1.5}Si_{14}B_9$ melt have been studied. The results of measurements were treated with the use of the Arrhenius law. The critical temperature is established above which the activation energy of the viscous flow of the melt changes. Comparison of the temperature dependences of the kinematic viscosity of the melts prepared from the initial ingot and an amorphous ribbon shows that the viscosity of the melt essentially depends on the initial structural state. In the amorphous ribbon obtained in the mode of the melt preparation with overheating above the critical temperature, an enhanced value of the crystallization enthalpy of is found. After the heat treatments of the cores, the mode with overheating the melt above the critical temperature results in higher magnetic properties.

Acknowlegements

This work was supported by the Scientific researches of higher education institutions within the State task of the Russian Federation No. 2014/236.

References

1. G. Herzer, *Handbook of Magnetic Materials*, edited by K.H.J. Buschow, **10**, 415 (1997)
2. V.P. Manov, S.I. Popel, P.I. Buler, A.B. Manukhin, D.G. Komlev, Mater. Sci. Eng. **A133**, 535 (1991)
3. G. Kumar, T. Ohkubo, K. Hono, J. Mater. Res. **24**, 2353 (2009)
4. P.S. Popel, V.E. Sidorov, Mater. Sci. Eng. **A226–228**, 237 (1997)
5. U. Dahlborg, M. Calvo-Dahlborg, P.S. Popel, V.E. Sidorov, Eur. Phys. J. **B14**, 639, (2000)
6. B.A. Baum, *Metal liquids* (Nauka, Moscow, 1979)
7. V. Tsepelev, V. Konashkov, Y. Starodubtsev, Y. Belozerov, D. Gaipisherov, IEEE Trans. Magn. **48**, 1327 (2012)
8. J. Frenkel, *Kinetic theory of liquids* (Claredon Press, 1946)
9. K. Hono, D.H. Ping, M. Onhuma, H. Ondera, Acta Mater. **47**, 997 (1999)
10. O. Kubaschewski, *Iron – binary phase diagrams* (Springer, 1982)
11. *Diagrams of binary metallic systems*, Vol. 2, Ed. N.P. Lyakishev (Moscow, 1997)

EPJ Web of Conferences **151**, 05001 (2017) DOI: 10.1051/epjconf/201715105001
LAM-16

Disorder trapping by rapidly moving phase interface in an undercooled liquid

Peter Galenko[1,*], *Denis* Danilov[2], *Irina* Nizovtseva[1,3], *Klemens* Reuther[1], and *Markus* Rettenmayr[1]

[1]Friedrich-Schiller-Universität Jena, Physikalisch-Astronomische Fakultät, D-07743 Jena, Germany
[2]Institute of Nanotechnology, Karlsruhe Institute of Technology, 76344 Eggenstein-Leopoldshafen, Germany
[3]Ural Federal University, Laboratory of Multi-Scale Mathematical Modeling, 620000 Ekaterinburg, Russian Federation

Abstract. Non-equilibrium phenomena such as the disappearance of solute drag, the origin of solute trapping and evolution of disorder trapping occur during fast transformations with originating metastable phases [D.M. Herlach, P.K. Galenko, D. Holland-Moritz, *Metastable solids from undrercooled melts* (Elsevier, Amsterdam, 2007)]. In the present work, a theoretical investigation of disorder trapping by a rapidly moving phase interface is presented. Using a model of fast phase transformations, a system of governing equations for the diffusion of atoms, and the evolution of both long-range order parameter and phase field variable is formulated. First numerical solutions are carried out for a congruently melting binary alloy system.

1 Introduction

Disorder trapping is a known non-equilibrium phenomenon occurring during transformations of congruently melting ordered intermetallic compounds [1, 2]. This phenomenon occurs especially during rapid crystallization of undercooled melts of intermetallics with a superlattice crystalline structure [3–5]. In such systems, the phase interface propagation is sluggish at small undercoolings. The attachment of atoms from the liquid to the phase interface needs short-range atomic diffusion, since atoms have to move to the proper lattice site in the superlattice structure. If the undercooling increases, the non-equilibrium effect of disorder trapping leads to the formation of a metastable disordered structure. Experimental evidence of disorder trapping has been demonstrated by *in situ* diffraction studies using synchrotron radiation on levitation-processed samples, in which a transition from ordered to disordered growth at a critical undercooling was unambiguously shown [6]. The phenomenon of order–disorder transition with distinct change of the order parameter is also important in structural transformations with non-monotonic relaxation processes of melts [8] and in liquid–liquid phase separation of undercooled metallic alloys [9, 10].

The disorder trapping was investigated from theoretic perspective by sharp interface models [2, 3, 6], a diffuse interface model [11] and methods of atomistic simulation [1, 12]. In the present work, using an existing approach for fast phase transformations [13], a diffuse

*e-mail: peter.galenko@uni-jena.de

interface model results in a system of hyperbolic equations to describe (i) the rapid interface motion and (ii) the transition from ordered to disordered structures formed from undercooled liquids. The pioneering atomistic model of Chernov [1] and thermodynamic model of Boettinger and Aziz [2] consider a process of ordering at the interface only. The present diffuse interface model introduces spatial variation of order parameter such that the order-disorder transition and relaxation of order parameter exist in the diffuse zone between phases as well as in a solid phase far behind the solid-liquid interface. In addition to previous works [1–3, 6, 11], the present model takes into account highly rapid interface propagation at which locally non-equilibrium states in chemical composition and in atomic ordering are freezing in solid phase with an appearance of metastable states.

In the numerical solutions of the model, we test the existence of pronounced disorder trapping by direct exchange of dissimilar atoms between sublattices at high interface velocities moving in an undercooled binary liquid. The main focus of the present work is to analyze numeric results and to formulate conditions for disorder trapping in rapidly transforming congruently melting alloy undercooled in the liquid state.

2 Definitions

2.1 Main functions

We consider a binary mixture consisting of A-atoms and B-atoms. For a given temperature T, the main functions of the model, which should be obtained by solutions, are defined as follows. *The first function* is φ the phase field with $\varphi = 1$ in the S-phase (solid) and $\varphi = 0$ in the L-phase (liquid). *The second function* is x_B the overall continuous concentration of B-atoms (which can be defined through the L-phase and S-phase concentrations). Concentrations in the S-phase are given by $x_i^j = n_i^j/(n_A^j + n_B^j)$, where n_i^j is the number of moles of atom i (i=A,B) on the sublattice j ($j = \alpha,\beta$). Finally, *the third function* is the long-range order parameter η, which is defined by the concentrations $x_i^j = (x_A^\alpha, x_A^\beta, x_B^\alpha, x_B^\beta)$ of the $\alpha-$ and $\beta-$sublattices as

$$\eta = x_A^\alpha - x_A^\beta = x_B^\beta - x_B^\alpha. \tag{1}$$

A completely disordered state in the S-phase occurs with $\eta = 0$. Atoms exhibit full order on sublattices if $\eta = 1$. The concentrations in the S-phase on sublattices are

$$x_A^\alpha = 1 - x_B - \frac{\eta}{2}, \quad x_B^\alpha = x_B - \frac{\eta}{2}, \tag{2}$$

$$x_A^\beta = 1 - x_B + \frac{\eta}{2}, \quad x_B^\beta = x_B + \frac{\eta}{2}. \tag{3}$$

2.2 Gibbs free energy

2.2.1 The entire system

The Gibbs free energy for the entire binary system is given by

$$G(\varphi, \nabla\varphi, \eta, \nabla\eta, x_B) = G_S(x_B, \eta)p(\varphi) + G_L(x_B)p(1-\varphi) + \frac{\varepsilon_\varphi}{2}(\nabla\varphi)^2 + \frac{\varepsilon_\eta}{2}(\nabla\eta)^2 + W_\varphi g_\varphi(\varphi) + W_\eta g_\eta(\eta). \tag{4}$$

The double-well functions in Eq. (4) for the phase field g_φ and for the long-range order parameter g_η are given by

$$g_\varphi(\varphi) = \varphi^2(1-\varphi)^2, \quad g_\eta(\eta) = \eta^2(1-\eta)^2. \tag{5}$$

The Gibbs free energy is interpolated between the phases using the following function

$$p(\varphi) = \varphi^2(3 - 2\varphi). \tag{6}$$

2.2.2 The phases

In the L-phase we assume the following approximation

$$G_L(T, x_B) = (1 - x_B)G_L^A(T) + x_B G_L^B(T), \tag{7}$$

$$G_L^A(T, x_B) = G_L^{0A}(T) + RT\ln(1 - x_B) + x_B\Omega_L, \tag{8}$$

$$G_L^B(T, x_B) = G_L^{0B}(T) + RT\ln(x_B) + (1 - x_B)\Omega_L, \tag{9}$$

which can be considered as contribution (7) from elements A and B, reference terms G_L^{0i}, ideal mixture terms $RT\ln x_i$, and excess terms $x_i\Omega_L$ (where again $i = A, B$).

In the S-phase, the Gibbs free energy has additional terms related to the disorder parameter

$$G_S(T, x_B, \eta) = G_S^{ref}(T, x_B) + G_S^{id}(T, x_B, \eta) + G_S^{ex}(x_B, \eta). \tag{10}$$

Assuming that the L-phase transforms into the body centered cubic modification of the S-phase, the following contributions into the free energy (10) are taken into account:
- reference contribution

$$G_S^{ref}(T, x_B) = (1 - x_B)G_A^{bcc}(T) + x_B G_B^{bcc}(T), \tag{11}$$

- ideal mixture contribution

$$G_S^{id}(T, x_B, \eta) = \frac{1}{2}RT\Bigg((x_B - \frac{\eta}{2})\ln(x_B - \frac{\eta}{2}) + (x_B + \frac{\eta}{2})\ln(x_B + \frac{\eta}{2}) + (1 - x_B - \frac{\eta}{2})\ln(1 - x_B - \frac{\eta}{2}) + (1 - x_B + \frac{\eta}{2})\ln(1 - x_B + \frac{\eta}{2})\Bigg), \tag{12}$$

- excess free energy

$$G_S^{ex}(x_B, \eta) = \Omega_2\left(x_B(1 - x_B) + \frac{\eta^2}{4}\right) + \Omega_3\eta^3 + \Omega_4\eta^4. \tag{13}$$

3 Governing equations

Following the model of fast phase transformations [13], one can derive governing equations represented by hyperbolic equations for non-conserved and conserved field variables. They were obtained by the condition of non-decreasing entropy in time that, for isothermal systems, is equivalent to the condition of non-increasing Helmholtz free energy in time [14]. Following these approaches [13, 14], one can derive the system of governing equations using the Gibbs free energies from Section 2.2 as thermodynamic potentials accessible from thermodynamic databases. Then, one can find governing equations for
- the phase field

$$\tau_\varphi\frac{\partial^2\varphi}{\partial t^2} + \frac{\partial\varphi}{\partial t} = -M_\varphi\frac{\delta G}{\delta\varphi}, \tag{14}$$

- the long-range order parameter

$$\tau_\eta\frac{\partial^2\eta}{\partial t^2} + \frac{\partial\eta}{\partial t} = -M_\eta\frac{\delta G}{\delta\eta}, \tag{15}$$

- the concentration and chemical potential

$$\tau_D \frac{\partial^2 x_B}{\partial t^2} + \frac{\partial x_B}{\partial t} = \nabla \cdot (M_x \nabla \mu_B), \quad \mu_B = \frac{\delta G}{\delta x_B}. \tag{16}$$

The hyperbolic equations (14) and (15) are damped wave equations, which extend known parabolic equation for the non-conserved order parameter derived first by Mandel'shtam–Leontovich [15, 16] and known in literature as the time–dependent Ginzburg-Landau equation [16, 17] or Allen-Cahn equation [18]. The hyperbolic equation (16) generalizes the known parabolic equation for the conserved order parameter derived and analyzed first by Khachaturyan [19] and Cahn and Hilliard [20]. Equations (14)-(16) were applied to non-monotonic relaxation [7, 8], phase segregation [9] and were analyzed in the context of fast dynamics during transitions from un(meta)stable to (meta)stable states [21]. The validity of hyperbolic type models in fast phase transition theory has been verified by comparison with experimental data [22], in molecular dynamics simulations of the solute trapping effect by rapidly moving interfaces [23] and by coarse graining derivations of phase field equations [24].

3.1 Phase-field

In its explicit form, the governing equation for the phase field comes from Eqs. (4)-(14). As a result one finds

$$\begin{aligned}
\frac{\tau_\varphi}{M_\varphi}\frac{\partial^2 \varphi}{\partial t^2} + \frac{1}{M_\varphi}\frac{\partial \varphi}{\partial t} &= \varepsilon_\varphi \nabla^2 \varphi - W_\varphi \frac{dg_\varphi(\varphi)}{d\varphi} - \Bigg[(1 - x_B)G_A^{bcc} + xG_B^{bcc} \\
&+ \frac{RT}{2}\Big((x_B - \eta/2)\ln(x_B - \eta/2) + (x_B + \eta/2)\ln(x_B + \eta/2) \\
&+ (1 - x_B - \eta/2)\ln(1 - x_B - \eta/2) + (1 - x_B + \eta/2)\ln(1 - x_B + \eta/2)\Big) \\
&+ \Omega_2(x_B(1 - x_B) + (1/4)\eta^2) + \Omega_3\eta^3 + \Omega_4\eta^4\Bigg]\frac{dp(\varphi)}{d\varphi} \\
&- \Big((1 - x_B)\big(G_L^{0A} + RT\ln(1 - x_B) + x_B\Omega_L\big) + x_B\big(G_L^{0B} + RT\ln(x_B) + (1 - x_B)\Omega_L\big)\Big)\frac{dp(\varphi)}{d\varphi}
\end{aligned} \tag{17}$$

Without the ordering process, i.e. if $\eta = 0$, Eq. (17) transforms into the one derived in Refs. [13, 14].

3.2 Order parameter

Atomic ordering in α- and β-sublattices is described by Eq. (15) together with the free energy (4)-(13) by the following equation:

$$\begin{aligned}
\frac{\tau_\eta}{M_\eta}\frac{\partial^2 \eta}{\partial t^2} + \frac{1}{M_\eta}\frac{\partial \eta}{\partial t} &= \varepsilon_\eta \nabla^2 \eta - W_\eta \frac{dg_\eta(\eta)}{d\eta} - \Bigg[\frac{RT}{2}\Big(-\frac{1}{2}\ln(x_B - \eta/2) + \frac{1}{2}\ln(x_B + \eta/2) \\
&- \frac{1}{2}\ln(1 - x_B - \eta/2) + \frac{1}{2}\ln(1 - x_B + \eta/2)\Big) \\
&+ \frac{1}{2}\Omega_2\eta + 3\Omega_3\eta^2 + 4\Omega_4\eta^3\Bigg]p(\varphi)
\end{aligned} \tag{18}$$

From Eq. (18) it follows that the process of atomic ordering is coupled with the phase field by the interpolation function $p(\varphi)$.

3.3 Chemical diffusion

The chemical potential from Eq. (16) is given by the variational derivative

$$\mu_B = \frac{\delta G}{\delta x_B} =$$

$$\left[G_B^{bcc} - G_A^{bcc} + \frac{RT}{2}\left(\ln(x_B - \frac{\eta}{2}) + \ln(x_B + \frac{\eta}{2}) - \ln(1 - x_B - \frac{\eta}{2}) - \ln(1 - x_B + \frac{\eta}{2})\right) + \Omega_2(1 - 2x_B)\right] p(\varphi)$$

$$+ \left[G_L^{0B} - G_L^{0A} + RT\{\ln(x_B) - \ln(1 - x_B)\} + 2(1 - 2x_B)\Omega_L\right] p(1 - \varphi). \quad (19)$$

The mobility M_x in the diffusion equation Eq. (16) is an interpolation between the bulk mobilities in the liquid M_L and in the solid M_S:

$$M_x = M_L(1 - p(\varphi)) + M_S(\eta)p(\varphi) = M_L(1 - p(\varphi)) + \left(M_S^{disorder}(1 - \eta) + M_S^{order}\eta\right) p(\varphi). \quad (20)$$

As follows from Eq. (20), the bulk mobility $M_S(\eta)$ in the solid is interpolated by atomic mobilities M_S^{order} and $M_S^{disorder}$ in the ordered state and disordered state, respectively.

Table 1. Physical parameters of the $A_{50}B_{50}$ alloy used for phase field modeling

Parameter	Value	Reference
Concentration of A(B), x_B	0.5 mole fraction	present work
η-rate relaxation time, τ_η	$4 \cdot 10^{-8}$ s	present work
φ-rate relaxation time, τ_φ	$4 \cdot 10^{-11}$ s	present work
x_B-rate relaxation time, τ_D	$4 \cdot 10^{-10}$ s	present work
Mobility of the η-field, M_η	$8 \cdot 10^4\ mole \cdot J^{-1} \cdot s^{-1}$	present work
Mobility of the φ-field, M_φ	$4 \cdot 10^5\ mole \cdot J^{-1} \cdot s^{-1}$	present work
Mobility of B-atoms in the L-phase, M_L	$2 \cdot 10^{-8}\ mole \cdot m^2 \cdot J^{-1} \cdot s^{-1}$	present work
Mobility of B-atoms in the disordered state, $M_S^{disorder}$	$7 \cdot 10^{-11}\ mole \cdot m^2 \cdot J^{-1} \cdot s^{-1}$	present work
Mobility of B-atoms in the ordered state, M_S^{order}	$9 \cdot 10^{-12}\ mole \cdot m^2 \cdot J^{-1} \cdot s^{-1}$	present work
Gradient factor for the η-field, ε_η	$1.4 \cdot 10^{-13}\ J \cdot m^2 \cdot mole^{-1}$	present work
Gradient factor for the φ-field, ε_φ	$5.5 \cdot 10^{-12}\ J \cdot m^2 \cdot mole^{-1}$	present work
Energy barrier between states in the η-field, W_η	$2.5 \cdot 10^3\ J \cdot mole^{-1}$	present work
Energy barrier between states in the φ-phase, W_φ	$4.5 \cdot 10^3\ J \cdot mole^{-1}$	present work
1st Thermodynamic parameter, Ω_2	$-2 \cdot 10^3 R$ J/mole	[2]
2nd Thermodynamic parameter, Ω_3	$-1.225 \cdot 10^6 R$ J/mole	present work
3rd Thermodynamic parameter, Ω_4	$-5.662 \cdot 10^2 R$ J/mole	[2]
4th Thermodynamic parameter, Ω_L	$-1.72 \cdot 10^4$ J/mole	[2]

4 Material parameters and modeling

The spatially inhomogeneous evolution of ordered and disordered states has been modeled by the numerical solution of the equations for diffusion (16), (19) and (20), phase field motion (17) and ordering (18). Material parameters have been chosen for a binary congruently melting $A_{50}B_{50}$ alloy. Functions $G_L^{0A}(T)$, $G_L^{0A}(T)$, $G_A^{bcc}(T)$, and $G_B^{bcc}(T)$ from Eqs. (7)-(9) and

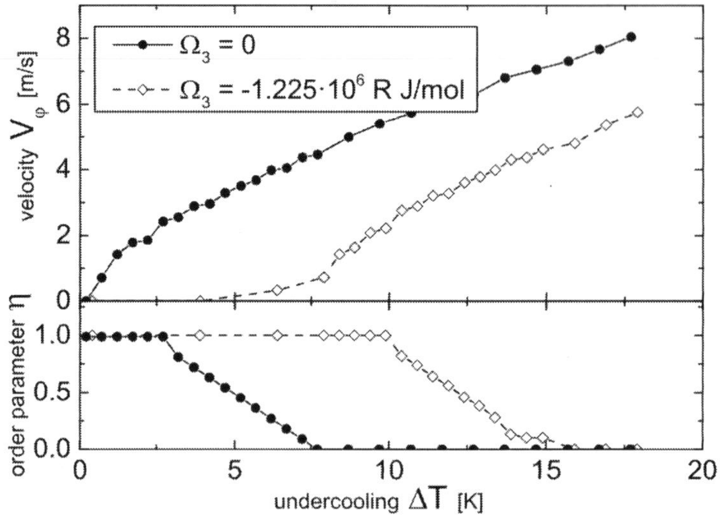

Figure 1. Behavior of the phase field interface velocity V_φ and long-range order parameter η as functions of liquid undercooling $\Delta T = T_L - T$, where T_L is the liquidus temperature (calculated as the temperature for "liquid–ordered phase" co-existence). Two cases show the change of the transformation kinetics depending on the thermodynamic parameter Ω_3 which regulates the barrier height for atomic diffusion jumps between sublattices α and β, see Eq. (13).

(11) are taken from the thermodynamical data base of elements (CALPHAD) for a $Ni_{50}Al_{50}$ alloy.

The numerical scheme for the solution of the model consists of a finite difference (FD) scheme of second order with explicit time stepping. The initial concentration distribution was taken as homogeneous, $x_B = 0.5$, while both phase field φ and order parameter η were initialized to unity in the solid and zero in the liquid, connected by a smooth transition at the interface, which was described by a Gaussian error function with a width of 1 micron, corresponding to 10 FD nodes. Boundary conditions were Dirichlet type conditions on the solidifying side and Neumann type (i.e., zero flux) conditions in the liquid far away from the interface.

Numerical solutions of Eqs. (16), (19), (20), 17), and (18) were achieved in a one-dimensional space. To obtain steady state results, average values of the interface velocity V_φ for the phase field and the long-range order parameter η where determined as follows. The phase-field has a step-like shape with $\varphi = 1$ in the solid and $\varphi = 0$ in the liquid. Because the solid-liquid interface is a diffuse interface, the conventional definition of the interface position is the point z_f where $\varphi(z_f) = 0.5$. In this work we adopt another definition of the interface position. $z_f = \int_{-\infty}^{+\infty} \varphi(z')dz'$ is exact for the case of a symmetric interface profile and allows us to find an exact interface velocity in case of a steady state profile. During the simulation run, we collect the values of the interface position z_f^n at times t^n after a fixed number (e.g. 2000) of time steps. The current interface velocity V_φ is obtained as $V_\varphi^n = (z_f^n - z_f^{n-1})/(t^n - t^{n-1})$. After the system reaches a steady state regime, we calculate *the steady state interface velocity* $V_\varphi^{(avg)}$ as the average of the last ten values V_φ^n. *The average value of the order parameter* η in the solid phase right beyond the interface is determined as the average of 11 FD nodes

 DOI: 10.1051/epjconf/201715105001

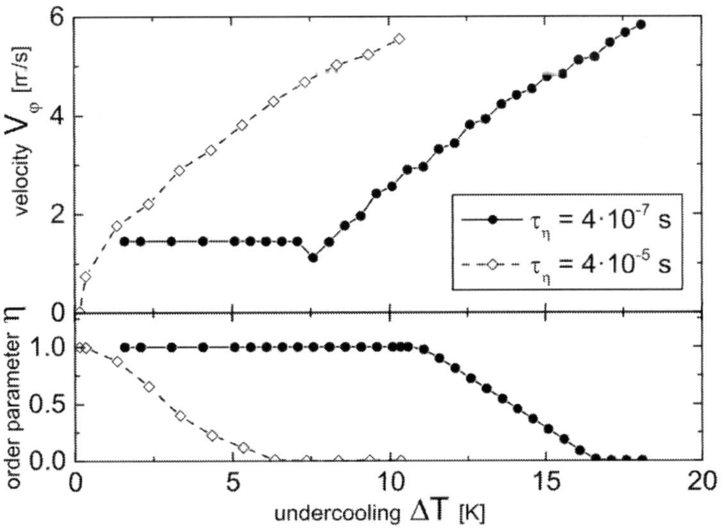

Figure 2. Behavior of the phase field interface velocity V_φ and long-range order parameter η as functions of liquid undercooling $\Delta T = T_L - T$, where T_L is the liquidus temperature (calculated as the temperature for "liquid–ordered phase" co-existence). Two cases show the change of the transformation kinetics depending of the time τ_η for relaxation of the ordering rate $\partial\eta/\partial t$ to change the regime from inertial (ballistic) to dissipative (diffusive) [21]. The η-gradient factor has been chosen as $\varepsilon_\eta = 1.4 \cdot 10^{-11}$ $(J \cdot m^2 \cdot mole^{-1})$.

at a distance of 10 FD spacings behind the interface, i.e. the FD nodes 10 to 20 behind the calculated interface position.

5 Results

Figures 1 and 2 show the influence of various parameters for the ordering process on both the transition from L-phase to S-phase and on the ordering process itself. Namely, Fig. 1 presents the kinetics change due to different values of Ω_3. It can be seen that with a non-zero value of the barrier Ω_3 [see Eq. (13)] the kinetics of transformation qualitatively changes with a decrease in V_φ and a shift in the disorder transformation velocity V_{DT} to its large values. Figure 2 illustrates the influence of gradient factor and relaxation time τ_η on the transformation kinetics. The time τ_η characterizes the relaxation to local equilibrium in the η-field, i.e., the relaxation of the rate $\partial\eta/\partial t$. An increase in the relaxation time τ_η moves the initiation of disorder trapping to smaller undercoolings.

6 Kinetics and temporal conditions for trapping of disorder

Finally, on the basis of our numerical results (see Section 5), we formulate a condition of disorder trapping by a moving diffuse interface. During fast transformations the disorder trapping effect occurs when a particular atom has no time to find its equilibrium position within the diffuse interface and remains in a higher energetic position. Instead of a phase with complete atomic order, a disordered metastable phase is appearing behind the diffuse interface. This effect becomes important at phase interface velocities V_φ larger than a characteristic

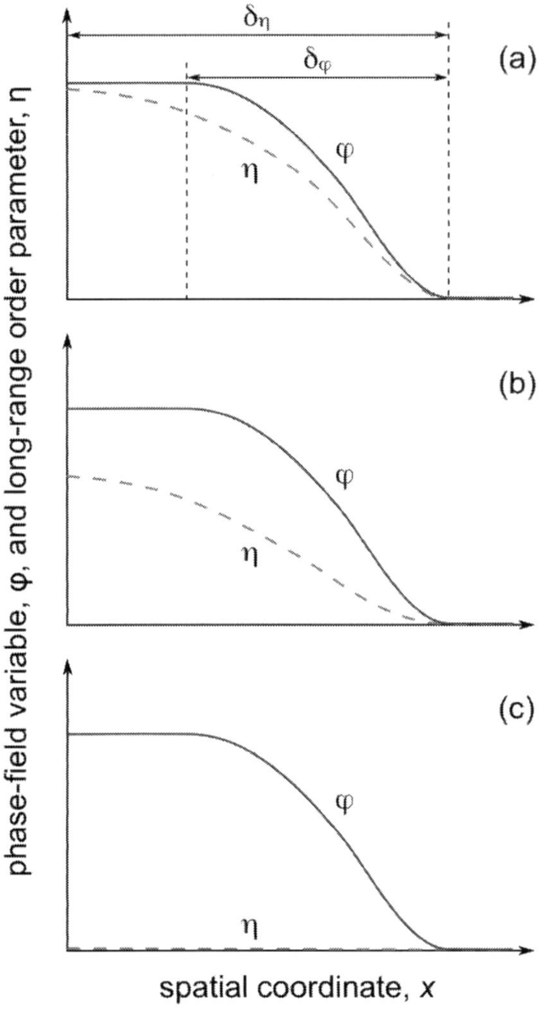

Figure 3. Profiles of the phase field ϕ and long-range order parameter η with their thickness δ_φ and δ_η, respectively. (a) Fully ordered state, $V_\varphi \ll V_{DT}$. (b) Partial disorder trapping, $V_\varphi \approx V_{DT}$. (c) Complete disorder trapping, $V_\varphi > V_{DT}$.

value of the order of $V_{DT} \propto D_\eta/\delta_\eta$ (with $D_\eta = M_\eta \varepsilon_\eta$ being the characteristic diffusion coefficient in solid phase for the ordering process, and δ_η the thickness of the interface for ordering). Indeed, when $V_\varphi > V_{DT}$, the characteristic time δ_φ/V_φ of the diffuse interface of thickness δ_φ becomes smaller than the typical time δ_η^2/D_η required for complete ordering within the diffuse interface. As a result, disorder trapping by the diffuse interface exists under the velocity condition $V > D_\eta/\delta_\eta$, or under the time scale condition $\delta_\varphi/V_\varphi < \delta_\eta^2/D_\eta = \delta_\eta/V_{DT}$. From these conditions it follows that the critical velocity for the disorder trapping, V_{DT}, becomes smaller as the ratio between the thickness of the ordering zone δ_η and the diffuse interface thickness δ_φ increases:

$$V_{DT} < \frac{\delta_\eta}{\delta_\varphi} V_\varphi. \tag{21}$$

As follows from the inequality (21), the disorder trapping becomes more pronounced with the increase of the diffuse interface velocity V_φ and the decrease of the diffuse interface thickness δ_φ. Furthermore, with a "sharper" ordering zone, i.e., with decreasing δ_η, the beginning of disorder trapping is shifted to larger phase interface velocities. These situations are illustrated in Fig. 3.

7 Conclusions

1. Using a phase field method in the theory of fast phase transformations, a new model for disorder trapping by a rapidly moving interface is formulated. The model consists of a system of hyperbolic equations for mass transport by atomic diffusion, dynamics of the phase field, and dynamics of atomic ordering.
2. The system of model equations has been solved numerically to reach steady-state from initially non-stationary states. For these steady-states, the interface velocity and long-range order parameter are analyzed depending on the type of free energy and relaxation time τ_η for the rate of long-range order parameter $\partial\eta/\partial t$.
3. It has been shown quantitatively that there are critical undercoolings at which the trapping of disorder begins and finishes with the formation of a completely disordered phase. Using the results of our computations, a criterion for disorder trapping by rapid diffuse interface has been formulated.

Acknowledgments

P. G. specially acknowledges the support the Russian Science Foundation (project no. 16-11-10095). I. N. acknowledges the support of Alexander von Humboldt Foundation (ID 1160779) and the Council on grants and scholarships (MK-3124.2015.5, SP-3122.2015.5). M.R. and K.R. acknowledge support from the Deutsche Forschungsgemeinschaft (DFG, grants no. Re 1261/8-2 and Re 1261/18-1).

References

[1] A.A. Chernov, Sov. Phys. JETP **26**, 1182 (1968)
[2] W.J. Boettinger, M.J. Aziz, Acta Metall. **37**, 3379 (1989)
[3] M. Barth, B. Wei, D.M. Herlach, Phys. Rev. B **51**, 3422 (1995)
[4] R. Ahmad, R.F. Cochrane, A.M. Mullis, J. Mater. Sci. **47**, 2411 (2012)
[5] C. Yang, J. Gao, J. Cryst. Growth **394**, 24 (2014).
[6] H. Hartmann, D. Holland-Moritz, P.K. Galenko, D.M. Herlach, Europhys. Lett. **87**, 40007 (2009)
[7] V.I. Lad'yanov, M.G. Vasin, S.V. Logunov, V.P. Bovin, Phys. Rev. B **62**, 12107 (2000)
[8] M.G. Vasin, V.I. Ladianov, Phys. Rev. E **68**, 051202 (2003)
[9] P. Galenko, Phys. Lett. A **287**, 190 (2001)
[10] E. Davidoff, P.K. Galenko, D.M. Herlach, M. Kolbe, N. Wanderka, Acta Mater. **61**, 1078 (2013)
[11] H. Assadi, Acta Mater. **55**, 5225 (2007)
[12] X.Q. Zhang, Y. Yang, Y.F. Gao, J.J. Hoyt, M. Asta, D.J. Sun, Phys. Rev. B **85**, 041601 (2012)
[13] P. Galenko, D. Jou, Phys. Rev. E **71**, 046125 (2005)
[14] V.G. Lebedev, A.V. Sysoeva, P.K. Galenko, Phys. Rev. E **83**, 026705 (2011)

[15] L.I. Mandel'shtam and M.A. Leontovich, JETP **7**, 438 (1937)
[16] A.Z. Patashinskii, V.L. Pokrovskii, *Fluctuation theory of phase transitions* (Pergamon, Oxford, 1979)
[17] E.M. Lifshitz, L.P. Pitaevskii, *Physical Kinetics. Course Theoretical Physics*, Vol 10, (Pergamon Press, Oxford, UK, 1981)
[18] S.M. Allen, J.W. Cahn, Acta Metall. **27**, 1085 (1979)
[19] A.G. Khachaturyan, *Theory of Structural Transformations in Solids* (Wiley, New York, 1983)
[20] J.W. Cahn, J.E. Hilliard, J. Chem. Phys. **28**, 258 (1958)
[21] P. Galenko, D. Danilov, V. Lebedev, Phys. Rev. E **79**, 051110 (2009)
[22] P.K. Galenko and D.M. Herlach, Phys. Rev. Lett. **96**, 150602 (2006)
[23] Y. Yang, H. Humadi, D. Buta D, B.B. Laird, D. Sun, J.J. Hoyt, M. Asta, Phys. Rev. Lett. **107**, 025505 (2011)
[24] D. Jou, P. Galenko, Phys. Rev. E **88**, 042151 (2013)

EPJ Web of Conferences **151**, 05002 (2017) DOI: 10.1051/epjconf/201715105002
LAM-16

Ab initio study of intrinsic profiles of liquid metals and their reflectivity.

B. G. del Rio[1,*], *J.* Souto[2], *M. M. G.* Alemany[2], and *L. E.* González[1]

[1]Departamento de Física Teórica, Universidad de Valladolid, Valladolid, Spain
[2]Departamento de Física de Partículas, Universidad de Santiago de Compostela, Spain

Abstract. The free surfaces of liquid metals are known to exhibit a stratified profile that, in favourable cases, shows up in experiments as a peak in the ratio between the reflectivity function and that of an ideal step-like profile. This peak is located at a wave-vector related to the distance between the layers of the profile. In fact the surface roughness produced by thermally induced capillary waves causes a depletion of the previous so called intrinsic reflectivity by a damping factor that may hinder the observation of the peak. The behaviour of the intrinsic reflectivity below the layering peak is however far from being universal, with systems as Ga or In where the reflectiviy falls uniformly towards the $q \to 0$ value, others like Sn or Bi where a shoulder appears at intermediate wavevectors, and others like Hg which show a minimum. We have performed extensive ab initio simulations of the free liquid surfaces of Bi, Pb and Hg, that yield direct information on the structure of the profiles and found that the macroscopic capillary wave theory usually employed in order to remove the capillary wave components fails badly in some cases for the typical sample sizes affordable in ab initio simulations. However, a microscopic method for the determination of the intrinsic profile is shown to be succesful in obtaining meaningful intrinsic profiles and corresponding reflectivities which reproduce correctly the qualitative behaviour observed experimentally.

1 Introduction.

Since the original prediction of Rice and coworkers that free liquid surfaces (FLS) of metals would exhibit surface layering in the density profile (DP) perpendicular to the interface [1], $\rho(z)$, almost a decade had to ellapse until this expectation was realized experimentally for the case of liquid Hg [2, 3], through the analysis of the x-ray reflectivity of the FLS, $R(q_z)$, which showed a peak related to the layering. Since then, up to 2009 [4], measurements have been performed for other five pure liquid metals, namely Ga, In, K, Sn and Bi [5]. All these studies were possible due to the advent of x-ray sources intense enough and reflectometers precise enough to measure reflectivities that decay by as much as 9/10 orders of magnitude as the wavevector increases above the position of the reflectivity peak. The experimental

*e-mail: beatriz@metodos.fam.cie.uva.es

requirements to make such experiments feasible, as discussed in [5], basically limit the set of pure liquid metals candidates to these studies to those already mentioned, plus liquid Pb and possibly liquid Ge. A particular source of complications due to the liquid nature of the samples is the presence of thermally induced capillary waves (CW) which blur the so-called intrinsic profile, $\rho_{\rm int}(z)$, leading to an average profile, $\rho_{\rm av}(z)$, and deplete even more the measured reflectivity, $R_m(q_z)$. The effects of CW are removed from $R_m(q_z)$ through the use of the macroscopic CW theory, that boils down to a division of $R_m(q_z)$ by a gaussian, Debye-Waller-like, factor that depends on a capillary roughness, $\sigma_{\rm cw}$, leading to the intrinsic reflectivity, $R_{\rm int}(q_z)$. In most cases, this is further divided by the Fresnel reflectivity, $R_F(q_z)$, that corresponds to a sharp flat surface and accounts for the overall decay of $R_{\rm int}(q_z)$. The resulting magnitude is called the surface structure factor, $|\Phi_{\rm int}(q_z)|^2$, which is directly related to the Fourier Transform of the electronic DP. Note that the capillary roughness depends on the temperature, T, and the surface tension, γ, of the liquid, on the atomic size, and on the lateral resolution of the detector.

The systems studied can be classified in three groups so far: those whose $|\Phi_{\rm int}(q_z)|^2$ decays monotonously from the layering peak towards $q_z \to 0$ (Ga, In and K), those that display a shoulder at an intermediate q_z (Sn and Bi) and those that display a minimum at an intermediate q_z (Hg). The origin of such different behaviors must lie in the different electronic DPs, but these are not directly available to experiments. Instead model profiles are proposed and the corresponding reflectivities are fitted to the measurements. The systems in the first group can be adequately described by a simple model named distorted crystal model (DCM), in which the DP is described by layers of equally dense and equispaced gaussians of linearly increasing variance as the bulk is approached. The other groups require some modifications to this model, changing either the positions or the density of some layer(s), in particular the outermost one. However the lack of uniqueness of model profiles that reproduce a given reflectivity poses a note to take into account in the interpretation.

An alternative route to study the FLS of metals is the use of computer simulations. These, however, rely on the availability of methods to describe properly the interaction among ions immersed in a valence electron sea, as adequate for a metal. In this respect, ab initio methods are certainly advantageous, since the underlying density functional theory (DFT) [6] shows a high degree of accuracy. Moreover, simulations give access to the atomic positions, so the DPs can be directly computed, and through them, the reflectivity. In fact, ab initio simulations have already been performed for the FLS of Si[7], Na[8], Sn[9], Hg[10] and Cd[11].

The sample size in simulations is utterly smaller than in experiments and consequently the abundance of CW is much reduced, but it is not null. It is therefore important to take into account this difference for proper comparison with experimental data. In previous studies the procedure adopted has been the same macroscopic CW theory as used in experiments, except for the aforementioned size differences. However a more direct access to the intrinsic profile and reflectivity is desirable, considering that the simulation samples are very far from being macroscopic. One such method was proposed by Chacon and Tarazona (CT) [12]. It relies on the calculation of a surface that separates the liquid and vapour phases, and sample the atomic positions with respect to this surface in order to obtain the intrinsic profile. The surface is computed so as to be the one that goes through the "surface atoms" and has "minimum area". Further methods have also been proposed, as the so called instaneous liquid-vapor interface [13], which has been used for the analysis of molecular dynamics simulations of the free surface of liquid water [14], but in this contribution we stick

Table 1. Simulation details.

	ρ_i (Å^{-3})	T (K)	N	$L_x = L_y$ (Å)	L_z (Å)	N_c
Hg	0.0403	300	116	13.03	17.10	8000
Bi	0.0288	800	160	13.57	38.69	27000
Pb	0.0309	850	160	13.57	37.33	27000

to the CT procedure, which has already proved fruitful in the analysis of the FLS of molten potasium [15].

In this contribution we report a reanalysis of previous ab initio molecular dynamics (MD) simulations of the FLS of Hg, as well as new calculations for Pb and Bi. We discuss the differences between the average and intrinsic profiles as obtained through the CT procedure, highlighting the advantages of the latter. Also, when available, we compare our reflectivities with the measured ones.

2 Theory.

The energy and the forces on the atoms for each MD configuration were computed through the use of DFT and pseudopotentials to describe the interaction between ions and valence electrons. The simulations for Hg [10] were performed using the Quantum Espresso (QE) code [16], and an ultrasoft pseudopotential with 12 valence electrons. In the cases of Pb and Bi we used the PARSEC code [17] along with norm-conserving pseudopotentials with 4 and 5 valence electrons respectively [18, 19]. The PARSEC code is very efficient, but it is not suited for ultrasoft pseudopotentials, which are a must for liquid Hg.

For the study of the FLSs we used N atoms in a slab geometry of initial width adequate to the experimental density, ρ_i, located in a tetragonal simulation box with $L_z > L_x = L_y$ and periodic boundary conditions. The vacuum region separating the replicas of each slab is taken large enough so that the inter-slab interaction can be safely neglected. After equilibration at the desired temperature N_c microcanonical configurations were generated from which the different properties were sampled. The particular values used for the different systems are shown in table 1.

The average electronic DP is obtained from the evaluation of a histogram of the z-coordinates of the particles, after representing each particle by the projection of the atomic electron density onto the z-direction.

The CT procedure to obtain the intrisic DP for a FLS relies on the computation, for each MD configuration, of a liquid-vapour separation surface, $z = \xi(x, y)$, so that the intrinsic DP is obtained from a histogram of the particles distances to this surface along the z-direction, i.e., according to their values of $z_{i;\mathrm{rel}} = z_i - \xi(x_i, y_i)$, where i runs over all the particles of the sample. The separation surface must comply with the periodic boundary conditions (PBC) of the simulation cell, so it is represented by its Fourier series. The separation surface should not oscillate within an atomic diameter. Therefore the series is terminated at q-values correponding to this size and compatible with the PBC, and consequently has a finite number of coefficients. These coefficients are determined by a constrained minimization of the surface area. The constraints imposed consist in the requirement that the separation surface passes through the surface atoms (pivots), i.e., $z_\ell = \xi(x_\ell, y_\ell)$, with ℓ running through the pivots. For each configuration 9 atoms are initially selected as pivots, by taking those whose z-coordinate is smallest among the atoms in each of the 3×3 prisms obtained by partitioning the simulation cell equally across the xy plane. After computing the minimal surface that passes through these pivots we locate the atom closest to this surface and include it as a new pivot. Further recalculations of minimal surfaces that pass through the pivots and inclusions of the closest atom as pivot are repeated until a specified number of pivots is reached, which is 21 atoms for Hg and 19 atoms for Pb and Bi. This number is self-consistently obtained from the method (and of course depends on the sample sizes), because the use of a different number of final pivots leads to intrinsic profiles with strange shapes that include shoulders at either the outermost layer or the first inner one. After the Fourier coefficients have been obtained one can optionally truncate the series, while keeping the coefficients, at lower cutoff wavevectors. This leads to a smoother separation surface which does not go exactly trough all the pivots, but close to them. Such separation surface is more physically appealing, since in the liquid there are not only capillary waves, but also other types of excitations that influence the vertical motion of the surface atoms, for instance transverse waves that propagate parallel to the surface. Due to the limited size of the samples used in ab initio simulations this further truncation of the series can only be performed one level, down to the next lower q compatible with the PBC.

Finally, after the DPs have been computed the reflectivities are obtained from the relations

$$\Phi(q_z) = \int_0^\infty dz\ \exp[-iq_z z]\frac{d\rho(z)/\rho_0}{dz} \qquad \frac{R(q_z)}{R_F(q_z)} = |\Phi(q_z)|^2. \tag{1}$$

where ρ_0 is the density of the bulk liquid. When $\rho(z)$ is the intrinsic profile these relations lead to the intrinsic reflectivity. However, when $\rho(z)$ is the average profile the resulting reflectivity function includes the CW present in the simulation. From this the corresponding intrinsic reflectivity, within the macroscopic CW theory, is obtained through the relation

$$\frac{R_{\text{int}}^{\text{CWT}}(q_z)}{R_F(q_z} = \frac{R_{\text{av}}(q_z)/R_F(q_z)}{\exp[-\sigma^2 q_z^2]} \tag{2}$$

where σ^2 includes the capillary roughness and also an intrinsic roughness, not really well defined but related to the atomic size [10].

3 Results and discussion.

3.1 Liquid Hg

The experiments on liquid Hg at several temperatures [3], without correcting for CW, yielded a $R_m(q_z)/R_F(q_z)$ with a layering peak at $q_z = 2.2$ Å^{-1} and a minimum at intermediate q between 0.6 and 1.2 Å^{-1} depending on the sample measured. A first attempt to fit the data with the DCM lead to an interlayer spacing of $\lambda = 2.72$ Å, but the fit was not good. Two modifications were proposed. The first one included a quite broad ad-layer of very small density (0.1 to $0.3\rho_0$) located a few Å into the vapour region. The second one corresponded to an interface layer broadened and a little depleted and with an expanded distance to the next layer. This second model was more in line with the one used in the first measurements [2], for which the first interlayer distance was $\lambda_1 = 3.05 \pm 0.15$ Å, and the rest were $\lambda = 2.76 \pm 0.20$ Å. The surface roughness was estimated as $\sigma \approx 1.0$ Å.

Our simulations [10] led to an average profile, shown in fig. 1, with $\lambda_1 \approx 2.95$ and $\lambda \approx 2.75$, in qualitative agreement with experiment. Moreover, the reflectivity,

 DOI: 10.1051/epjconf/201715105002

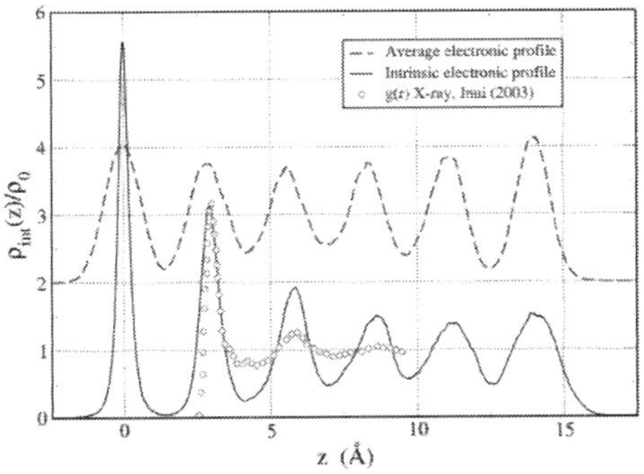

Figure 1. Intrinsic and average electronic density profiles for Hg at 300 K. Also shown is the experimental pair correlation function of the bulk liquid.

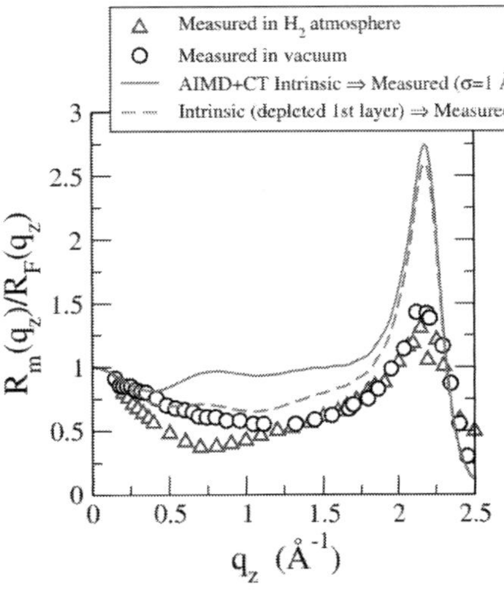

Figure 2. Reflectivities for Hg at 300 K.

modified through the CWT to be directly comparable to the measured one, showed a peak at a similar q_z as the experiment but did not reproduce the intermediate minimum. The intrinsic electronic profile computed through the CT procedure is also shown in fig. 1 and leads to the same λ_1 as the average profile, whereas the inner interlayer distance increases a little to $\lambda \approx 2.84$ Å. The layers are however narrower and higher than in $\rho_{\rm av}(z)$. The layer positions correlate very well with the available pair distribution function of the bulk liquid [20]. It is interesting to note that $\rho_{\rm int}(z)$ cannot be fit by the models used by experimentalists, since the first two layers don't have a gaussian shape, although they can be reproduced as the sum of two gaussians for each layer. The resulting intrinsic reflectivity has been modified through the same procedure and using the same parameters as in experiment [3] and is shown in fig. 2. It keeps the correct peak position, although the height is overestimated, but it still does not produce the dip found in the measurements. Nevertheless, the form of the reflectivity is very sensitive to small details in the profile, and a 10 % reduction of the height of the first layer does produce a correct minimum in the q_z region where it was experimentally observed. Therefore we can conclude that the expanded distance and a small depletion of the fist layer are behind the appearance of the minimum, in line with the second experimental model.

3.2 Liquid Pb and Bi

In table 1 one can observe a much higher number of configurations for these two systems than those for Hg. The main reason was the lack of convergence of the average DP as obtained directly from the simulations. We observed that further increase in N_c did not lead to any substantial improvement in the convergence. We performed several tests, using different simulation cells, different pseudopotentials (ultrasoft vs norm-conserving) and/or different computer codes (QE and VASP vs PARSEC), but the lack of convergence persisted in all cases. We consider that such problems in the simulation, not present in Hg, are due to the combination of a high temperature and low surface tension, both favouring large fluctuations in the height

DOI: 10.1051/epjconf/201715105002

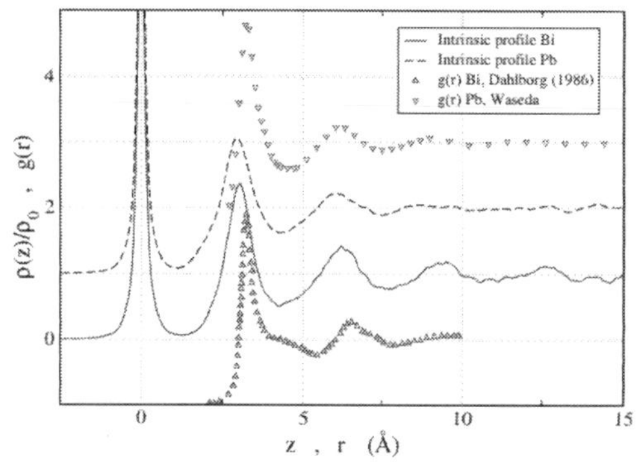

Figure 3. Intrinsic electronic density profiles for Bi and Pb. Also shown are the experimental pair correlation functions of the bulk liquids. The different functions are shifted vertically for clarity.

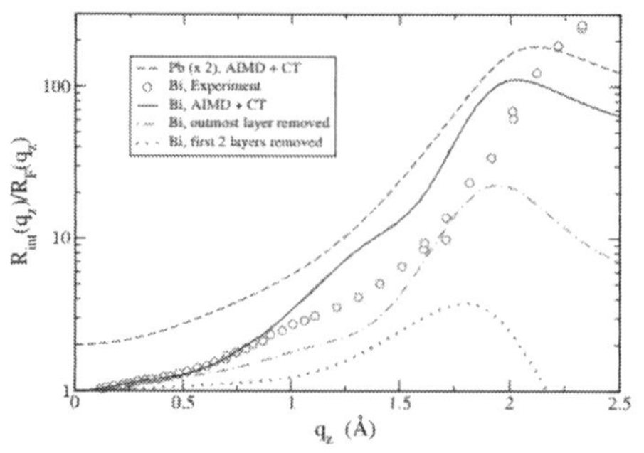

Figure 4. Reflectivities for liquid Bi and Pb (multiplied by two for clarity).

of the surface atoms, which are enhanced even more by the small number of them at the surface layer.

No experiments have been performed for the FLS of Pb. On the contrary, Bi (together with Sn) is one of the systems whose measured reflectivity, when transformed into the intrinsic one, shows a shoulder at intermediate q values [4] between 0.6 and 1.4 $\mathrm{\AA}^{-1}$. The authors constructed several models that could fit the measured reflectivity, with an adlayer on top of a DCM. The main conclusion was that the adlayer should be denser than the bulk in order to produce the shoulder, but different values of the width, position and density of the adlayer could do the job. The inner interlayer distance, as obtained from the DCM, was $\lambda \approx 2.83 \pm 0.05$ Å.

The intrinsic electronic DP obtained for Bi and Pb are shown in fig. 3, along with the corresponding experimental $g(r)$ [21, 22]. The shape of the layers is not DCM-like, requiring several gaussians for a proper fit. The layer positions in general agree with the maxima of $g(r)$ although some relaxation towards smaller distances is observed in the profile. Moreover, specific features as the shoulder of $g(r)$ in Bi, are reflected also in the shape of $\rho_{\text{int}}(z)$. The inner interlayer distance in Bi is found to be ≈ 3.12 Å, somewhat larger than the one experimentally determined through the fit to $R_m(q_z)$. Despite the many non-DCM features of the profile, its form is in line with one of the possible models put forward by Pershan and coworkers, with a very narrow outmost layer, at a slightly contracted distance to the inner layer ($\lambda_1 = 3.02$ Å) and with a slightly enhanced integrated density.

The corresponding intrinsic reflectivities are displayed in fig. 4 along with the experimental one for Bi [4]. In Pb we find a monotonous function, akin to those of Ga, In and K. For Bi we find, in agreement with experiments, a shoulder at intermediate q_z, although the magnitude of the reflectivity is rather overshot. To check the experimentalists suggestion that an enhanced first layer is responsible for the shoulder, we decreased the density of this layer, obtaining still a reflectivity with a shoulder. However, including one by one layers from the bulk towards the surface for computing $R_{\text{int}}(q_z)$ we find that the shouldered structure appears after the first inner layer is included. Consequently we can conclude that the two first layers are responsible for this kind of behaviour.

4 Conclusions.

The first important point to note is that the CT procedure leads to converged DPs, which is not always the case for the average profile, possibly due to a combination of high temperature and low surface tension.

The combination of AIMD simulations and the CT procedure leads to reflectivities which, either directly or with small modifications, agree qualitatively with experiments.

The profiles obtained are not in general well described by the DCM or variations thereof, but the general characteristics of the intrinsic profiles are similar to the models proposed by the experimentalists. We observe a high degree of correlation between the bulk $g(r)$ and the characteristics of the intrinsic profile, although with some relaxation in the distances, which become somewhat smaller. The particular behaviour of the reflectivities has been traced back to the depletion of the first layer in the case of Hg, and to a combined effect of the first two layers in the case of Bi.

We acknowledge the support of MECD (FIS2012-33126 and FIS2014-59279-P), Xunta de Galicia (AEFIS: AGRUP2015/11) and JCyL (VA104A11-2). BGR acknowledges the financial support of the Universidad de Valladolid.

References

[1] M. P. d'Evelyn and S. A. Rice, Phys. Rev. Lett. **47**, 1844 (1981).
[2] O. M. Magnussen *et al*, Phys. Rev. Lett. **74**, 4444 (1995).
[3] E. DiMasi *et al*, Phys. Rev. B **58**, R13419 (1998).
[4] P. S. Pershan *et al*, Phys. Rev. B **79**, 115417 (2009).
[5] P. S. Pershan and M. L. Schlossman, *Liquid Surfaces and Interfaces. Synchrotron X-ray Methods*, (New York: Cambridge, 2012), and references therein.
[6] P. Hohenberg and W. Kohn, Phys. Rev. **136**, B864 (1964); W. Kohn and L.J. Sham, Phys. Rev. **140**, A1133 (1965).
[7] G. Fabricius *et al*, Phys. Rev. B **60**, R16283 (1999).
[8] B. G. Walker, C. Molteni and N. Marzari, J. Phys.: Condens. Matter **26**, S2575 (2004).
[9] L. Calderín, L. E. González and D. J. González, Phys. Rev. B **80**, 115403 (2009).
[10] L. Calderín, L. E. González and D. J. González, Phys. Rev. B **87**, 014201 (2013).
[11] L. Calderín, L. E. González and D. J. González, J. Phys.: Condens. Matter **25**, 065102 (2013).
[12] E. Chacón and P. Tarazona, Phys. Rev. Lett. **91**, 166103 (2003).
[13] A. P. Willard and D. Chandler, J. Phys. Chem. B **114**, 1954 (2010).
[14] J. Kessler *et al*, J. Phys. Chem. B **119**, 10079 (2015)
[15] E. Chacón, P. Tarazona and L. E. González, Phys. Rev. B **74**, 224201 (2006).
[16] P. Gianozzi *et al*, J. Phys.: Condens. Matter **21**, 395502 (2009).
[17] M. M. G. Alemany *et al*, Phys. Rev. B **69**, 075101 (2004); L. Kronik *et al*, Phys. Status Solidi B **243**, 1063 (2006); M. M. G. Alemany *et al*, Comput. Phys. Comm. **177**, 339 (2007).
[18] M. M. G. Alemany *et al*, Phys. Rev. B **76**, 214203 (2007).

[19] J. Souto *et al*, Phys. Rev. B **81**, 134201 (2010).
[20] M. Inui, X. Hong and K. Tamura, Phys. Rev. B **68**, 094108 (2003).
[21] U. Dahlborg and M. Davidovič, Phys. Chem. Liq. **15**, 243 (1986).
[22] Y. Waseda, http://res.tagen.tohoku.ac.jp/~waseda/scm/LIQ/all/pb.html

Statistics and Thermodynamics of Fe-Cu Alloys at High Temperatures.

Leonid Son[1,*], *Valery* Sidorov[1,2] , and *Nikolay* Katkov[1,2]

[1]Ural Federal University, 19 Myra st., Ekaterinburg, Russia
[2]Ural State Pedagogical University, 26 Cosmonavtov ave., Ekaterinburg, Russia

Abstract. Statistics and thermodynamics of binary Fe-Cu system is described in terms of local state representation combined with the account of nonergodic microheterogeneity in terms of frozen fluctuations of chemical potential. It is shown that the microheterogeneity in this system is an equilibrium property up to 1700^0 C.

1 Introduction

Fe-Cu system is a binary system with limited miscibility of the components. Most of such systems correspond to the eutectic or monotectic type of the concentration – temperature phase diagram, but Fe-Cu system demonstrates peritectic triple point [1] and strong microheterogeneity in the liquid state [2].

In [3], we suggested the description of binary systems with limited miscibility, which is based on the local state representation of condensed matter statistics and thermodynamics. The description arises from the melting theory [4] based on the Potts model [5].
In the theory, it is suggested that small local volume (first coordination shell) of condensed matter may be characterized by strict local order with local anisotropy. The interaction between neighboring local volumes makes the coinciding mutual orientation of the local anisotropy axis preferable, i.e. corresponding to the energy minima. The disorientation leads to the energy rising. Thus, one can divide all possible orientations into n orientation shells, and suppose that the energy of interaction of nearest neighbors is $-J$, if their orientations get into one and the same shell, and 0 otherwise. The parameters J,n characterize the depth and the width of the orientation interaction, respectively. The system is obviously being described by the n - states Potts model, which demonstrates first order phase transition from the orientationally ordered state at low temperatures (all local volumes get into one and the same orientation shell) to the disordered one (all orientations have equal probabilities) at high temperatures. It is natural to interpret the transition as a melting. In [3], we applied this approach to binary systems and show that systems with limited miscibility may be successfully described.

In [6], we introduced a way to account nonergodic microheterogeneity in binary systems, which arises from the dynamical limitations imposed on particles diffusion. It was accounted by frozen fluctuations of chemical potential. Their mean square value serves as

* Corresponding author: ldson@yandex.ru

an additional thermodynamic parameter. As a result, one gets a possibility to calculate boundaries of microheterogeneity at the phase diagram.

In the present work, we combine the approaches [3,6] to describe the Fe – Cu system.

2 The model

2.1 Local state modeling

The state of small volume in binary system located at point r may be described by 2 – index local state vector $\sigma_k^i(r)$, where the upper index $i = \overline{1,2}$ corresponds to the component, and the lower numerates the orientation. Here, the overbar denotes the limits of the index. In general case, $k = \overline{1,n}$ when $i = 1$, and $k = \overline{1,m}$ when $i = 2$. Thus, the numbers n, m characterize the width of orientation interaction for the first and second components respectively. The vector is defined by the rule

$$\sigma_k^i(r) = \begin{cases} 1, \textit{ if } r \textit{ is occupied by component } i \textit{ in } k^{th} \textit{ orientation} \\ 0, \textit{ otherwise} \end{cases} \quad (1)$$

The Hamiltonian of system in terms of these variables is

$$H\{\sigma\} = \frac{1}{2}\sum_{r,r'} \sigma_k^i(r) M_{kl}^{ij}(r-r') \sigma_l^j(r') + \sum_r \mu_k^i(r)\sigma_k^i(r),$$
$$\mu_k^i = \begin{cases} \mu, i = 1 \\ 0, i = 2 \end{cases} \quad (2)$$

Here, μ is the chemical potential, and kernel $M_{kl}^{ij}(r-r')$ differs from zero only for nearest neighbors and realizes the idea of orientation interaction

$$M_{kl}^{11} = -J_1\delta_{kl}, \quad M_{kl}^{22} = -J_2\delta_{kl}. \quad (3)$$

In [3], inter - component terms of the kernel were taken in the form

$$M_{kl}^{12} = M_{kl}^{21} = \varepsilon, \ \forall k,l \quad (4)$$

This form is appropriate, if the two components differ sharply from each other (have different local structures or interatomic distances) – then one can suppose that if two different local structures share a boundary, then the energy of their interaction is ε regardless of their orientation. For the Fe-Cu system, the interaction kernel should be modified. First of all, the two metals have similar local structures (corresponding to FCC close packed lattice), so that we keep the form of eq.(3) and choose $m = n$. Besides, if the two local structures are similar to each other, then coinciding orientations are preferable for nearest neighbors in the case of different components also, and instead of (4) one writes

$$M_{kl}^{12} = M_{kl}^{21} = \varepsilon + J_0\delta_{kl} \tag{5}$$

This form of interaction contains itercomponent tension ε and energy benefit J_0 for coinciding orientations. The next step is to calculate the partition and thermodynamic potential

$$Z = \sum_{\{\sigma\}} \exp\left[-\frac{H\{\sigma\}}{T}\right], \quad F = -T\ln Z \tag{6}$$

This can be made in the mean – field approximation, which implies introduction of the mean value

$$<\sigma_j^i(r)>= \omega_j^i \ , \tag{7}$$

in terms of which the unequilibrium thermodynamic potential should be written out. As a result, one gets the Landau theory, where (7) plays the role of order parameter. In detail, the procedure is presented in [3]. The result is as follows. The nonequilibrium Landau potential is given by

$$\begin{gathered} F = \frac{1}{2}\omega_k^i M_{kl}^{ij}\omega_l^j - T\ln Z_{MF} \ , \\ Z_{MF} = \sum_{i,k} \exp\left(\frac{1}{T}M_{kl}^{ij}\omega_l^j + \frac{1}{T}\mu_k^i\right). \end{gathered} \tag{8}$$

The structure of the order parameter (7) is

$$\begin{gathered} \omega_1^1 = x\sigma_1, \quad \omega_{k\neq 1}^1 = x\frac{1-\sigma_1}{n-1}, \\ \omega_1^2 = (1-x)\sigma_2, \quad \omega_{k\neq 1}^2 = (1-x)\frac{1-\sigma_2}{n-1} \end{gathered} \tag{9}$$

Here, x is the first component concentration. We assigned number 1 to the orientation that is preferable in the crystal state, so that σ_1, σ_2 are the parameters of crystal order (for the first and second components respectively): in the crystal state, these values are close to unity, while in the liquid $\sigma_{1,2} = \frac{1}{n}$. In terms of (3-5, 9), quadratic and logarithmic parts of this expression take the form

$$F_{quad} = \frac{J_1}{2}x^2\sigma_1^2 + \frac{J_1}{2}x^2\frac{(1-\sigma_1)^2}{n-1} +$$
$$+\frac{J_2}{2}(1-x)^2\sigma_2^2 + \frac{J_2}{2}(1-x)^2\frac{(1-\sigma_2)^2}{n-1} + \tag{10}$$
$$+J_0x(1-x)\left(\sigma_1\sigma_2 + \frac{(1-\sigma_1)(1-\sigma_2)}{n-1}\right) + \varepsilon x(1-x),$$
$$Z_{MF} = E_1^{(1)} + (n-1)E_2^{(1)} + E_1^{(2)} + (n-1)E_2^{(2)},$$

where

$$E_1^{(1)} = \exp\left[\frac{J_1}{T}x\sigma_1 + \frac{J_0}{T}(1-x)\sigma_2 + \frac{\varepsilon}{T}(1-x) + \frac{\mu}{T}\right]$$
$$E_2^{(1)} = \exp\left[\frac{J_1}{T}x\frac{1-\sigma_1}{n-1} + \frac{J_0}{T}(1-x)\frac{1-\sigma_2}{n-1} + \frac{\varepsilon}{T}(1-x) + \frac{\mu}{T}\right]$$
$$E_1^{(2)} = \exp\left[\frac{J_2}{T}(1-x)\sigma_2 + \frac{J_0}{T}x\sigma_1 + \frac{\varepsilon}{T}x\right] \tag{11}$$
$$E_2^{(2)} = \exp\left[\frac{J_2}{T}(1-x)\frac{1-\sigma_2}{n-1} + \frac{J_0}{T}x\frac{1-\sigma_1}{n-1} + \frac{\varepsilon}{T}x\right]$$

Minimizing F with respect to ω_k^i at fixed temperature T, one gets mean field equations on the crystal order parameters

$$\sigma_1 = \frac{E_1^{(1)}}{E_1^{(1)} + (n-1)E_2^{(1)}}, \quad \sigma_2 = \frac{E_1^{(2)}}{E_1^{(2)} + (n-1)E_2^{(2)}} \tag{12}$$

and equation on the phase separation

$$x = \frac{E_1^{(1)} + (n-1)E_2^{(1)}}{Z_{MF}}. \tag{13}$$

The scheme of the phase diagram calculation is as follows. First, at fixed temperature T and concentration x one has to solve equations (12). Each of these equations may have two stable solutions $\sigma_{1,2} \approx 1$, and $\sigma_{1,2} = \frac{1}{n}$. The first corresponds to the crystal, and second – to the liquid.

Then, for all solutions of (12), which represent possible phases, the thermodynamic potential (8) has to be plotted as a function of (x,T), afterwards the phase diagram may calculated by standard shared tangent method. Equation (13) serves to indicate phase separation in liquid: when this equation has single solution at all values of chemical potential μ , then the phase separation is impossible, and it takes place if multiple solutions realize while changing μ

2.2 Account of nonergodic diffusivity

In [6], it was shown that nonergodic diffusivity may be accounted in terms of random frozen spatial fluctuations of chemical potential. The corresponding corrections may be derived by substituting

$$\frac{\mu}{T} \Rightarrow \frac{\mu}{T} + \frac{\Delta\mu}{T}, \tag{14}$$

where $\Delta\mu$ is random spatial fluctuation. Then one averages (10, 12, 13) over these fluctuations. Since equations (12) contain no μ , they do not experience any changes. Correction to the thermodynamic potential is

$$\begin{aligned} F_{corr} &= F + \frac{\partial^2 F}{\partial(\mu/T)^2} C = F_{quad} - T\ln Z_{MF} - \\ &- TC\left(\frac{E_1^{(1)} + (n-1)E_2^{(1)}}{Z_{MF}} - \frac{(E_1^{(1)} + E_2^{(1)})^2}{Z_{MF}^2} \right) \end{aligned} \tag{15}$$

where the dimensionless constant C is the mean square of $\Delta\mu/T$ fluctuations and should be treated as an additional phenomenological parameter. Corrected form of eq.(13) is

$$\begin{aligned} x &= \frac{E_1^{(1)} + (n-1)E_2^{(1)}}{Z_{MF}} + C\frac{E_1^{(1)} + (n-1)E_2^{(1)}}{Z_{MF}^3} \times \\ &\times \left(1 - 3\frac{E_1^{(1)} + (n-1)E_2^{(1)}}{Z_{MF}} + 2\frac{(E_1^{(1)} + (n-1)E_2^{(1)})^2}{Z_{MF}^2} \right) \end{aligned} \tag{16}$$

In the next section we calculate the Fe-Cu phase diagram using (12, 15, 16).

3 Phase diagram calculation.

We start with determination of model parameters, which allow one to fit the experimental diagram presented in fig.1. We did not take into account the phase transformation in iron near melting point, and treated iron having FCC local order all over the diagram. Also, we do not consider the low – temperature $\gamma \leftrightarrow \beta$ transformation.

First, for the close – packed systems, number of orientation shells $n = 12$ is appropriate [4]. Parameters J_1, J_2 were chosen to reproduce melting temperatures of the pure components:

$$J = T_{melt}\frac{2(n-1)}{(n-2)}\ln(n-1)\ . \tag{17}$$

DOI: 10.1051/epjconf/201715105003

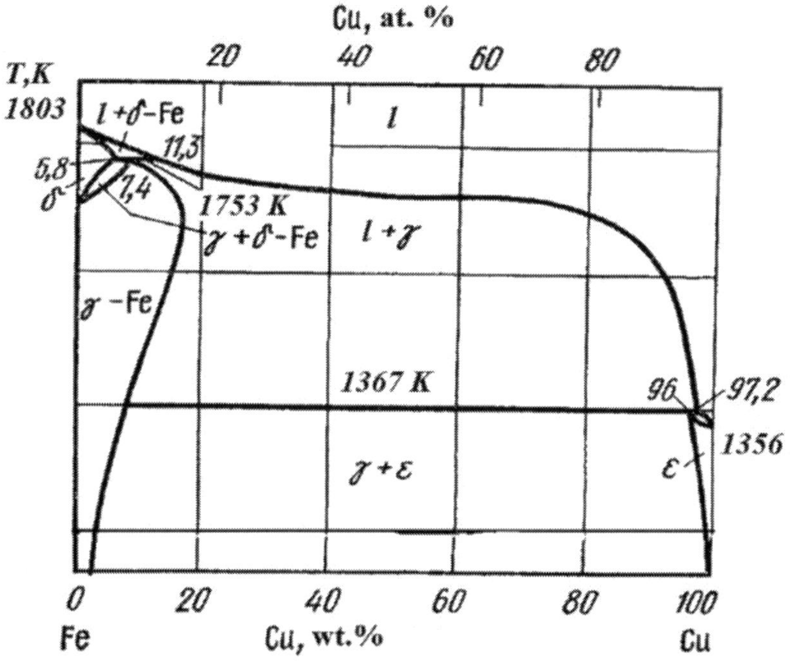

Fig. 1. Experimental Fe-Cu phase diagram according to [1].

We used energy scale in Kelvin degrees, so that $J_1 = 9443$ and $J_2 = 7153$. Parameter J_0 was chosen to reproduce the slope of liquidus lines near pure components, $J_0 = 8400$. Parameters

ε, C were determined from peritectic temperature and critical point in the undercooled illiquid. The later is $T_c = 1670^0 K$ [8]. We used the relation

$$T_c = (1 - \frac{C}{2})\left(\frac{J_1 + J_2 - 2J_0}{4n} - \varepsilon\right), \tag{18}$$

which may be derived from (16). We varied ε to reproduce the peritectic triple point, and got $\varepsilon = -4010$, $C = 0.32$.

The results of calculation are presented in Fig.2. Equilibrium diagram is drawn by solid lines. One can see that the diagram is reproduced rather well. The lower dashed line corresponds to the phase separation [8] in undercooled liquid. The upper dashed curve eliminates area, where nonergodic microheterogeneity [6] takes place (shadowed area in fig.2). It was calculated as a miscibility cupola according to eq. (13). Nonergodic microheterogeneity in Fe-Cu liquid system was discovered in [2] by viscosity measurements, and justified in [8]. Note, that dashed lines do not correspond to any line of equilibrium phase transition. Thus, we can to fit the phase diagram by 6 – parameter model, including melting and phase separation in undercooled liquid, and to explain the features of system behavior in the liquid state. Thus, the thermodynamics of melting and dissolution in the Fe-Cu system is described rather well.

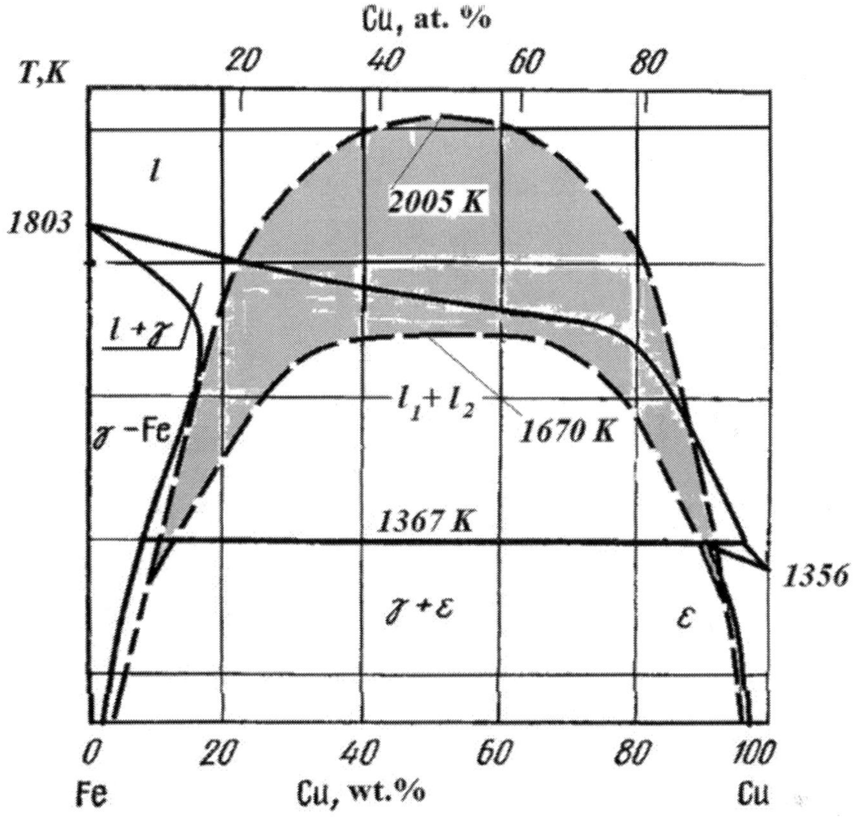

Fig.2. Phase diagram, calculated in the present paper.

The work was supported by Russian Ministry of Science and Education (Government job 2014/392 projects 2391 and 1177), and by Government of the Russian Federation, act 211 contract № 02.A03.21.0006. L.Son is also grateful to RFBR (grant 14-02-00359)

References

1. Lyakishev N.P., Bannykh O.A., Rokhlin L.L., et. al. *Phase Diagrams of Binary Metallic Systems.* Handbook. Moscow, Mashinostrojenie, 1996 (in Russian)A.
2. V. S. Guschin, D. B. Shulgin, B. A. Baum, and T.P.Buhler, Metallofizika **11** (6), 33 (1989)
3. Son L., Patashinski A., Rusakov G., Ratner M., Physica A, **248** (3-4), 305-322 (1998)
4. A.C.Mitus, A.Z.Patashinski, Sov.Phys. JETP, **53**, 798 (1981); Phys. Lett. **87A**, 179 (1982)
5. F.Y.Wu. Rev.Mod.Phys, **54**, 239 (1982).
6. L.D.Son .Physica A, **449C**, 395-400 (2016)
7. Turchanin M.A., Agraval P.G., Nikolaenko I.V. J. Phase Equilibria, **24,** (4) 307-319 (2003)
8. O. A. Chikova, M. A. Vityunin, V. P. Chentsov, and G.V. Sakun, Colloidnyi Zhurnal **72** (2), 1–7 (2010)

New potentials for multiscale simulations of liquid metals

Andrey Sobolev[1,*], *Viktor* Starukhin[1], *Ivan* Buldashev[1], and *Alexander* Mirzoev[1]

[1]South Ural State University, Chelyabinsk, Russia

Abstract. We present the technique and results for finding norm-conserving pseudopotentials and EAM potentials that can be used to recover atomic and electronic structure of liquid iron at and above the melting point. Pseudopotentials were found by minimizing the energy differences of our results with all-electron reference methods; EAM potentials — by the modified hybridization method proposed earlier by Belashchenko. We show that these potentials are at least as accurate in describing liquid iron as the established potentials in the field.

1 Introduction

In the last decade multiscale computer simulations have transitioned from the category of curiosities to the toolbox of practicing material scientists. Increasing computational resources and no need for setting of expensive experiments make such simulations an essential step in the development of new materials. Multiscale modelling is typically understood [1] as the modelling from the smallest scale (i.e., electrons and atoms) to the full system level (e.g., autos). Every step in this multiscale hierarchy requires its own simulation method — from quantum mechanics to continuum modelling — which depends on a set of parameters obtainable from previous steps. Each method, in its turn, has its own assumptions and limitations, that can greatly influence the obtained results.

The first two steps in the multiscale modelling hierarchy usually comprise ab initio electronic simulations, done with methods based on the density functional theory (DFT) [2], and atomic scale simulations with classical molecular dynamics. Ab initio simulation methods are based on the solution of Schrödinger equation for a system of particles. The DFT methods suggest searching for the solution of the equation in the form of one-electron wavefunction, which can be expanded in terms of some basis — typically either atomic orbitals or plane waves are used. If the electron density is close to the atomic nucleus, large gradients occur in it, that can be described properly only by enlarging the expansion basis substantially. For such electrons pseudopotentials are used, that replace the Coulomb repulsion usually present in Schödinger equation. The choice of pseudopotentials and basis sets are two main sources of errors within the DFT framework. Classical molecular dynamics, in its turn, relies heavily on the choice of interatomic potentials, which are usually constructed largely basing on the author's intuition, as the choice of the fitting parameters is very wide. We tried to tackle the problem sequentially: first, we estimated the error of ab initio calculations of liquid iron; then, we tried to minimize it; and finally on the basis of ab initio data we built the series of EAM potentials and chose the most adequate one. In this paper we present the findings.

*e-mail: sobolevan@susu.ru

2 Computational details

A prerequisite for the error analysis is the repeatability of results — different software implementations of the same DFT formalism must give identical predictions for identical systems. Recently Lejaeghere et al [3] developed a technique to consistently determine the Δ-factor — the single number that describes the difference in the predictions of different methods. The technique was applied to a variety of DFT methods, however, the majority of them use plane waves as a basis. We used the technique to find Δ-factor for SIESTA software [4], that uses atomic orbitals, and wrote the PseudoGenerator code [5] to minimize Δ-factor in comparison with de-facto standard Wien2K and VASP software packages.

Later, atomic and electronic structure of crystalline and liquid iron was found by ab initio molecular dynamics (AIMD). AIMD simulations were made with SIESTA software package for supercells of 200 atoms of iron at 4 temperatures from the range of liquid phase existence: 1833K, 1873K, 1923K and 2023K. Siesta implements Car–Parrinello molecular dynamics in the basis of numerical atomic orbitals, with core electrons represented by non-local norm-conserving pseudopotentials [6]. Standard DZP basis was employed, with GGA as the approximation of exchange and correlation. As local magnetic moments on iron atoms exist also above Curie temperature, we employed spin polarization for all AIMD runs. Previously we performed the study of influence of local magnetic moments on the structure of liquid iron [7] and found that it is necessary to properly account for magnetic effects to get the structure RDF close to experimental. The same conclusion was made by Ganesh and Widom in [8].

Based on ab initio atomic structure, we constructed a set of EAM potentials using a modification of the method developed by Belashchenko in [9] for recovering an EAM potential from an RDF. The details of algorithm will be published elsewhere; here we will give only an outline.

To reconstruct EAM potentials, first we needed to establish a set of hybrid effective pair potentials (EPPs). It is known that one can approximately represent an EAM potential as the effective pair potential corresponding to the average electron density in a system under fixed thermodynamic conditions:

$$\phi_{EAM}(r) \approx \phi(r) + 2\psi(r)\left.\frac{dF(\rho)}{d\rho}\right|_{\rho=\langle\rho\rangle}. \tag{1}$$

Here $\phi_{EAM}(r)$ is the effective pair representation of EAM potential, and $\langle\rho\rangle$ is the mean electron density in the system at the given conditions. The key point of the Belashchenko method is the choice of the pair part recovered from RDF as $\phi(r)$. He proved, that if some pair potentials reconstruct the RDF for the liquid, then the linear combination of the found potentials also reconstructs the target RDF:

$$\phi_{hyb}(r) = \lambda_1\phi_1(r) + \lambda_2\phi_2(r) + \lambda_3\phi_3(r), \tag{2}$$

given that

$$\lambda_1 + \lambda_2 + \lambda_3 = 1. \tag{3}$$

$\phi_i(r)$ here are the basis effective pair potentials, and λ_i are arbitrary real numbers. We employed a method by Schommers [10], which is used to generate pair potentials from a known RDF, to obtain three linearly independent pair potentials. The potentials are exploited later to get a set of hybrid pair potentials from Eq. (2), which differ by the choice of λ_i. Any set of λ_i coefficients from the known range (range bounds are determined by the thermodynamical stability of the system) does not distort significantly the model RDF curve in NVT ensemble. λ_i can be selected to fit several physical properties of the modelled system. Namely,

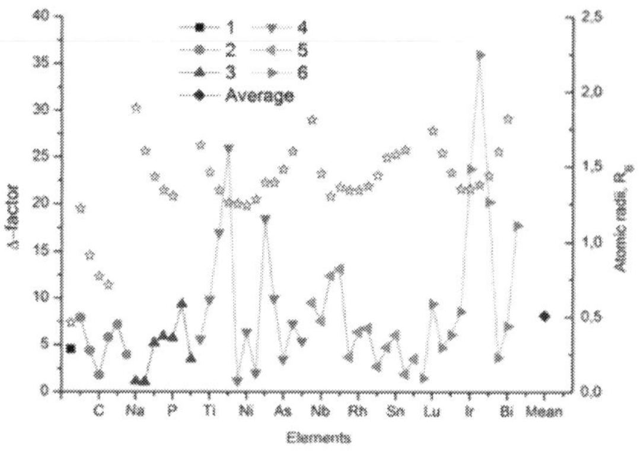

Figure 1. Δ-factor for SIESTA software package compared to all-electron calculations. The coloured symbols distinguish the elements from different periods of the Periodic Table (see legend for correspondence between symbols and period numbers); atomic radii are also shown (empty star symbols).

Table 1. The properties of BCC iron at 0K, acquired by different methods

	Magnetic moment, μ_B	Lattice constant, Å	Δ-factor
Experiment [13]	2.22	2.866	–
Wien2K	2.27	2.87	–
VASP [14]	2.24	2.87	–
SIESTA [15]	2.31	2.91	25.3
SIESTA, present work	2.21	2.862	15.3

we use one of these coefficients to optimize hybrid EPPs to experimental atomic density in NPT ensemble. With condition (3), we have one free parameter λ left. In this paper, we used this parameter to fit the potential to a number of physical properties of the system, both experimentally measured and found by ab initio simulations.

With such a pair part, EAM potential also reproduces target RDF, provided the derivative of embedding function in (1) is equal to or sufficiently close to zero. Embedding function was computed from the universal equation of state for metals by Rose et al [11]. Effective electron density $\psi(r)$ was chosen in the form of decaying exponent.

3 Results

3.1 Norm-conserving pseudopotential for iron

Employing the technique described in [3], we calculated Δ-factors for more than 50 different elements with SIESTA software package, using default pseudopotentials (taken from [12]); the results are shown in Fig. 1 along with atomic radii. It can be seen, that the difference between SIESTA and all-electron calculation results grows with the period number. Apart from this, for each period Δ-factor is larger for the elements with the smallest atomic radius, i.e. situated in the middle of the period. Two notable outliers are iron and platinum.

To decrease Δ-factor for iron we used two pseudopotentials for iron — the first proposed by Izquierdo et al [15], with 7 valence electrons on the 3d orbital and 1 valence electron on the 4s orbital, with cut-off radii for each orbital equal to 2 Bohr (formula

$4s^1(2.0)4p^0(2.0)3d^7(2.0)4f^0(2.0)$), and the second was fitted to minimize Δ-factor (we used PseudoGenerator code to do the fitting). The resultant pseudopotential has the formula $4s^2(2.0)4p^0(2.18)3d^6(2.0)4f^0(2.0)$; the calculated properties of BCC iron at 0K are shown in Tab. 1, along with reference data.

It can be seen from the table, that the change of pseudopotential improves both the lattice constant and the magnetic moment of iron; the resulting magnetic moment appearing even closer to the experimental data than that from Wien2K and VASP calculations. To check if the pseudopotential can be applied to the calculations of structure in the liquid state, we have simulated the structure of liquid iron, which is shown in Fig. 2 (along with experimental data and RDFs made from classical MD calculations). It can be seen, that the structure recovered with the new pseudopotential is at least as close to the X-ray diffraction data as the old one.

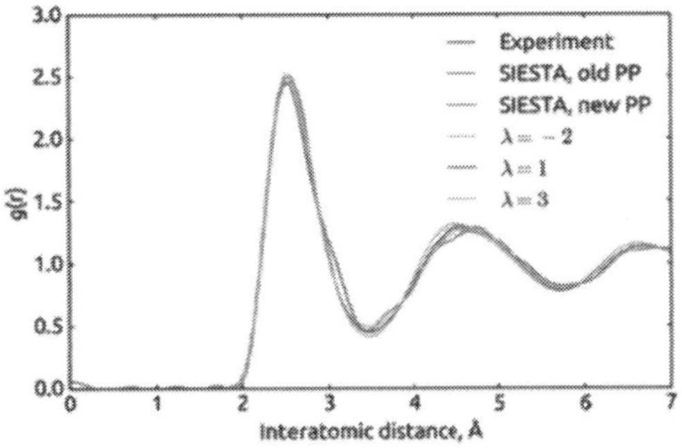

Figure 2. Radial distribution functions for liquid iron at T=1833K: experimental data [16], AIMD calculations and classical MD calculations with different EAM potentials from the hybrid set.

3.2 EAM potential for iron

To check the influence of EAM potential on the structure of the liquid we have calculated RDFs of liquid iron for potentials corresponding to several λ parameters ranging from -2 to 3. The results are presented in Fig. 2. We can see, that the hybridization principle stands — the differences in RDFs are very subtle, that is, different potentials resulted in the same liquid structure.

To determine how model properties depend on a hybrid EAM potential choice, we calculated several iron properties both in liquid state and during melting, versus hybridization parameter λ. We examined the following liquid iron characteristics at $T = 1833K$: self-diffusion coefficient D, dynamic viscosity coefficient η and volume thermal-expansion coefficient β. Melting temperature T_{melt}, enthalpy ΔH_{melt} and changing the volume at melting ΔV_{melt} have been studied during melting process. Table 2 presents all mentioned characteristics, which were calculated for the potentials with $\lambda = 1$ (the best fit to properties), compared to the potentials from λ range boundaries ($\lambda = -2$ and $\lambda = 3$). The reference data are also presented for comparison. As a reference we used experimental and ab initio data; besides, we compared data obtained with our potentials to that from EAM potentials for iron reported by other authors. We used several EAM potentials: potentials No.2 and No.5 by Mendelev et al [17] (below referenced to as M2 and M5), potential by Belashchenko [9] (referenced to as B) and potential reported by Chamati et al [18] (referenced to as C).

When comparing the results for different values of λ, given in Tab. 2, we can see that every physical property is dependent on λ; however, different properties depend on the parameter

DOI: 10.1051/epjconf/201715105004

Table 2. The calculated properties of liquid iron at T = 1833K and at the melting point. See text for explanations.

	Properties at T=1833K			Properties at melting		
	D, 10^{-5} cm·sec^{-2}	η, mPa·sec	β, 10^{-5} K^{-1}	T_{melt}, K	ΔH_{melt}, eV·atom^{-1}	ΔV_{melt}, Å^3·atom^{-1}
$\lambda = -2$	2.68	5.9	10.4	1892	0.192	0.90
$\lambda = 1$	3.05	5.1	5.4	1815	0.141	0.51
$\lambda = 3$	3.36	4.5	4.4	1738	0.119	0.38
M2	3.44	4.6	7.9	1793	0.162	0.62
M5	3.55	4.5	9.1	1827	0.188	0.88
B	3.57	4.4	6.5	1624	0.108	-0.01
C	3.51	4.3	9.0	1967	0.181	0.90
Experiment		5.6 [19]	13.2 [19]	1812 [17]	0.143 [17]	0.38 [17]
Ab initio	2.9					

to a different extent. The most sensitive property appears to be the volume change at melting ΔV_{melt}, which changes by 237% across the λ interval; the least sensitive property is T_{melt}, with 9% change. We optimized λ against the set of properties using least squares, and found $\lambda = 1$ to be the parameter that leads to the most optimal potential.

Comparing the results for $\lambda = 1$ with reference data, we can find that the relative deviation from the target values for D and η is comparatively small (D — 5%, η — 9%). The value of β, on the contrary, deviates greatly from the experimental value, with relative error of 59% with EAM and 29% with the pair potential. As for the characteristics at melting point, the calculated values of T_{melt} and ΔH_{melt} are very close to reference for EAM with $\lambda = 1.0$; relative errors are just 0.2% and 1.2% respectively. ΔV_{melt} was recovered with sufficiently larger error of 34%; however, it is the best result compared to other authors.

4 Conclusions

In the paper we have found the Δ-factor for SIESTA software package and compared it with other packages implementing DFT. We showed, that the Δ-factor depends on the choice of pseudopotential, and suggested a procedure to fit pseudopotential parameters taking the data on Δ-factor into account. Using proposed procedure, we modified the pseudopotential and showed that the new pseudopotential performs better in describing structure and properties of both crystalline and liquid iron. After that, taking structural data from ab initio calculations as reference, we built a set of EAM potentials that differ by the only parameter λ and each of which accurately reproduces the RDF of liquid iron from ab initio calculations; most of the physical characteristics of liquid iron, as well as at melting, appeared to have strong dependence on the parameter. We chose one of the potentials (corresponding to $\lambda = 1$) as the most adequate based on the RDF recovery error and found it to recover the whole set of properties at least as accurate as the majority of existing EAM potentials for iron.

Acknowledgements

The authors would like to acknowledge the support of the Russian Science Foundation (project no. 16–19–10252; ab initio part of the work) and the Russian Foundation for Basic Research (project no. 16–03–00486; EAM potential recovery).

References

[1] M.F. Horstemeyer, *Integrated Computational Materials Engineering (ICME) for metals: using multiscale modeling to invigorate engineering design with science* (John Wiley & Sons, 2012)
[2] P. Hohenberg, W. Kohn, Phys. Rev. **136**, B864 (1964)
[3] K. Lejaeghere, V. Van Speybroeck, G. Van Oost, S. Cottenier, Critical Reviews in Solid State and Materials Sciences **39**, 1 (2014)
[4] J.J.M. Soler, E. Artacho, J.D.J. Gale, A. García, J. Junquera, P. Ordejón, D. Sánchez-Portal, Journal of Physics: Condensed Matter **14**, 2745 (2002)
[5] *Pseudogenerator*, URL: https://github.com/ansobolev/PseudoGenerator, accessed: 2016-09-28
[6] N. Troullier, J.L. Martins, Physical Review B **43**, 1993 (1991)
[7] A. Mirzoev, A. Sobolev, EPJ Web of Conferences **15**, 01012 (2011)
[8] P. Ganesh, M. Widom, Physical Review B **77**, 014205 (2008)
[9] D. Belashchenko, Russian Journal of Physical Chemistry A **80**, 758 (2006)
[10] W. Schommers, Physical Review A **28**, 3599 (1983)
[11] J.H. Rose, J.R. Smith, F. Guinea, J. Ferrante, Physical Review B **29**, 2963 (1984)
[12] *Siesta pseudopotential database*, URL: http://icmab.es/siesta/Databases/Pseudopotentials/, accessed: 2016-09-28
[13] C. Kittel, *Introduction to solid state physics* (John Wiley & Sons, 1996)
[14] E. Moroni, G. Kresse, J. Hafner, J. Furthmüller, Physical Review B **56**, 15629 (1997)
[15] J. Izquierdo, A. Vega, L. Balbás, D. Sánchez-Portal, J. Junquera, E. Artacho, J. Soler, P. Ordejón, Physical Review B **61**, 13639 (2000)
[16] A. Il'inskii, S. Slyusarenko, O. Slukhovskii, I. Kaban, W. Hoyer, Materials Science and Engineering: A **325**, 98 (2002)
[17] M. Mendelev, S. Han, D. Srolovitz, G. Ackland, D. Sun, M. Asta, Philosophical magazine **83**, 3977 (2003)
[18] H. Chamati, N. Papanicolaou, Y. Mishin, D. Papaconstantopoulos, Surface Science **600**, 1793 (2006)
[19] M.J. Assael, K. Kakosimos, R.M. Banish, J. Brillo, I. Egry, R. Brooks, P.N. Quested, K.C. Mills, A. Nagashima, Y. Sato et al., Journal of Physical and Chemical Reference Data **35**, 285 (2006)

Asymmetrical bonding in liquid Bi disentangled by inelastic X-ray scattering

Masanori Inui[1,*], *Yukio* Kajihara[1], *Shuji* Munejiri[1], *Shinya* Hosokawa[2], *Ayano* Chiba[3], *Koji* Ohara[4], *Satoshi* Tsutsui[4], and *Alfred Q. R.* Baron[5,4]

[1]Graduate School of integrated Arts and Sciences, Hiroshima University, Higashi-Hiroshima, Hiroshima 739-8521, Japan
[2]Department of Physics, Graduate School of Science and Technology, Kumamoto University, Kumamoto 860-8555, Japan
[3]Department of Physics, Keio University, Yokohama 223-8522, Japan
[4]Japan Synchrotron Radiation Research Institute, SPring-8, Sayo-gun, Hyogo 679-5198, Japan
[5]Materials Dynamics Laboratory, RIKEN SPring-8 Center, Sayo-gun, Hyogo 679-5148, Japan

Abstract. The structure of liquid Bi has been debated in relationship with the Peierls distortion, as crystalline Bi takes A7 structure. A recent *ab initio* molecular dynamics simulation for liquid Bi predicted a flat-topped profile of the acoustic dispersion curve. To confirm the prediction, we have carried out inelastic x-ray scattering (IXS) for liquid Bi. The dynamic structure factor obtained by the IXS exhibits a distinct inelastic excitation of the longitudinal acoustic mode up to 14 nm^{-1} and the dispersion curve of the excitation energy obtained by the memory function analysis becomes a flat-topped one. We found that a linear chain model including the interatomic interaction with the second nearest neighbors can explain the flat-topped profile. The result suggests that the anomalous dispersion curve in liquid Bi arises from local anisotropy related to the Peierls distortion in the crystalline phase. .

1 Introduction

Bi belongs to the same group as As in the periodic table of elements, and the local structure of liquid Bi has been discussed from a view-point of the Peierls distortion [1]. The Peierls distortion is realized in a A7 crystalline structure type comprising three short and three long bonds in a distorted simple cubic lattice to gain the electronic energy.

The structure factor $S(Q)$ of liquid Bi, where Q is momentum transfer, exhibits a shoulder at high momentum transfer side of the first maximum (see Fig. 3) [2–4], in contrast to a symmetrical profile of simple liquid metals. The origin of the shoulder in $S(Q)$ of liquid Bi was theoretically studied and it was found that an effective pair potential with a ridge in the repulsive component could reproduce the shoulder [5, 6]. Later, an *ab initio* molecular dynamics (AIMD) simulation for liquid Bi [7] revealed that longer and shorter bonds produce a distinct shoulder at the first peak in $S(Q)$.

Atomic dynamics in liquid Bi had been studied by inelastic neutron scattering (INS) experiments. Most recently, Sani *et al.* [8] carried out INS experiments at $Q \leq 6$ nm^{-1} and

*e-mail: masinui@hiroshima-u.ac.jp

reported the dynamic structure factor $S(Q, E)$, where E is energy transfer, with good statistics. The $S(Q, E)$ exhibits the inelastic excitations of the acoustic mode dispersing with Q. However, this result did not fix conflicting INS results obtained in 1980's: Dahlborg and Olsson [9, 10] reported no distinct inelastic excitation of the acoustic mode in $S(Q, E)$ at $Q > 6$ nm^{-1}, while Shibata *et al.* [11] reported experimental observation of the acoustic excitations up to 12 nm^{-1}. Later Dzugutov and Dahlborg [5] reported that there appears no distinct acoustic excitation at $Q > 6$ nm^{-1} with a classical molecular dynamics simulation (CMD), consistent with the INS results by Dahlborg and Olsson.

Contrary to the CMD result, Souto *et al.* [7] reported that inelastic excitations of the acoustic mode survive up to 14 nm^{-1} in $S(Q, E)$ obtained by an *ab initio* molecular dynamics (AIMD) simulation. Furthermore the AIMD simulation for liquid Bi predicted an intriguing dispersion curve of the acoustic mode with a flat-topped profile, which is much different from a sinusoidal profile in simple liquid metals [12]. To investigate the behavior of the acoustic excitation in liquid Bi, we carried out inelastic x-ray scattering (IXS) experiments. A high-resolution IXS technique allows us to avoid the kinematic constraints of INS and to obtain good spectra out to large energy transfer. The technique is also advantageous for measurements at extreme conditions such as high temperatures and high pressures owing to a small beam size of approximately 0.1 mm in diameter. Details of the IXS results have already been reported in Ref.[13]. In this article, we report the results of the acoustic mode in liquid Bi briefly and present results obtained by $S(Q, E)$ near the first peak position of $S(Q)$ that are not published.

2 Experimental procedure

The IXS experiments were conducted at the high-resolution IXS beamline (BL35XU) at SPring-8 in Japan [14]. Backscattering at the Si (11 11 11) reflection provided a beam of approximately 10^{10} photons/sec in a 0.8 meV bandwidth onto the sample. The energy of the incident beam and the Bragg angle of the backscattering were 21.747 keV and approximately 89.98°, respectively. The spectrometer resolution was approximately 1.5 meV depending on the analyzer crystal, and it was experimentally determined by scattering from polymethyl methacrylate (PMMA). Q resolution, ΔQ, was set to be 0.45 and 1.0 nm^{-1} (full width) for $Q \leq 11$ and $Q > 11$ nm^{-1}, respectively.

X-ray diffraction measurements using monochromatized high-energy x-ray were carried out at BL04B2/SPring-8 [15], to obtain $S(Q)$. Details of the data processing are described in Ref. [13].

Liquid Bi of 99.999% purity was mounted in a single-crystalline sapphire cell of Tamura-type [16] that was carefully machined to provide a 0.04 mm sample thickness. The cell was placed in "Marburg" chamber [17]. IXS spectra were measured at 573 and 1023 K [18] in pure He atmosphere at 0.1 MPa. The background spectra were measured at 573 K with an empty cell. After the absorption correction, the backgrounds were subtracted from the data, and $S(Q, E)/S(Q)$ were obtained. More details were described in Ref. [13].

3 Computer simulations

AIMD simulations were performed at 573 K using Quantum ESPRESSO package [19], which is based on the density functional theory, plane waves and pseudopotentials. The generalized gradient approximation(GGA) [20] was adopted for the exchange-correlation energy. We used the norm-conserving pseudopotential Bi.pbe-hgh.UPF [21] for the electron-ion interaction with five valence electrons, $6s^2 6p^3$. The plane-wave cutoff energy for the electronic

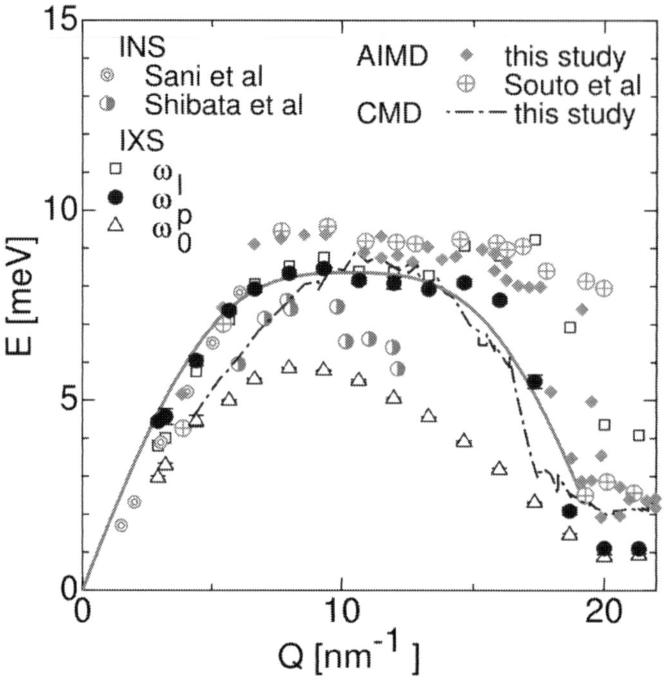

Figure 1. $\omega_\ell(Q)$ (open squares), $\omega_p(Q)$ (black circles) and $\omega_0(Q)$ (open triangles) obtained by IXS results at 573 K. $\omega_p(Q)$ by AIMD simulations of this study and that at 600 K [7] is indicated by diamonds and circles with cross, respectively. Also shown are the excitation energy obtained by INS by Sani *et al.* (double circles) and by Shibata *et al.* (circles with half moon), and CMD (dash-dotted line). A bold red curve denotes the dispersion relation obtained by a regular chain model with two force constants.

wavefunctions was 15 Ry. The Γ point was only used to sample the Brillouin zone of the MD supercell. We used 128 Bi atoms in a cubic MD cell with periodic boundary conditions. The length of the side of the MD cell was 1.6422 nm. The MD simulation was performed on the ground-state Born-Oppenheimer surface at 573 K for 32 000 steps with a time step of 4.8 fs, and reproduced $S(Q, E)$ at $Q \geq 3.8$ nm^{-1} with quality as good as that by Souto *et al* [7]. We also carried out a CMD simulation at 573 K using the model potential reported in [5] for a system with 256 particles. To obtain $S(Q, E)$, the simulation was performed for 60 000 steps with a time step of 2.4 fs. More details are described in Ref. [13].

4 Results and discussion

4.1 The acoustic dispersion

The integration of $S(Q, E)$ obtained by IXS was carefully compared with $S(Q)$ obtained independently by our own x-ray scattering experiment. By adjusting the integration at 8 nm^{-1} with $S(Q)$, we obtained $S(Q, E)$ of the absolute scale. Then $S(Q, E)$ obtained by IXS was compared with that by INS. Our data nicely agree with those by Sani *el al.* [8]. The difference is that INS spectra are truncated at an energy where inelastic excitations still exist, probably owing to the kinematic constraint of INS. Our results exhibit that the inelastic excitation is

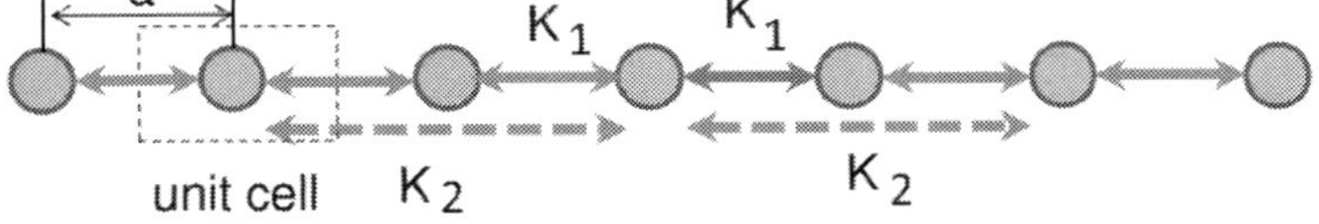

Figure 2. A schematic illustration of a linear chain connected with springs between the first and second nearest neighbors.

visible at $Q \geq 6$ nm^{-1} and it disperses with increasing Q. Hence the prediction by Souto *et al.* [7] was experimentally confirmed. Details are shown in Ref. [13].

To determine the excitation energy of the acoustic mode, $S(Q, E)/S(Q)$ experimentally obtained should be deconvoluted with the resolution function. We tried three model functions; (a) the memory function formalism with two viscous decay channels [22–25], (b) a model function consisting of single Lorentzian and single damped harmonic oscillator (DHO) [26], and (c) a model function consisting of single Lorentzian and two DHO's. Hereafter, we present the results using (a). Details of these model functions are described in Ref. [13]. Using the memory function formalism, the excitation energy of the acoustic mode, $\omega_p(Q)$, was defined as the peak position of the current-current correlation function deduced from the deconvoluted $S(Q, E)/S(Q)$.

Figure 1 shows $\omega_p(Q)$ of liquid Bi at 573 K as a function of Q, with the normalized second frequency moment, $\omega_0(Q)$, and the normalized fourth frequency moment $\omega_\ell(Q)$. $\omega_0(Q)$ and $\omega_\ell(Q)$ are fitting parameters in the memory function analysis. The figure also depicts the excitation energies obtained by the AIMD and CMD simulations, and those of INS measurements by Sibata *et al.* [11] and Sani *et al.* [8]. As shown in the figure, $\omega_p(Q)$ obtained by IXS linearly disperses up to 7 nm^{-1} This dispersion curve is consistent with that obtained by INS by Sani *et al.* $\omega_p(Q)$ deviates from the linear dispersion and stays at approximately 8.5 meV at $10 \leq Q \leq 15$ nm^{-1}. A flat region in the dispersion curve is consistent with the prediction by the AIMD simulations by Souto *et al.* [7] although the AIMD results lie at energies slightly higher than black circles. The IXS results surely confirm the theoretical prediction of a flat-topped dispersion curve.

As shown in Fig.1, the peak positions of the current-current correlation function obtained by CMD (a dash-dotted curve) agree with $\omega_p(Q)$ (black circles) at approximately 10 nm^{-1}. With decreasing Q, however, real excitation energies of the acoustic modes deviate much from the CMD results at approximately 5 nm^{-1}. This result suggests that the effective pair potential, able to reproduce characteristics in S (Q), cannot exactly simulate the atomic dynamics in liquid Bi at low Q.

We could related the flat-topped dispersion curve of the acoustic mode in liquid Bi, using a linear chain model. We describe it using the illustration in Fig. 2. A simple linear chain connected by a spring with force constant K_1 gives a sinusoidal dispersion curve expressed by $M\omega^2 = 2K_1(1-\cos Qa)$, where M, ω and a are atomic mass, an excitation energy and a lattice constant, respectively. When the second springs of a force constant K_2 are fixed between the central atom and the second nearest neighbors as shown in the figure, the equation of the dispersion curve is modified as $M\omega^2 = 2K_1(1-\cos Qa)+2K_2(1-\cos 2Qa)$. This equation can reproduce a flat-topped dispersion curve by adjusting parameters K_1, K_2 and a [13]. Although the model is based on a regular structure, the concept of a pseudo-Brillouin zone [27] may support applicability of our model to a disordered structure. A linear chain connected by

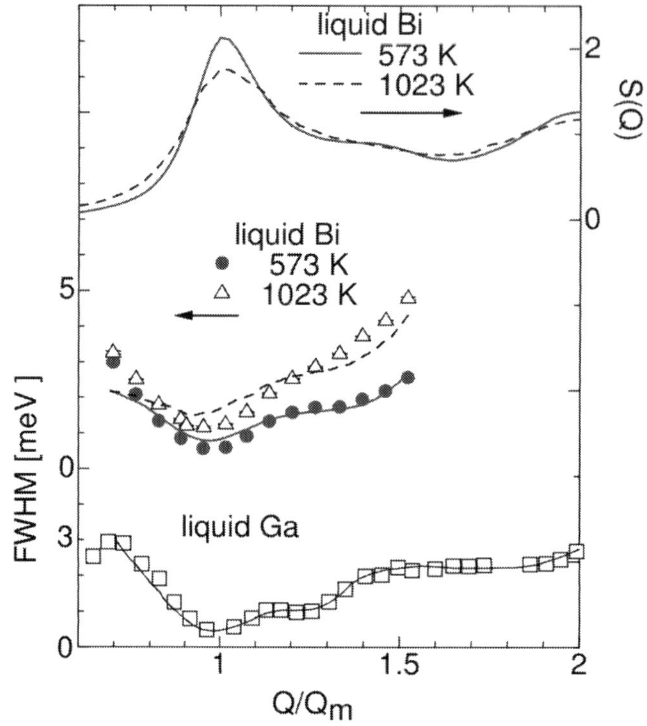

Figure 3. $S(Q)$ (upper curves) and the FWHM of the dynamics structure factor (middle curves and symbols) of liquid Bi at 573 and 1023 K in this work, and the FWHM of liquid Ga (a bottom curve and symbols) reported in [28]. The FWHM denoted by symbols is the IXS results and that denoted by lines is the theoretical optimization by Eq. (1).

strong and weak springs hints the one-dimensional Peierls distortion. Hence, we could relate the flat-topped dispersion curve with the Peierls distortion in liquid Bi.

4.2 Hard-sphere-like dynamics

$S(Q)$ of liquid Bi exhibits the first maximum with a shoulder at higher Q. $S(Q)$ of liquid Ga exhibits a similar profile. Scopigno *et al.* [28] carried out IXS measurements of liquid Ga near the melting point, and analyzed the linewidth of $S(Q, E)$ in high Q region including the $S(Q)$ maximum using an Enskog's theory. The equation given by de Schepper and co-workers [29] expresses the linewidth $z_h(Q)$ as,

$$z_h(Q) \quad = \quad \frac{1}{S(Q)} \frac{D_E Q^2}{1 - j_0(Q\sigma) + 2j_2(Q\sigma)}, \tag{1}$$

where D_E, $j_0(x)$ and $j_2(x)$ are Enskog's diffusion coefficient, zero-th and second order spherical Bessel functions, respectively. D_E is related to Boltzmann diffusion coefficient D_0 as $D_E = D_0/g(\sigma)$, where $g(\sigma)$ is the pair distribution function at a hard sphere diameter σ. By using free parameters of D_E and σ, and $S(Q)$ experimentally obtained, Scopigno *et al.* could reproduce Q dependence of the linewidth well as shown at the bottom of Fig.3. To evaluate

 DOI: 10.1051/epjconf/201715106001

Table 1. The optimized Enskog's diffusion coefficient D_E and hard sphere diameter σ with a packing fraction φ and a number density ρ.

T [K]	D_E [m^2/s]	$\hbar D_E$ [meV nm^2]	σ [nm]	φ	ρ [nm^{-3}]
573	3.6×10^{-9}	2.4×10^{-3}	0.36	0.71	28.9
1023	6.1×10^{-9}	4.0×10^{-3}	0.35	0.63	27.2

the effective mass, D_E expressed by the packing fraction φ [28, 30]

$$D_E = \frac{1}{16}\sqrt{\frac{\pi k_B T}{m}}\left(\frac{6}{\pi\rho\varphi^2}\right)^{\frac{1}{3}}\frac{(1-\varphi)^3}{(1-\varphi/2)}, \tag{2}$$

was used. Here m and ρ are the effective mass and number density, respectively. In a hard sphere model, φ, ρ and σ are related as $\varphi = \pi\rho\sigma^3/6$. In the case of liquid Ga, σ slightly larger than the atomic diameter of Ga, and an effective mass sightly larger than the atomic mass were obtained. These results obtained for liquid Ga were discussed as an indication of existence of dimeric clusters in Ref. [28].

To apply the Enskog's theory, we deconvoluted $S(Q,E)/S(Q)$ with a Lorentzian and the resolution function. Figure 3 shows a full width at half maximum (FWHM) obtained for liquid Bi at 573 (blue circles) and 1023 (open triangles) K as a function of Q/Q_m, where Q_m denotes the first peak position of $S(Q)$. Q_m values of liquid Bi and liquid Ga are 0.21 and 0.25 nm^{-1}, respectively. In our analysis, $S(Q)$ of liquid Bi was our own data obtained by x-ray diffraction measurements shown at the top of the figure. Bold and broken curves at the middle position of the figure denote the optimized fits using eq.(1), where D_E and σ are free parameters as the previous study [28]. As shown in the figure, the best fits reproduce FWHM at 573 K fairly well but does not at 1023 K. Although FWHM denoted by symbols seems to follow the profile of $S(Q)$, the fitting profiles are clearly worse than the results of liquid Ga. The optimized D_E and σ are tabulated in Table 1. σ of liquid Bi is larger than a cutoff distance in $g(r)$, which is approximately 0.29 nm. Asymmetrical bonding consisting of longer and shorter bonds may be inferred in liquid Bi from the optimized σ. However it was not a reasonable result that the effective mass obtained using eq. (2) was much smaller than the atomic mass.

5 Conclusion and future perspective

The present results suggest that a flat-topped dispersion curve observed in liquid Bi can be explained by taking the interatomic force between the second nearest neighbors into account. This result strongly hints asymmetrical bonding in liquid Bi. Symmetry breaking of bonding related to electronic energy levels in a molecule is well known as the Jahn-Teller effect. Hence, local structure may also be distorted in a liquid if a distortion makes the electronic energy lowered, as the case of liquid Bi.

Peierls distortion in liquid Bi has been expected by detailed analysis of the average structure [1]. Such a picture was strongly supported by atomic configurations obtained by AIMD simulations AIMD [7], and primitive analysis for the flat-topped dispersion curve confirmed by the present IXS results. It is very important to study the profiles of acoustic dispersion curves in liquid metals and semiconductors where asymmetrical bonding is expected, in order to understand the present results deeply.

Acknowledgements

The authors would like to thank the Ministry of Education, Culture, Sports, Science and Technology, Japan, for a Grant-in-Aid for Scientific Research. The synchrotron radiation experiments were performed at the SPring-8 with the approval of the Japan Synchrotron Radiation Research Institute (JASRI) (Proposals No. 2011B1314, No. 2010B1173, and No. 2009B1283). M. I. would like to thank The Japan Science Society for Sasakawa Grants for Science Fellows (SGSF).

References

[1] M. Mayo *et al.*, J.Phys:Condens. Matter **25**, 505102 (2013).
[2] M. North *et al.*, J. Phys. C **1**, 1075-1087 (1968).
[3] Y. Waseda and K. Suzuki, Phys. Stat. Sol. B **49**, 339-347 (1972).
[4] Y. Greenberg *et al.*, Euro. Phys. Lett. **86**, 36004 (2009).
[5] M. Dzugutov M and U. Dahlborg, Phys. Rev. A **40**, 4103-4106 (1989).
[6] W. Jank W and J. Hafner, Phys. Rev. B **45**, 2739-2749 (1992).
[7] J. Souto *et al.*, Phys. Rev. B **81**, 134201 (2010).
[8] L. Sani *et al.*, J. Non-Cryst. Solids **353**, 3139-3144 (2007).
[9] U. Dahlborg and L. G. Olsson, Phys. Rev. A **25**, 2712-2720 (1982).
[10] U. Dahlborg and L. G. Olsson, J. Phys. F **13**, 555-561 (1983).
[11] K. Shibata *et al.*, J. Phys. Soc. Jpn. **53**, 899-902 (1984).
[12] T. Scopigno *et al.*, Rev. Mod. Phys. **77**, 881-933 (2005).
[13] M. Inui *et al.*, Phys. Rev. B **92**, 054206 (2015).
[14] A. Q. R. Baron *et al.*, J. Phys. Chem. Solids **61**, 461-465 (2000).
[15] S. Kohara *et al.*, J. Phys.: Condens. Matter **19**, 506101 (2007).
[16] K. Tamura *et al.*, Rev. Sci. Instrum. **70**, 144-152 (1999).
[17] S. Hosokawa and W.-C. Pilgrim, Rev. Sci. Instrum. **72**, 1721-1728(2001).
[18] $S(Q, E)$ at 1023 K was consistent with that obtained by AIMD simulations using 512 atoms by M. Ropo *et al.* [J. Chem. Phys. **145**184502, (2016).]
[19] P. Giannozzi *et al.*, J. Phys.:Condens. Matter **21**, 395502 (2009).
[20] J. P. Perdew *et al.*, Phys. Rev. Lett. **77**, 3865-3868 (1996).
[21] *http://www.quantum-espresso.org.*
[22] D. Levesque *et al.*, Phys. Rev. A **7**, 1690-1700 (1973)
[23] U. Balucani and M. Zoppi *Dynamics of liquid State* (Clarendon press. Oxford) (1994)
[24] T. Scopigno *et al.*, J. Phys.: Condens. Matter **12**, 8009-8034(2000).
[25] F. Bencivenga *et al.*, Phys. Rev. E **75**, 051202 (2007).
[26] B. Fåk B and B. Dorner, Physica B **234-236**, 1107 (1997).
[27] T. Scopigno *et al.*, Phys. Rev. B **64**, 012301 (2001).
[28] T. Scopigno *et al.*, Phys. Rev. Lett. **94**, 155301 (2005).
[29] E. G. D. Cohen *et al.*, Phys.Rev. Lett. **59**, 2872 (1987).
[30] T. E. Faber *Introduction to the theory of liquid metals* (Cambridge university press.) 162-165 (1972).

Dispersion relations of the acoustic modes in divalent liquid metals

Masanori Inui[1,*], *Yukio* Kajihara[1], *Koji* Kimura[2,**], *Kazuhiro* Matsuda[2], *Tetsu* Miyatake[2], *Ayano* Chiba[3], *Shinya* Hosokawa[4], *Satoshi* Tsutsui[5], and *Alfred Q. R.* Baron[6,5]

[1]Graduate School of integrated Arts and Sciences, Hiroshima University, Higashi-Hiroshima, Hiroshima 739-8521, Japan
[2]Graduate School of Science, Kyoto University, Kyoto 606-8502, Japan
[3]Department of Physics, Keio University, Yokohama 223-8522, Japan
[4]Department of Physics, Kumamoto University, Kumamoto 860-8555, Japan
[5]Japan Synchrotron Radiation Research Institute, SPring-8, Sayo-gun, Hyogo 679-5198, Japan
[6]Materials Dynamics Laboratory, RIKEN SPring-8 Center, Sayo-gun, Hyogo 679-5148, Japan

Abstract. Collective dynamics in liquid Ca and liquid Cd was studied by inelastic x-ray scattering (IXS). Using our experimental technique to prepare proper sample cells and high performance of an IXS beamline (BL35XU) at SPring-8 in Japan, the dynamic structure factor with reasonable statistics was obtained for these divalent liquid metals. For both liquids, the dynamic structure factor at low Q exhibits a central peak with a shoulder or small hump clearly visible on each side, and the inelastic excitation energy determined using the model function composed of Lorentzian and the damped harmonic oscillator function disperses with increasing Q. The dispersion curves of these liquids were compared with that of the longitudinal acoustic phonon in each crystalline phase. From these results, clear difference in the interatomic interaction between liquid Ca and liquid Cd was inferred.

1 Introduction

Atomic dynamics in a liquid has intensively been investigated since inelastic neutron scattering (INS) was found to probe atomic motions in a material [1]. After an INS study on liquid Rb [2], liquid metals are known to exhibit a distinct inelastic excitation of the collective motion of atoms in the dynamic structure factor $S(Q, E)$, where Q and E are momentum and energy transfer, respectively. Since high-resolution inelastic x-ray scattering (IXS) was available at the end of last century, collective and single particle dynamics in many liquid metals have been studied using these methods [3].

Among liquid metals, *sp*-valent metals are classified as a simple liquid metal and they are considered typical liquids. In fact, inelastic scattering measurements of liquid alkali metals were carried out by many researchers [3]. However, those on alkaline earth metals exist only in a few studies. Although IXS results of liquid Mg [4] and INS results of liquid Ba and Liquid Ca [5] were reported, the INS results presented excitation energies only at a few Q

*e-mail: masinui@hiroshima-u.ac.jp
**present address:Nagoya Institute of Technology, Nagoya 466-8555, Japan

 DOI: 10.1051/epjconf/201715106002

points. We could not find dispersion curves of the acoustic mode in liquid alkaline earth metals besides liquid Mg.

Recent development of computer resource has made application of *ab initio* molecular dynamics (AIMD) simulations more popular. Del Rio and Gonzalez [6] reported a result of systematic studies on atomic dynamics in liquid alkaline earth metals using AIMD. Their results seem consistent with the previous experimental results. However, to compare the simulation results with the experimental ones in detail, it is needed to accumulate experimental data of divalent liquid metals.

We carried out IXS measurements of liquid Ca using synchrotron radiation at SPring-8 in Japan. Although liquid Ca has a strong chemical reactivity as well as a much higher melting point compared to alkali metals, we could obtain $S(Q,E)$ with reasonable statistics at $Q < 12$ nm^{-1}, using a special cell made of stainless steel. A high-brilliant IXS beamline (BL35XU) at SPring-8 was also needed to obtain good data. As another type of divalent metals, we investigated atomic dynamics in liquid Cd using IXS. INS measurements for liquid Cd of natural isotopes must be very difficult because the ^{113}Cd isotope has a very high absorption coefficient for neutron. Using IXS, we could obtain reasonable $S(Q,E)$ and compare our results with AIMD ones of liquid Cd [7]. In this article, we report our IXS results of liquid Ca and liquid Cd, and discuss the interatomic interaction in these liquids.

2 Experimental procedure

The experiments were conducted at the high-resolution IXS beamline (BL35XU) at SPring-8 in Japan [8]. Backscattering at the Si (11 11 11) reflection provided a beam of approximately 10^{10} photons/sec in a 0.8 meV bandwidth onto the sample. The energy of the incident beam and the Bragg angle of the backscattering were 21.747 keV and approximately 89.98°, respectively. The spectrometer resolution was approximately 1.5 meV depending on the analyzer crystal, as was experimentally determined by scattering from polymethyl methacrylate (PMMA). The Q resolution, ΔQ, was set to be 0.45 and 1.0 nm^{-1} (full width) for $Q \leq 11$ and $Q > 11$ nm^{-1}, respectively.

Liquid Ca of 99.99% purity was mounted in a stainless-steel cell of Tamura-type [9] that was machined to provide a 1.5 mm sample thickness. The cell was placed in the Marburg chamber [10]. IXS spectra were measured at 1128 K in pure He atmosphere at 0.1 MPa. The background spectra were measured at the same temperature using the same cell without sample. After the absorption correction, the backgrounds were subtracted from the data, and the normalized dynamic structure factor $S(Q,E)/S(Q)$ were obtained by normalizing the signal spectrum with the integration with respect to E. Because x-ray windows of the cell were made of polycrystalline foil, the background was larger than a single crystalline sapphire cell. However, the same cell was used for the empty and sample measurements, phonon excitations from stainless-steel foil were safely subtracted.

Liquid Cd of 99.999 % purity was mounted in a single crystalline sapphire cell of the Tamura type [11] that was carefully machined to provide a 0.2mm sample thickness. The cell was placed in a high pressure vessel equiped with large Be windows [12]. IXS spectra from 1.9 to 32 nm^{-1} were measured at 623 and 1173 K in pure He atmosphere at 1 MPa. The background spectra were measured at 623 K using an empty sapphire cell and $S(Q,E)/S(Q)$ were obtained by the way described in the previous paragraph.

3 Results

Figure 1 shows $S(Q,E)/S(Q)$ of liquid Ca at 1128 K. The Q maximum of 10.6 nm^{-1} was limited by the aperture of a scattering window of the cell. $S(Q,E)/S(Q)$ at 1.7 nm^{-1} exhibits

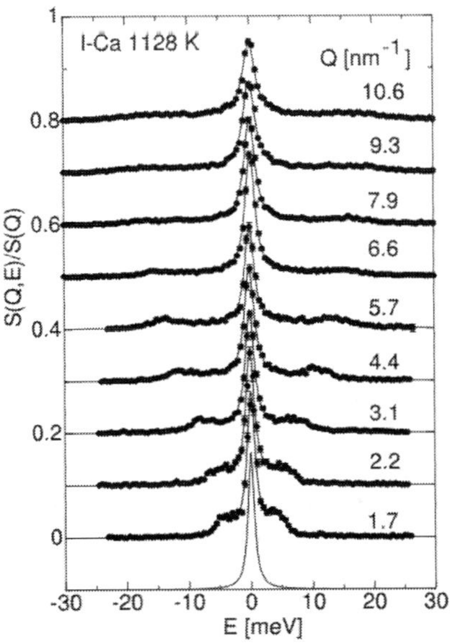

Figure 1. $S(Q,E)/S(Q)$ of liquid Ca at 1128 K. Black circles are the experimental results and lines are optimized fits. A sharp peak at the bottom is the resolution function.

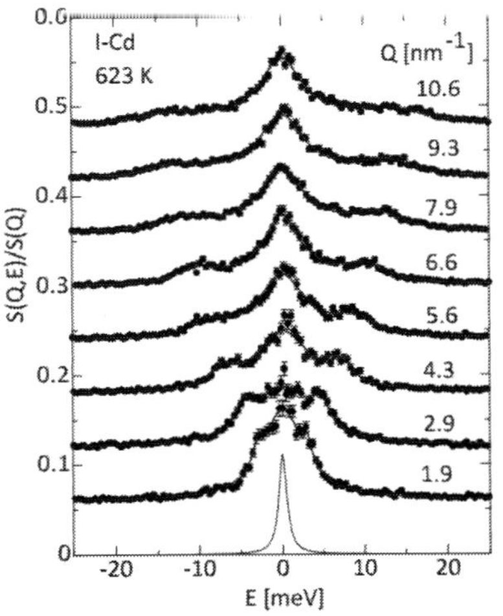

Figure 2. $S(Q,E)/S(Q)$ of liquid Cd at 623 K. Black circles are the experimental results and lines are optimized fits. A sharp peak at the bottom is the resolution function.

the inelastic excitation as a shoulder of the quasielastic peak on each side. The inelastic excitation disperses with increasing Q and an excitation peak profile is visible at 4.4 and 5.7 nm^{-1}.

Figure 2 shows $S(Q,E)/S(Q)$ of liquid Cd at 623 K at $Q \leq 10.6$ nm^{-1}. As shown in the figure, $S(Q,E)/S(Q)$ at 1.9 nm^{-1} exhibit the inelastic excitation as a shoulder of the quasielastic peak similarly to liquid Ca. The inelastic excitation disperses with increasing Q up to 14.6 nm^{-1}. As the first peak position of $S(Q)$ in liquid Cd is approached, the inelastic excitation on each side in $S(Q,E)/S(Q)$ is merged into the central peak, and de-Gennes narrowing profile of $S(Q,E)/S(Q)$ is observed at 21.3 nm^{-1} (the data are not shown in Fig.2).

We deconvoluted the experimental $S(Q,E)/S(Q)$ of liquid Ca and liquid Cd using the resolution function $R(E)$ and a model function $F_{\rm model}(Q,E)$ consisting of a Lorentzian for the quasi-elastic peak and a single damped harmonic oscillator (DHO) function [13] for the inelastic excitation, as following equations.

$$
\begin{aligned}
S(Q,E)/S(Q) &= \int dE' \left[B(E')F_{\rm model}(Q,E')\right] R(E-E'), \qquad (1)\\
F_{\rm model}(Q,E) &= \frac{A_0}{\pi}\frac{\Gamma_L}{\Gamma_L^2+E^2} + \frac{A_1}{\pi\beta}\frac{4\Gamma_1\sqrt{\Omega_1^2-\Gamma_1{}^2}}{(E^2-\Omega_1^2)^2+4E^2\Gamma_1^2},\\
B(E) &= \beta E/\left[1-\exp(-\beta E)\right],
\end{aligned}
$$

 DOI: 10.1051/epjconf/201715106002

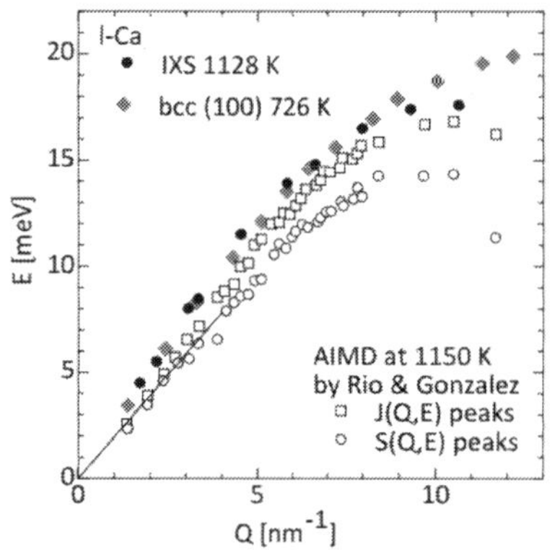

Figure 3. Dispersion relation of the acoustic mode in liquid Ca. Black circles, open squares and open circles denote the IXS results, peak positions of the current-current correlation function and the dynamic structure factor obtained by the AIMD [6], respectively. Also shown are the excitation energy of the longitudinal acoustic phonons in crystalline Ca of a bcc phase (Blue diamonds) and the ultrasonic sound speed in liquid Ca (solid line).

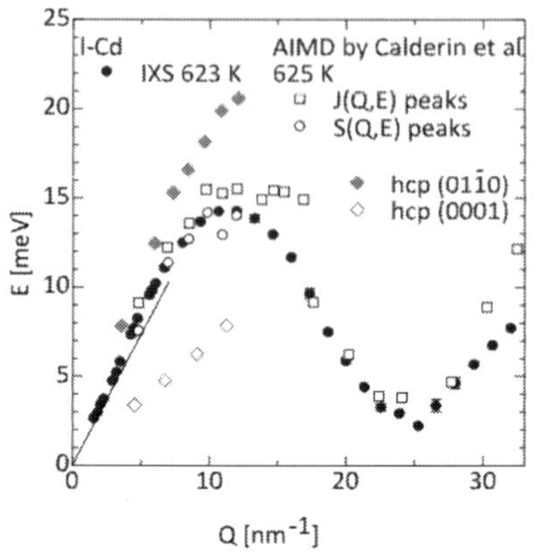

Figure 4. Dispersion relation of the acoustic mode in liquid Cd. Black circles, open squares and open circles denote the IXS results, peak positions of the current-current correlation function and the dynamic structure factor obtained by the AIMD [7], respectively. Also shown are the excitation energies of the longitudinal acoustic phonon to (01$\bar{1}$0) direction (solid blue diamonds) and (0001) direction (open blue diamonds) in crystalline Cd of a hcp phase, and the ultrasonic sound speed in liquid Cd (solid line).

where $\beta = (k_B T)^{-1}$. A_0 and Γ_L are the amplitude and the linewidth of Lorentzian, respectively. A_1, Γ_1 and Ω_1 are the amplitude, the linewidth and the excitation energy for the inelastic excitation. Although a transverse acoustic excitation at lower E is often discussed in collective dynamics in a liquid, a simple model function of a Lorentzian and a DHO gave a good fit to the experimental $S(Q,E)/S(Q)$, as indicated through solid lines in Figs.1 and 2.

The excitation energy of the DHO component Ω_1 (black circles) for liquid Ca is shown in Fig. 3. It is known that Ω_1 does not coincide with the energy of the inelastic peak in $S(Q,E)/S(Q)$ but approximate the peak energy of the current-current correlation function $J(Q,E)$. The results of liquid Ca show that Ω_1 (black circles) at $Q < 5$ nm^{-1} disperses along with approximately 3900 m/s, which is 34 % faster than the ultrasonic sound speed of 2950 m/s [14]. This panel also depicts the excitation energy of the longitudinal acoustic mode predicted by the AIMD simulation. The simulation gave the energy of the peak positions in $S(Q,E)$ and $J(Q,E)$. As shown in the figure, the AIMD simulation sightly underestimated the IXS results. Interestingly, the IXS results are very similar to a phonon dispersion curve to (100) direction of high-temperature bcc phase of crystalline Ca [15] as indicated in the figure.

The Q dependence of Ω_1 (black circles) in liquid Cd is shown in Fig.4. Ω_1 at $Q \sim 5$ nm^{-1} disperses along approximately 2600 m/s, which is 19 % faster than the ultrasonic sound speed of 2200 m/s [14]. The excitation energies of the longitudinal acoustic mode predicted by the AIMD simulation are plotted in the figure, using the same symbols in Fig. 3. The AIMD results agree with the IXS ones although open squares give a slightly higher energies than black circles at the top of the dispersion curve. We plot the dispersion curve of crystalline Cd with hcp form [16]. Probably owing to directional dependence of the elastic property, the

 DOI: 10.1051/epjconf/201715106002

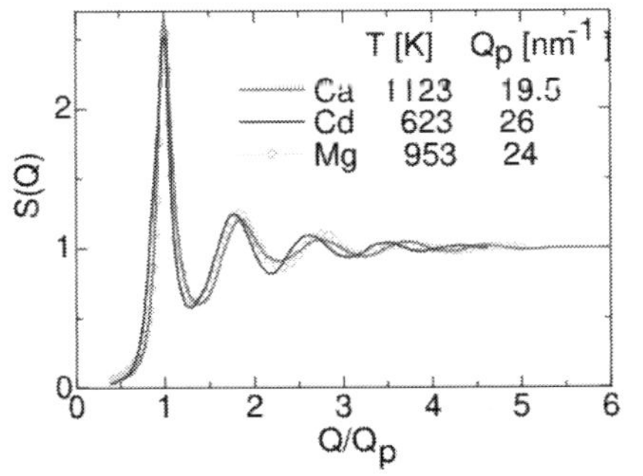

Figure 5. $S(Q)$ of liquid Mg, Ca and Cd obtained by x-ray diffraction [17] on the normalized Q scale.

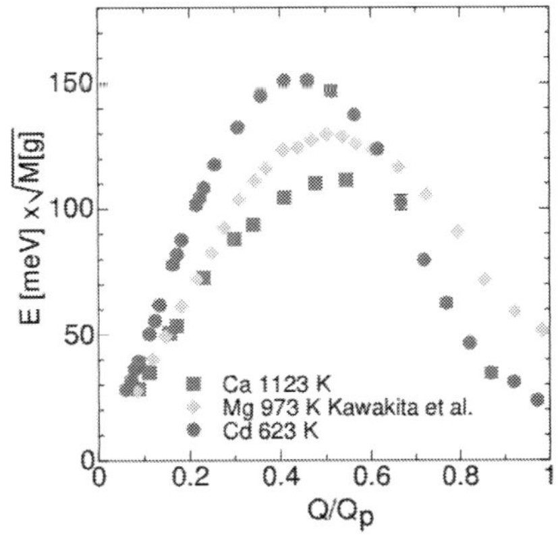

Figure 6. Excitation energies of the acoustic mode normalized by the atomic mass for liquid Ca and Cd as a function of the normalized Q scale. Also shown is that of liquid Mg [4].

excitation energy in hcp form largely depends on the crystalline axes as shown in the figure. The IXS results at $Q \leq 5$ nm^{-1} are located near the phonon dispersion curve to $(01\bar{1}0)$.

4 Discussion

Figure 5 shows $S(Q)$ of liquid Ca, Mg and Cd reported by Waseda [17], as a function of Q/Q_p, where Q_p is the first peak position of $S(Q)$ indicated in the figure. $S(Q)$ profiles on the normalized Q scale are similar between the alkaline earth elements but that of Cd is different from them. The oscillatory phase of liquid Cd is faster than those of liquid Ca and Mg. These results hint that the interatomic interaction may be similar between liquid Ca and liquid Mg but it is different from that of liquid Cd.

The effective pair potential of divalent liquid metals were reported by Jank and Hafner [18]. They showed that the density of states (DOS) of liquid Mg is similar to DOS of an free electron gas but DOS of liquid Ca is different from that because d-states in Ca is closer to the Fermi level compared to Mg. However, the effective pair potential of liquid Mg and Ca that they deduced is similar; the potential exhibits a clear minimum near the first nearest neighbor distance in the liquids. In contrast, the effective pair potential of liquid Cd exhibits a profile that a repulsive component is dominant when a relativistic effect is included. Their results are qualitatively consistent with the results inferred from the normalized $S(Q)$ profiles shown in Fig. 5.

We compare the dispersion curves of liquid Ca, Mg and Cd in Fig. 6 on the normalized Q and E scales, where the E is normalized by the square root of atomic mass. As shown in the figure, the dispersion curves of liquid Ca and Mg agree with each other up to $Q/Q_\mathrm{p} = 0.2$. Although the maximum position in the dispersion curves appears at approximately $Q/Q_\mathrm{p} = 0.5$, the maximum energy of liquid Ca is smaller than that of liquid Mg. These results indicate that the interatomic interaction of liquid Ca is slightly different from that of liquid Mg.

As shown in Fig. 6, the dispersion curve of liquid Cd is much different from those of the alkaline earth elements. The result suggests that the interatomic interaction in liquid Cd is much different from that in liquid Ca and liquid Mg, as predicted by Jank and Hafner [18].

The maximum energy is larger than those of the alkaline earth elements. It is noteworthy that the maximum position of liquid Cd shifts to $Q/Q_p = 0.4$ smaller than those of liquid Ca and Mg. This behavior may be correlated with the profile of the first maximum of $S(Q)$ in liquid Cd that is broadened toward lower Q than those in liquid Ca and Mg, as shown in Fig. 5.

5 Summary

IXS measurements of liquid Ca and liquid Cd provide the dynamic structure factors of these liquids that exhibit distinct acoustic excitations at low Q. The dispersion curves are compared with each other including that of liquid Mg. We found that although $S(Q)$ of liquid Ca and Mg is similar after the atomic size is normalized, the dispersion curves of these elements are not perfectly consistent after atomic size and mass are normalized. Meanwhile, the structure and dynamics of liquid Cd must be much different from these liquid alkaline earth elements. The results are qualitatively consistent with the difference in the effective pair potential between these liquids reported by Jank and Hafner [18].

Acknowledgements

The authors would like to thank the Ministry of Education, Culture, Sports, Science and Technology, Japan, for a Grant-in-Aid for Scientific Research. The synchrotron radiation experiments were performed at the SPring-8 with the approval of the Japan Synchrotron Radiation Research Institute (JASRI) (Proposals No. 2013A1165 and No. 2014A1154). M. I. would like to thank to The Japan Science Society for Sasakawa Grants for Science Fellows (SGSF).

References

[1] L. Van Hove, Phys. Rev. **95**, 249-262 (1954).
[2] J. R. D. Copley and J. M. Rowe, Phys. Rev. Lett. **32**, 49-52 (1974).
[3] T. Scopigno *et al.*, Rev. Mod. Phys. **77**, 881-933 (2005).
[4] Y. Kawakita et al., J. Phys. Soc. Jpn., 72, 1603-1606 (2003).
[5] L. Sani et al, 14th Liquid and Amorphous Metals conference, Rome (11 褰・6 July 2010).
[6] Beatriz G del Rio and Luis E Gonzalez, J.Phys.:Condens Matter **26**, 465102 (2014).
[7] L Calderín *et al.*, J.Phys.:Condens Matter **25**, 065102 (2013).
[8] A. Q. R. Baron *et al.*, J. Phys. Chem. Solids **61**, 461-465 (2000).
[9] K. Matsuda *et al.*, Rev. Sci. Instrum. **75**, 709-712 (2004).
[10] S. Hosokawa and W.-C. Pilgrim, Rev. Sci. Instrum. **72**, 1721-1728 (2001).
[11] K. Tamura *et al.*, Rev. Sci. Instrum. **70**, 144-152 (1999).
[12] M. Inui *et al.*, Eur. Phys. J. Special Topics **196**, 167-174 (2011).
[13] B. Fåk B and B. Dorner, Physica B **234-236**, 1107-1108 (1997).
[14] S. Blairs, Int. Mat. Rev. **52**, 321-342 (2007).
[15] M. Heiroth *et al.*, Phys. Rev. B **34**, 6681-6688 (1986).
[16] H. P. Sinha and J. C. Upadhyaya, J. Phys. Chem. Sol. **38**, 41-44 (1977)
[17] Y Waseda *The structure of non-crystalline materials* (McGraw-Hill, Inc.) (1980).
[18] W. Jank W and J. Hafner, Phys. Rev. B **42**, 6926-6938 (1990).

EPJ Web of Conferences **151**, 06003 (2017)
LAM-16
DOI: 10.1051/epjconf/201715106003

Inelastic x-ray scattering measurements of liquid water-glycerol mixtures

Yukio Kajihara[1,*], *Nanako* Shibata[1], *Masanori* Inui[1], *Kazuhiro* Matsuda[2], and *Satoshi* Tsutsui[3]

[1]Graduate School of Integrated Arts and Sciences, Hiroshima University, Higashi-Hiroshima 739-8521, Japan
[2]Graduate School of Science, Kyoto University, Kyoto 606-8502, Japan
[3]Japan Synchrotron Radiation Research Institute, SPring-8, Sayo, Hyogo 679-5198, Japan

Abstract. We have carried out inelastic x-ray scattering measurements on liquid water-glycerol mixtures. The data are analyzed by a damped harmonic oscillator model with two excitations, longitudinal and transverse modes. The sound velocity of the longitudinal mode is almost constant (about 3.1 km/s), being independent of the composition. Thus the strength of 'fast sound' which we define the ratio of this IXS sound velocity to ultrasonic one is largest at pure water and gradually decrease with increasing mole fraction of glycerol. This result indicates that the relaxation phenomena of pure water gradually reduce with increasing the fraction of the solute: the scenario which we proposed for water-monohydric alcohol mixtures hold true for this water-trihydric alcohol mixtures.

1 Introduction

Liquid water-monohydric alcohol mixtures exhibit many thermodynamic anomalies [1]. For example, a maximum appears at around 10-20% alcohol mole fraction in the concentration dependence of the ultrasonic sound velocity (v_{US}, frequency in MHz) [2–4] in water-ethanol and water-methanol mixtures. The anomalies have been believed to be of structural origin [1] and the relation between clathrate structure such like Ethanol-17H_2O and Ethanol-5.75H_2O [2, 5] or the special cage structure has been discussed [2, 6]. Anyway, there has been no direct experimental evidence and the discussion has not been concluded.

On the other hand, it is well known that water itself has many thermodynamic anomalies: The density shows a maximum at 4 °C and the ultrasonic sound velocity exhibits anomalous increase below 80 °C. There is an opinion that these anomalies of pure water should affect the properties of water-alcohol mixtures [7]. The so-called 'fast sound' problem is one of such anomalies of pure water [9]. The sound velocity of water at ambient conditions which is obtained by microscopic methods (molecular dynamics and inelastic neutron / x-ray scattering) is much faster (about by a factor of two) than the one obtained by a macroscopic method like ultrasonic measurements. The advent of inelastic X-ray (IXS) and ultraviolet scattering technique lead to the conclusion that a relaxation phenomena, which is assigned to making and breaking hydrogen bond, is the essence of the 'fast sound' of water [10–13].

*e-mail: kajihara@hiroshima-u.ac.jp

The strength of the relaxation is so large as to slow down the ultrasonic sound velocity by half compared to the IXS sound velocity. From this fact, it is natural that the question of how this strength will change when alcohol is added to water. We speculated that it should be the origin of the ultrasonic and also the thermodynamics anomalies of water-(monohydric) alcohol mixtures, which is the same scenario to [7]. We recently carried out IXS experiment on water-ethanol mixtures at ambient condition [8] and on water-methanol mixtures from ambient to low temperature conditions [4]. The concentration dependence of the strength of the relaxation exhibits a maximum for pure water and gradually decreases with increasing alcohol concentration in both water-ethanol and water-methanol mixtures. From these experimental results, we concluded that the ultrasonic anomalies of the water-monohydric alcohol mixtures are derived from the pure water itself, not from the complex mixing state of water and alcohols.

In the present study, we carried out IXS measurement of water-glycerol mixtures, to verify whether the same scenario as observed for water-monohydric alcohol mixtures holds also true for water-trihydric alcohol mixtures.

2 Experimental details and Results

The IXS experiment is performed at SPring-8/BL35XU in Japan [14] with standard beamline conditions for liquids. Si (11 11 11) reflection is used as a backscattering monochromator and the energy of the incident beam is 21.747 keV. The energy resolution of the spectrometer is about 1.5 meV and the scan ranges of energy transfer (E) is $\pm$ 35 meV. There are 12 analyzers (and corresponding 12 detectors) in total, four in the horizontal direction and three in the vertical direction, which enables us to measure 12 data at once whose momentum transfer (Q) are different. We set 2 horizontal angles of the analyzers to cover Q range from 1.5 to 11 nm^{-1}. An aluminum cell with two sapphire single crystal windows of diameter 8 mm and thickness 0.5 mm each is used to hold the liquid samples. The flat samples have a thickness of 2 mm.

Figure 1 shows $S(Q, E)$ of the water-10 mol% glycerol sample recorded at a temperature of 25 °C. Closed circles, open circles, closed triangles and open triangles correspond to Q equal to 1.56, 4.24, 6.56 and 9.25 nm^{-1}, respectively. In the spectra, we can see the shoulders correspond to the longitudinal acoustic mode at around the energy of ± 2, ± 8, ± 14 and ± 20 for Q equals to 1.58, 4.24, 6.56 and 9.25 nm^{-1}, respectively. In pure water, it was concluded that IXS spectrum has two collective modes (longitudinal and transverse acoustic excitations) [10, 15]. Thus we analyzed the spectrum by a damped harmonic oscillator (DHO) model which has two DHO components correspond to these two modes in addition to a Lorentzian component corresponds to quasi-elastic scattering as shown in the following equation.

$$\frac{I(Q,E)}{I(Q)} = \int \frac{S(Q,E)}{S(Q)} R(E-E')dE'$$

$$\left(\frac{S(Q,E)}{S(Q)}\right) = \left(\frac{E/k_BT}{1-e^{-E/k_BT}}\right) \times \left((1-A_1-A_2)f_0 + \sum_i^{1-2} A_i f_i\right)$$

$$f_0 = \frac{1}{\pi\Gamma_0}\frac{\Gamma_0}{E^2+\Gamma_0^2}$$

$$f_{i=1,2} = \frac{k_BT}{\pi}\frac{4\Gamma_i\sqrt{E_i^2-\Gamma_i^2}}{(E^2-E_i^2)^2+4\Gamma_i^2E^2} \tag{1}$$

Here Γ_0 is the width of the central Lorentzian and A_i, Γ_i and E_i are the amplitude, width and energy of the DHO for longitudinal (i = 1) and transverse (i = 2) modes, respectively.

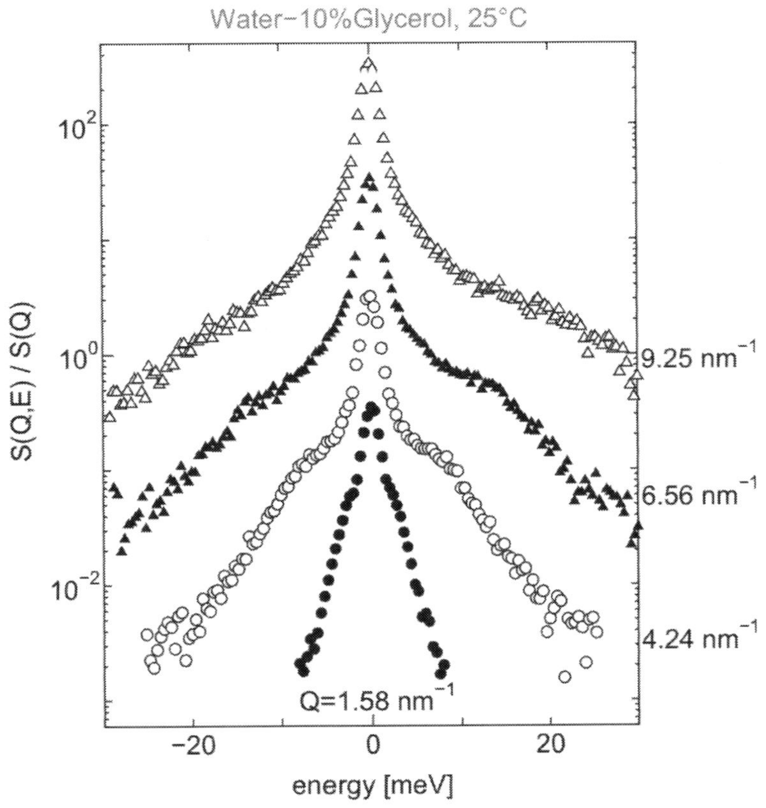

Figure 1. $S(Q,E)$ of water-10 mole % glycerol mixture at 25 °C. Closed circles, open circles, closed triangles and open triangles correspond to Q equal to 1.58, 4.24, 6.56 and 9.25 nm^{-1}, respectively. Each spectra is shifted by a factor of 5 for clarity.

k_B is Boltzmann constant and T is the temperature of the sample. R(E) represents the resolution function experimentally determined from the measurements of PMMA (polymethyl methacrylate). We fitted the experimentally obtained spectrum by this model function $I(Q,E)$ and optimized these parameters by the non-linear least squares method. In figure 2, the calculated model spectra are plotted by solid lines; the spectra with two-DHO components are plotted by the red line and that with one-DHO (only longitudinal mode) by the blue one. The shoulder of the spectrum at around 14 meV, which corresponds to the longitudinal mode, is better described by the red line than the blue one, which means that there exists a transverse mode in addition to a longitudinal one in this mixture. The estimated energy (E_1=13.81 meV) correspond to the longitudinal mode of 2-DHO model is slightly higher than that (11.95 meV) of 1-DHO model. Dashed lines are the components of the calculated spectrum (without considering resolution function); the green, red and pink lines correspond to the Lorentzian component, DHO for the longitudinal mode and DHO for the transverse mode, respectively

Figure 3(a) shows dispersion curves for water-10 mole % glycerol mixture at 25 °C. Closed and open circles indicate the estimated energies of the DHO function for longitudinal ($E_1(Q)$) and transverse modes ($E_2(Q)$), respectively. From the curve of longitudinal mode, we estimate the Q- dependent sound velocity $v(Q) = E_1(Q)/Q$ as shown in figure 3(b). $v(Q)$ is almost constant in the higher-Q region above 4 nm^{-1} and shows a decrease approaching

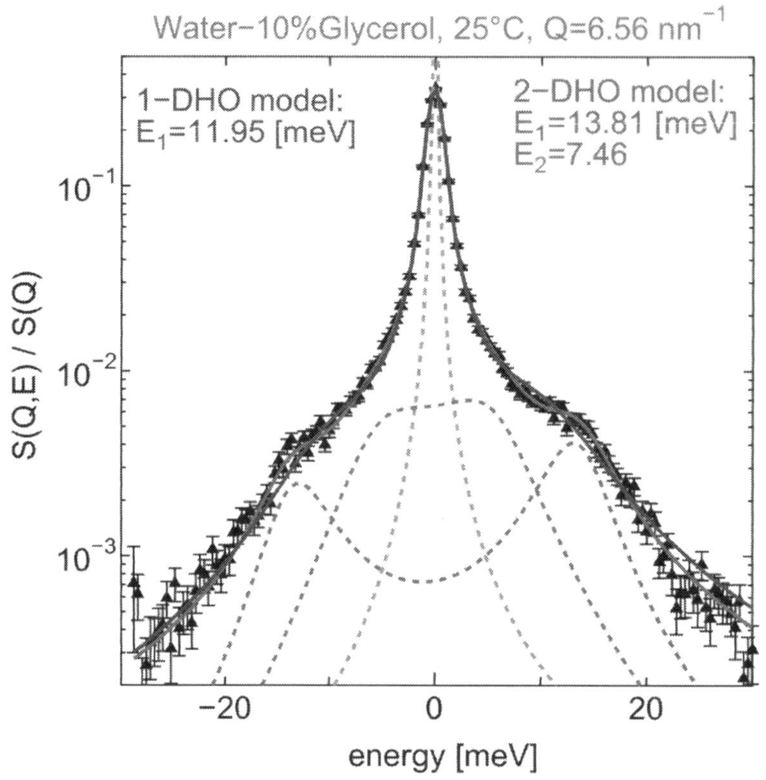

Figure 2. $S(Q, E)$ of water-10 mole % glycerol mixture at temperature of 25 °C and Q equal to 6.56 nm^{-1}. Open circles are experimental values. The blue and red solid lines represent the calculated model functions of 1-DHO and 2-DHO models, respectively. Dashed lines are the components for 2-DHO model function without convolution of the resolution function; the green, red and pink lines correspond to a Lorentzian component, DHO for longitudinal mode and DHO for transverse one, respectively.

v_{US} (indicated by triangle in the figure) at lower Q- values. But for mixtures containing more than 60 mole % glycerol $v(Q)$ gradually decrease with increasing Q above about 8 nm^{-1}. Thus we determined IXS sound velocity v_{IXS} by only averaging $v(Q)$ in the range of 4 to 7 nm^{-1} for all the mixtures as indicated by the red open circles in figure 3.

Figure 4 (a) shows the concentration dependence of v_{IXS} at 25 °C. v_{US} [18] is also plotted in the figure represented by a solid line. Irrespective of concentration, the present v_{IXS}, indicated by closed circles, is almost constant with a value of 3.1 km/s which compares well with literature data [10, 16, 17] denoted by open marks. This result suggests that there does not exist special structure or characteristic mixing state in this system. To characterize the relaxation phenomena in this system, we define the simple parameter as the strength of 'fast sound' $S_f \equiv v_{\mathrm{IXS}}/v_{\mathrm{US}}$. This is a measure for the strength of the relaxation processes that exist in the frequency range between that of US (MHz) and that of IXS (THz) measurements as followings. Sound velocity v can be related with the apparent longitudinal elastic modulus M as $M = \rho v^2$. Frequency (ω) dependence of the elastic modulus is expressed by the Debye approximation as,

$$M(\omega) = M_\infty + \frac{M_0 - M_\infty}{1 + i\omega\tau}. \tag{2}$$

DOI: 10.1051/epjconf/201715106003

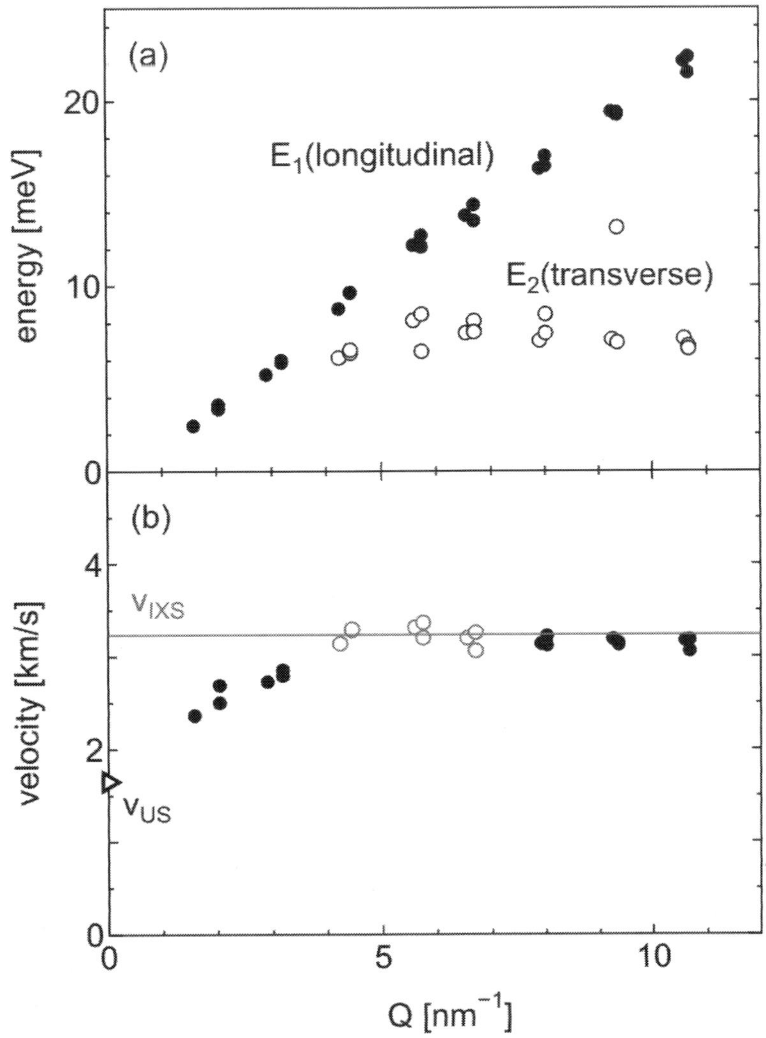

Figure 3. (a) Dispersion curve of water-10%glycerol mixture at 25 °C. Closed and open circles indicate the estimated energies of DHO function for longitudinal and transverse modes, respectively. (b) Q-dependent sound velocity of the longitudinal mode $v(Q)$. IXS velocity $v_{\rm IXS}$ is determined by the average of $v(Q)$ in the Q range between 4 and 7 nm^{-1} denoted by red open circles. In the low-Q region, $v(Q)$ approaches to the ultrasonic sound velocity $v_{\rm US}$.

where the M_0 and M_∞ are bulk modulus for ω=0 and $\omega \to \infty$, respectively, and τ is the characteristic time of the relaxation. Similar formalism was adopted and the characteristic time and other parameters were discussed for liquid water [11]. Here we focus on the strength of the relaxation ϵ, which is related with the strength of fast sound as,

$$\epsilon \equiv \frac{M_\infty - M_0}{M_0} = \frac{v_\infty^2 - v_0^2}{v_0^2} \approx \frac{v_{\rm IXS}^2 - v_{\rm US}^2}{v_{\rm US}^2} = S_f^2 - 1. \tag{3}$$

In this equation, we assume that the frequency of IXS (THz) and that of US (MHz) is enough fast and slow compared to the characteristic frequency of the system $1/\tau$. Thus the parameter

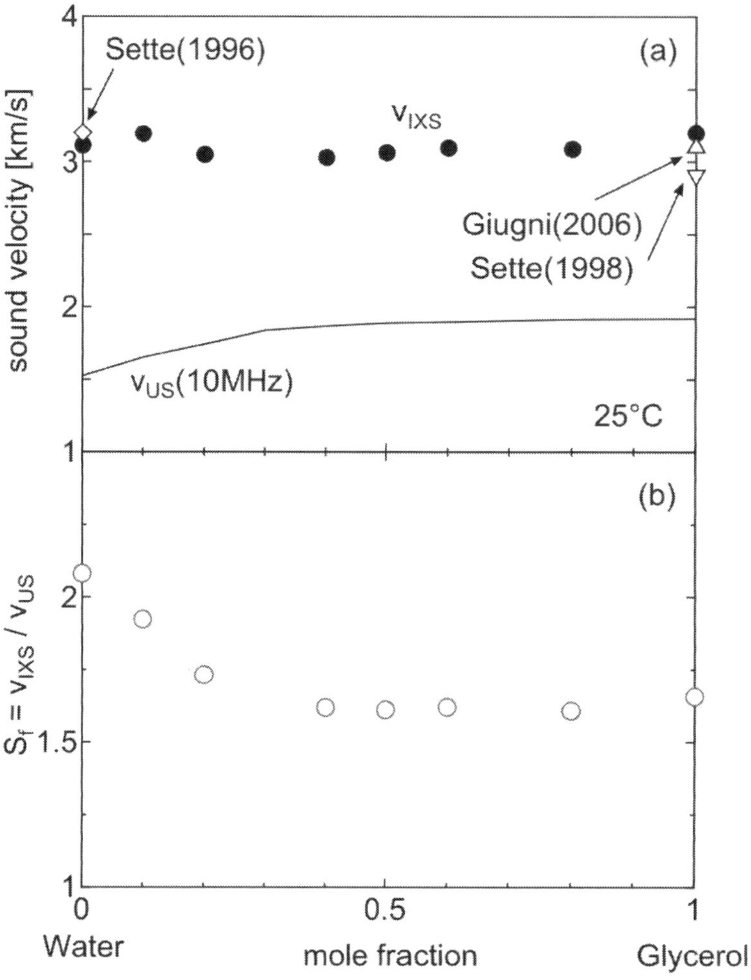

Figure 4. (a) Concentration dependence of sound velocity of water-glycerol mixtures. Marks are obtained by IXS measurement: closed circles are by us and open marks are by others [10, 16, 17]. Solid line shows ultrasonic sound velocity [18]. (b) Ratio of the two sound velocities.

S_f is a good measure for the strength of the relaxation processes that exist in the frequency range between that of US and that of IXS measurements. S_f is plotted in figure 4 (b) by open circles, which is largest for pure water and gradually decrease with increasing glycerol content in the water-rich region (mole fraction of glycerol x_g<0.5). In the glycerol rich region (x_g>0.5), S_f is almost constant. We conclude that the concentration dependence of v_{US} is mainly characterized by the relaxation, not by the characteristic mixing state or special structure of these two molecule (water and glycerol). In the low-alcohol mixtures, the relaxation should be caused by pure water because the strength of the relaxation is largest. These conclusion is same as that for the water-monohydric alcohol mixtures [4, 8] and thus should be universal in water-alcohol mixtures. We believe that the origin of this relaxation of pure water is the critical fluctuations for liquid-liquid phase transition whose critical point is suggested to locate in the deep supercooled region (critical temperature is about -50 °C and pressure is from 100 MPa to 150 MPa) [19, 20]. Because S_f increases with decreasing temperature in water-rich region of water-methanol mixtures. There is one difference between these monohydric- and trihydric- alcohol solutions : S_f is from 1.6 to 1.7 for glycerol-rich mixtures which is not so small as compared to that of high monohydric-alcohol mixtures (S_f is about 1.2 to 1.3). This is because the glycerol is a well-known glass forming liquid and has a substantial value of the strength of relaxation [17], which should affect the relaxation of glycerol-rich samples.

3 Summary

We have carried out IXS measurements on liquid water-glycerol mixtures. The IXS determined sound velocity is almost constant irrespective of the glycerol concentration. The strength of the relaxation for the mixtures, which is estimated by taking the ratio of this IXS velocity and the one measured by the ultrasonic method, is largest at pure water and gradually decreases with increasing the concentration of glycerol. We conclude that our scenario, that the ultrasonic feature (and we believe all the other thermodynamic anomalies) in water-alcohol mixtures is characterized by the relaxation phenomena, not by the special mixing state of the two materials (water and alcohol). We believe that the fluctuations of pure water corresponding to liquid-liquid phase transition should be dominant in the low alcohol (both monohydric-alcohol and glycerol) mixtures. In the glycerol-rich samples, the relaxation is thought to be characterized by the glass forming ability of pure glycerol.

Acknowledgments

This work was partially supported by JSPS (Japan Society for the Promotion of Science) KAKENHI Grant Number JP26610116. The synchrotron radiation experiments were performed at the BL35XU of SPring-8 with the approval of the Japan Synchrotron Radiation Research Institute (JASRI) (Proposal No. 2015B1788).

References

[1] F. Franks and D. J. G. Ives, Quart. Rev. Chem. Soc. **20**, 1 (1966)
[2] G. D'Arrigo and A. Paparelli, J. Chem. Phys. **88**, 405 (1988)
[3] G. Onori, J. Chem. Phys. **87**, 1251 (1987);
[4] N. Shibata, Y. Kajihara et al, in preparation
[5] D. N. Glew, Nature **195**, 698 (1962)
[6] H. Endo, Bull. Chem. Soc. Jpn **46**, 1106 (1973)
[7] O. Conde, J. Teixeira and P. Papon, J. Chem. Phys. **76**, 3747 (1982)
[8] Y. Kajihara et al, in preparation
[9] G. Ruocco, F.Sette, Condens. Matt. Phys. **11**, 29 (2008)
[10] F. Sette, G. Ruocco, M. Krisch, C. Masciovecchio, R. Verbeni, and U. Bergmann, Phys. Rev. Lett. **77**, 83 (1996)
[11] A. Cunsolo, G. Ruocco, F. Sette, C. Masciovecchio, A. Mermet, G. Monaco, M. Sampoli and R. Verbeni Phys. Rev. Lett. **82**, 775 (1999)
[12] G. Ruocco and F. Sette, J. Phys.: Condens. Matter **11**, R259 (1999)
[13] S. C. Santucci, D. Fioretto, L. Comez, A. Gessini, and C. Masciovecchio, Phys. Rev. Lett. **97**, 225701 (2006)
[14] A. Q. R. Baron et al, J. Phys. Chem. Solids **61**, 461-465 (2000)

[15] A. Cimatoribus, S. Saccani, F. Bencivenga, A. Gessini, M. G. Izzo and C. Masciovecchio, New J. Phys. **12**, 053008 (2010)

[16] F. Sette, Michael H. KrischC. Masciovecchio, G. Ruocco, G. Monaco, Science **280**, 1550 (1998)

[17] A Giugni and A Cunsolo, J. Phys.: Condens. Matter **18**, 889 (2006)

[18] W. M. Slie, A. R. Donfor Jr., and T. A. Litovitz, J. Chem. Phys. **44**, 3712 (1966)

[19] P. H. Poole, F. Sciortino, U. Essmann and H. E. Stanley, Nature **360**, 324 (1992)

[20] O. Mishima, H. E. Stanley, Nature **396**, 329 (1998)

EPJ Web of Conferences **151**, 07001 (2017) DOI: 10.1051/epjconf/201715107001
LAM-16

Effect of annealing and cobalt content on relaxation and crystallization behavior of zirconium based bulk metallic glasses

Yue Dong[1*], *Rainer* Wunderlich[1], and *Hans-Jörg* Fecht[1]

[1]Institute of Micro and Nanomaterials, University of Ulm, Albert-Einstein Allee 47, 89081 Ulm, Germany

Abstract. The effects of annealing and cobalt content on relaxation and the crystallization process of $Zr_{64}Ni_{10}Al_7Cu_{19}$ bulk metallic glasses were investigated. β-relaxation occurs during annealing, leading to increased endotherm before crystallization. α-relaxation during high temperature annealing (higher than Tg) affects the crystallization process. The introduction of cobalt leads to an inhomogeneous amorphous structure and two-step crystallization due to the positive mixing enthalpy between cobalt and copper. Non-affine thermal strain arising from low temperature annealing of heterogeneous structure leads to a reduced endotherm phenomenon during relaxation on the DSC curves and a reduction in hardness.

1 Introduction

Metallic glasses are amorphous alloys, meaning they are disordered at the atomic scale. Different from standard metals, metallic glasses have no long-range order. Due to their unique structures, including a lack of grain boundaries or dislocations, bulk metallic glasses often exhibit excellent properties, including high strength, high hardness, low Young's modules and good corrosion resistance. However, metallic glasses are still not widely used. Although metallic glasses have been investigated for many years, the amorphous structure is still not well understood [1].

To produce a metallic glass, the metallic liquid needs to be cooled down without any nucleation. Thus, the properties of the supercooled liquid are correlated with the structure and properties of the formed metallic glass [2,3]. Due to their metastability, the relaxation and crystallization behaviors of the supercooled liquids and glassy solids are related to their structures and properties. It has been shown that activation energy of the relaxation is nearly equivalent to the activation energy of shear transformation zones, and the temperature difference between crystallization and relaxation has been shown to be related to the metal's glass forming ability [4]. Furthermore, the study of metallic glasses is vital to understanding the nature of glass. However, the factors that govern relaxation and

* Corresponding author: yue.dong@uni-ulm.de

crystallization are still unclear and are thus actively and systematically being investigated [5,6,7].

Zr-based bulk metallic glasses have been the subject of intensive research because of their good glass forming ability and excellent mechanical properties [8]. W Wang et al. [9] investigated the Zr-Cu-Ni-Al system and found that for some compositions, the metallic glasses show good plasticity. In this work, we will focus on effects of annealing and composition on the relaxation and crystallization processes of this alloy series.

2 Experimental methods

Ingots of different compositions ($Zr_{64-x}Ni_{10}Al_7Cu_{19}Co_x$, x=0-8.3 at.%) were prepared by arc-melting a mixture of high purity Zr, Ni, Al, Cu and Co elements in a Ti-gettered argon atmosphere. The as-obtained ingots were then melted and sucked into a water-cooled copper mold. The samples were glassy rods 4mm in diameter and 40mm in length. The structure of the as-casted samples were analyzed by X-ray diffraction (XRD) with Cu-Kα radiation. Thermal analysis was carried out in a differential scanning calorimeter under a high purity Ar flow. In continuous heating, the scan rate was 20K/min from 373K to 830K. Each sample was heated twice so that the baseline could be obtained from the second run and subtracted to get the final curve. For the isothermal annealing, the samples were first heated to a fixed temperature with a fast heating rate (160K/min) from room temperature and then held. The results can also be used to calculate the heat flow as a function of time. To monitor changes in the structure of the BMGs, selected samples were analyzed with a transmission electron microscope (TEM), a Philips CM20 operating at 200kV. Electron transparent TEM lamellae were produced by mechanical grinding and polishing followed by low angle Ar-ion milling. To investigate the change in hardness, nanoindentation measurements were completed with a Berkovich indenter on the polished samples.

3 Results

Fig. 1 shows the XRD results for samples with different cobalt contents. All of the samples have fully amorphous structures. With increasing cobalt content, the diffraction maxima shift to the right, indicating that the sample with the highest cobalt content has the smallest interatomic distance due to cobalt having smaller atom diameter in comparison to zirconium.

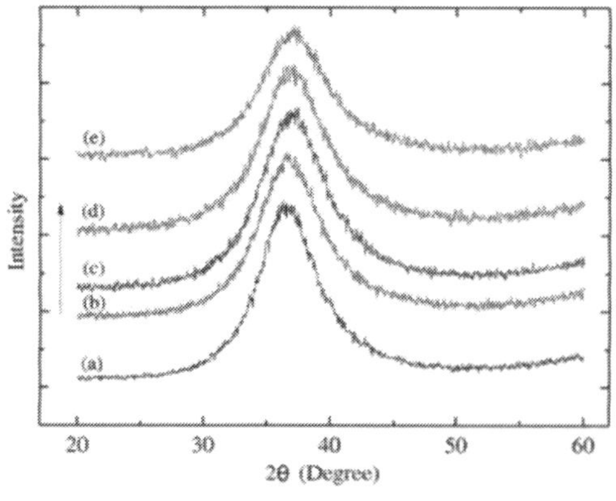

Fig. 1. X-ray diffraction patterns for $Zr_{64-x}Ni_{10}Al_7Cu_{19}Co_x$ as a function of the Co content after casting: (a) 0%, (b) 2.7%, (c) 4%, (d) 6%, (e) 8.3%

To study the effect of annealing on the relaxation process, the $Zr_{64}Ni_{10}Al_7Cu_{19}$ alloys were annealed at different temperature 606 K-658 K (Tg for this alloy is 648K.). Fig. 2(a) shows that annealing introduces an endothermic peak at the beginning of the relaxation on the DSC curves. The peak intensity increases at first and then reduces with the rising annealing temperature. This peak also shifts to a higher temperature with the rising annealing temperature. Furthermore, in Fig. 2(b), it shows that with the rising annealing temperature, the peak intensity is reduced, the peak width increases and the crystallization peak shifts to a lower temperature when the annealing temperature is higher than Tg.

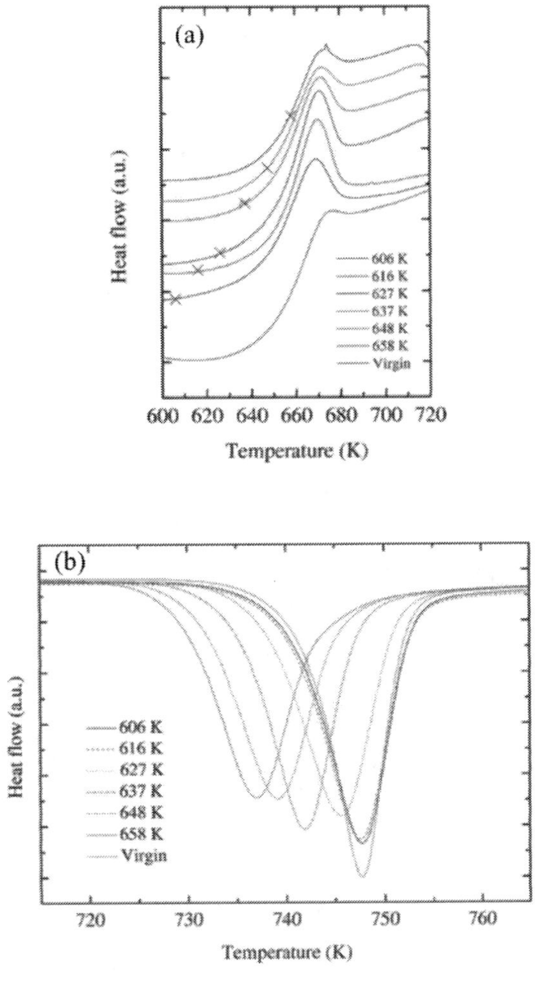

Fig. 2. (a) Changes in the heat flow around Tg on the DSC curves of $Zr_{64}Ni_{10}Al_7Cu_{19}$ BMGs annealed at different temperatures. The cross on the curve is the annealing temperature before the DSC measurement. (b) Changes in the crystallization peak on the DSC curves for $Zr_{64}Ni_{10}Al_7Cu_{19}$ BMGs annealed at different temperature.

Fig. 3(a) shows the continuous heating DSC curves for alloys with different cobalt contents. The DSC curves show that the cobalt composition affects the thermal properties of the alloys so dramatically that the crystallization peak splits into two peaks. With the increasing cobalt content, the distance between the two peaks becomes larger, the intensity of the first peak increases, and the second peak exhibits a corresponding decrease so that the total enthalpy remains consistent. In Fig. 3 (b), it can be seen that the glass transition temperature Tg shifts to a higher temperature due to the increase in cobalt content. The temperature range of the relaxation (between Tg and Tx) also reduces with increasing cobalt content.

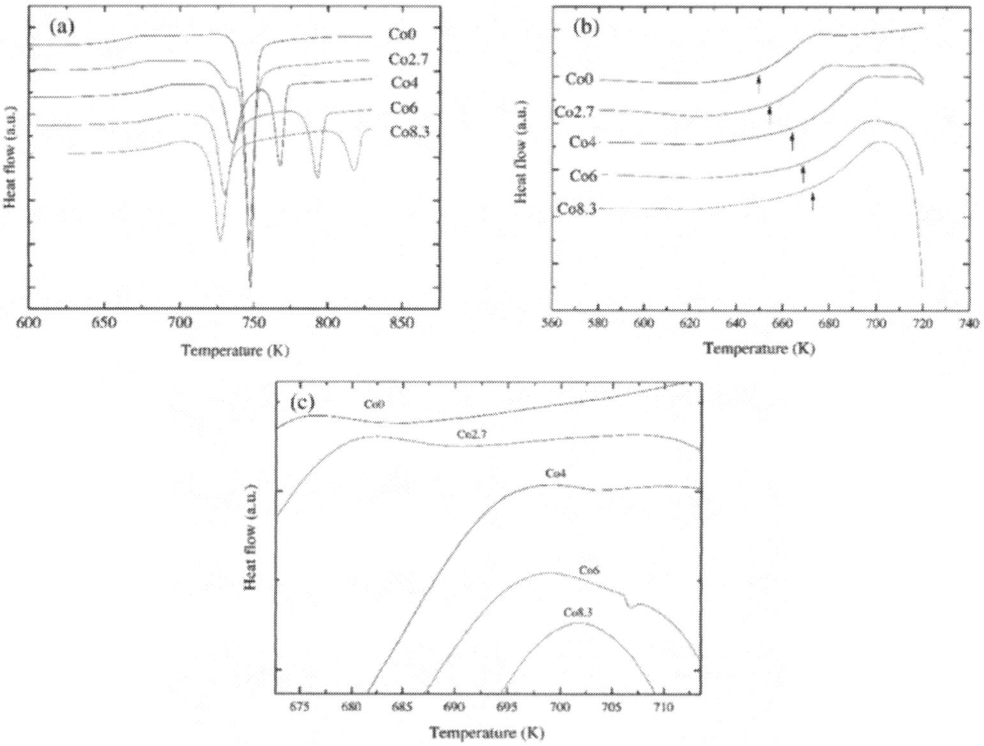

Fig. 3. (a) DSC curves of the BMGs with different cobalt content. (b) Change of Tg as a function of Co content (0%, 2.7%, 4%, 6%, 8.3%). (c) Change in the relaxation properties as a function of Co content as measured by DSC

Fig. 4 shows the DSC curves of samples with various Co contents annealed at different temperatures. When the samples that contained Co were annealed at a lower temperature (563 K), a special endotherm reducing event occurs before relaxation, as shown in the DSC measurement Fig. 4 (d). Furthermore, with increasing of Co content, the presence of the endotherm reducing event becomes more obvious. Whereas, when the samples were annealed at a higher temperature (583 K), there was no substantial change from the results of samples without Co. However, the height of the endothermic peak before relaxation is reduced with the addition of Co.

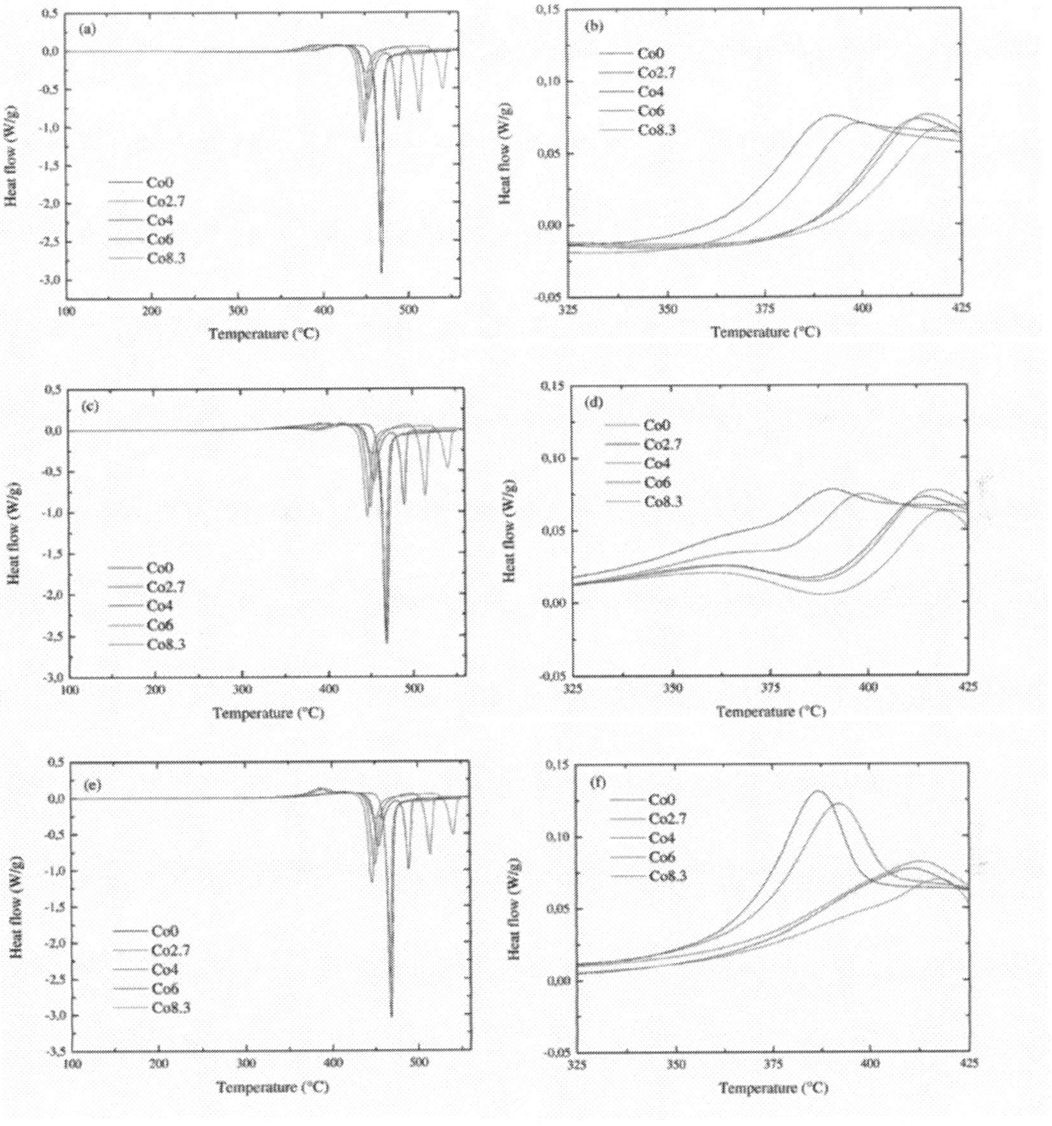

Fig. 4. (a) DSC curves of the as-cast BMGs with different cobalt contents. (b) The relaxation peak as a function of cobalt content for the as-cast BMGs. (c) DSC curves of the BMGs with different cobalt contents annealed at 563 K. (d) Relaxation of the BMGs with different cobalt contents annealed at 563 K. (e) DSC curves of the BMGs with different cobalt contents annealed at 583 K. (f) Relaxation of the BMGs with different cobalt contents annealed at 583 K.

4 Discussion

The XRD results show that the introduction of cobalt does not dramatically reduce the glass forming ability. However, if the Co content is too high (more than 10%), the BMG will be partially crystallized. Due to the brittle nature of the intermetallic compound formed at high cobalt contents, we focus only on the composition range that results in a fully amorphous structure for this BMG (i.e., with Co content between 0 and 8.3%). From the shift in the diffraction maxima it can be inferred that with increasing cobalt content, the average interatomic distance is reduced, resulting in a stronger atomic interaction.

To investigate the effect of annealing on the relaxation process, $Zr_{64}Ni_{10}Al_7Cu_{19}$ BMGs were first annealed at a temperature around Tg. An endothermic peak appears on the DSC curves prior to relaxation. The peak intensity increases with elevation of the annealing temperature while it is below Tg; however, the peak intensity starts to decrease at annealing temperatures higher than Tg, as shown in Fig. 2(a). This phenomenon has been previously observed in the literature [10]. The energy landscape model can be used to explain this phenomenon [11]. The β-relaxation determines the features of the α-relaxation by acting as a precursor to the glass transition. When the temperature is below Tg, β-relaxation is dominant (Fig. 5(b)). As shown in Fig. 5(a), with higher annealing temperatures, the high energy barriers present can be surmounted and thus, the system arrives at a configuration with a lower energy (deeper valley on energy landscape). Then, after annealing, the α-relaxation in the following DSC measurement needs higher activation energy to surmount the initial energy barriers. In this way, the heat flow increases and the peak shifts to a higher temperature. If the annealing temperature is higher than Tg, α-relaxation will occur, accompanied by β-relaxation during annealing, such that the endothermic peak intensity on DSC curve is reduced. In another words, the low temperature β-relaxation results in a structure with lower energy, which indicates denser atomic packing. Thus, the low temperature β-relaxation leads to a higher diffusion energy for the following α-relaxation, resulting in the appearance of the endothermic peak prior to relaxation on the DSC curves; alternatively, when the annealing temperature is higher than Tg, the diffusion energy is reduced.

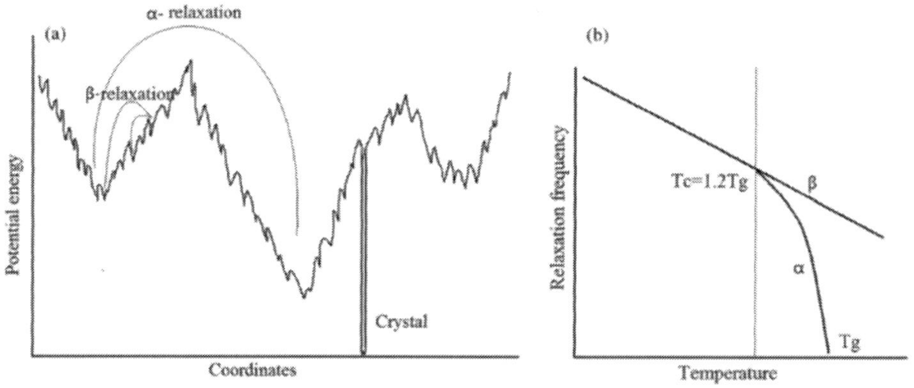

Fig. 5. Schematic illustration of (a) an energy landscape, (b) relaxation frequency change as a function of temperature [11]

The difference between low and high temperature annealing (lower or higher than Tg, respectively) can also be found in Fig. 6. There is an apparent difference in heat flow during annealing. The curves can be well fitted with the equation $W(t) = W_0 (1\text{-}e^{t/\tau})$, where W_0 is the increase in heat flow $W(t)$ after an infinite time and τ is the relaxation time. The fit parameters are given in Table 1.

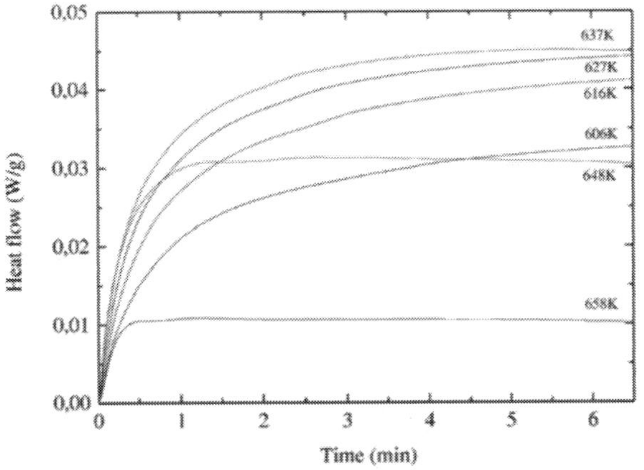

Fig. 6. Heat flow on $Zr_{64}Ni_{10}Al_7Cu_{19}$ BMGs as a function of time during annealing at different temperatures

Table 1. Fitting parameters for heat flow during annealing

Annealing temperature	W_0(W/g)	τ (s)
606K	0.031	0.943
616K	0.039	0.904
627K	0.043	0.765
637K	0.044	0.635
648K	0.031	0.296
658K	0.012	0.164

The difference in heat flow can also be explained by the energy landscape model. When the annealing temperature is below Tg, the final heat flow, δ_0, increases with increasing temperature. This means that additional energy barriers can be surmounted, allowing the system to arrive at the structure with the lowest energy. When the annealing temperature is above Tg, the relaxation mechanism changes. Due to the occurrence of α-relaxation, the ergodic configuration on energy landscape is smaller, resulting in a much shorter relaxation time, τ, compared to the relaxation time during the annealing process below Tg, where β-relaxation is dominant.

In Fig. 2(b), the crystallization peak shifts to lower temperatures after annealing, and at higher temperatures, this is even more apparent. This is due to significant relaxation occurring during the annealing process, especially from the α-relaxation at higher temperatures. This relaxation represents a major challenge in glassy physics. Perturbations from some external vital fields can markedly accelerate crystallization. [12] Furthermore, diffusion control, instead of nucleation control, becomes more important in the crystallization process, which leads an increase in the peak width and a reduction in the peak intensity.

Thermal properties of $Zr_{64-x}Ni_{10}Al_7Cu_{19}Co_x$ BMGs have been analyzed by DSC to investigate the effect of composition on the relaxation process. The DSC results show a split peak as a result of the introduction of cobalt. By calculating the enthalpy, one can determine the two peaks correspond to a two-step crystallization process. To determine origin of the two-step crystallization process, we completed electron diffraction and high-resolution TEM investigations for two BMG samples, $Zr_{64}Ni_{10}Al_7Cu_{19}$ and

 DOI: 10.1051/epjconf/201715107001

$Zr_{55.7}Ni_{10}Al_7Cu_{19}Co_{8.3}$ (see Fig. 7). Electron diffraction patterns show that both samples are amorphous. However, HRTEM images revealed a nanometer-scaled inhomogeneous structure on the $Zr_{55.7}Ni_{10}Al_7Cu_{19}Co_{8.3}$ BMG sample that was not present in the $Zr_{64}Ni_{10}Al_7Cu_{19}$ BMG sample. Phase separation on the scale of 5-10 nm occurs in the $Zr_{55.7}Ni_{10}Al_7Cu_{19}Co_{8.3}$ BMG sample, which could be caused by the positive enthalpy of mixing between copper and cobalt. Thus, the normal one step crystallization process splits into two because of the introduction of Co. With increasing of cobalt content, the second crystallization peak is stabilized, whereas the first crystallization peak shifts to lower temperatures. Thus, the relaxation processes occurs over a smaller temperature range as shown in Fig. 3(c).

Fig. 7. TEM images of (a) $Zr_{64}Ni_{10}Al_7Cu_{19}$ and (b) $Zr_{55.7}Ni_{10}Al_7Cu_{19}Co_{8.3}$ BMG samples.

As shown in Fig. 4, annealed BMGs with Co show an obvious difference in relaxation behavior. After low temperature annealing (563 K), the DSC curve for the BMG with Co shows an endotherm reducing event prior to relaxation that is not present for the annealed BMGs without Co (Fig. 4(d)). This change in ΔH can be used to characterize the effects of thermal and mechanical treatments on metallic glasses. For example, heavy plastic deformation or room temperature-liquid nitrogen temperature cycling which introduce non-affine strains in amorphous structure can lead to a exothermic phenomenon before relaxation, that is termed rejuvenation [13]. In this case, it can be deduced that due to the difference in thermal expansion, non-affine thermal strains arise from the pre-existing heterogeneity in BMGs with Co (see Fig. 7(b)), which leads to the change in ΔH visible on DSC curves. It is worth noting that this kind of "rejuvenation" is only in reference to the annealed homogeneous BMGs without Co (Fig. 7(b)), not the as-cast state of the BMGs. This phenomenon disappears when the annealing temperature reaches 583 K due to the easier diffusion of atoms at higher temperatures. Stress relief annealing occurs. It can also be confirmed by the nanoindentation hardness measurement. It has been reported that hardness of BMGs will increase after annealing due to free volume annihilation [14]. However, Fig. 8 shows that the hardness is instead reduced by the annealing process when the cobalt content reaches 8.3%. The BMGs without cobalt or with low cobalt content have a relatively homogeneous structure and thus show a decrease in free volume during annealing that increases their hardness, as previously reported. The inhomogeneous amorphous structure forms when the cobalt content is at least 8.3% (see Fig. 7(b)). As discussed earlier, non-affine thermal strains arise during annealing of this structure which is beneficial for the formation free volume, and the hardness decreases.

DOI: 10.1051/epjconf/201715107001

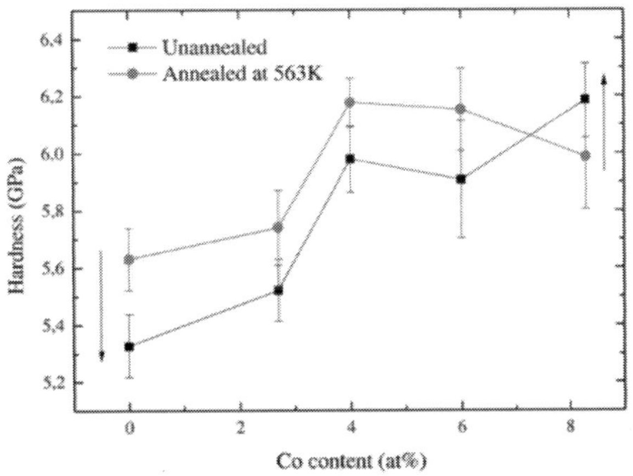

Fig. 8. Hardness of as-cast samples and samples annealed at 563K

5 Conclusion

The effects of annealing and composition on the relaxation and crystallization processes of $Zr_{64-x}Ni_{10}Al_7Cu_{19}Co_x$ bulk metallic glasses were studied. The peak at the beginning of relaxation in the DSC curve demonstrates the occurrence of β-relaxation during annealing; the α-relaxation then affects the following crystallization. The introduction of cobalt reduces the temperature range of relaxation and results in a splitting of the crystallization peak into two peaks due to its inhomogeneous amorphous structure. Non-affine thermal strains arise from this structure during low temperature annealing, which leads to an endotherm reducing event before relaxation and a reduction in hardness.

References

1. A.L. Greer, Y.Q. Cheng, E. Ma, Materials Science and Engineering R **74**, 71-132 (2013)
2. L.C. Chen, F. Spaepen, Nature **336**, 366-368 (1988)
3. K. Biswas, S. Ram, L. Schultz, J. Eckert, Journal of Alloys and Compounds **397**, 104-109 (2005)
4. K. Biswas, S. Venkataraman, W.Y. Zhang, S. Ram, J. Eckert, Applied Physics Letter **100**, 023501 (2006)
5. S. Ram, G.P. Johari, Philosophical Magazine B **61**, 299-310 (1990)
6. T. Matsuyama, N.Terada, T. Baba, T. Sawada, S. Tsuge, K. Wakisaka, S. Tsuda, Journal of Non-Crystalline Solids **198-200**, 940-944 (1996)
7. P.S. Frankwicz, S. Ram, H.-J. Fecht, Materials Letters **28,** 77-82 (1996)
8. U. Koester, J. Meinhardt, S. Roos, H. Liebertz, Applied Physics Letter **69**, 179-181 (1996)
9. YH Liu, G Wang, RJ Wang, DQ Zhao, MX Pan, WH Wang, Science **315,** 3185-3188 (2007)
10. G.J. Fan, J.F. Loeffler, R.K. Wunderlich, H.-J. Fecht, Acta Materialia **52**, 667-674 (2004)
11. PG Debenedetti, FH Stillinger, Nature **410**, 259-267 (2001)
12. HB Yu, WH Wang, K Samwer, Materials Today **16**, 183-191 (2013)

13. S.V. Ketov, Y.H. Sun, S. Nachum, Z. Lu, A. Checchi, A.R. Beraldin, H.Y. Wang, D.V. Louzguine-Luzgin, M.A. Carpenter, A.L. Greer, Nature **524**, 200(2015)
14. J Gu, M Song, S Ni, S Guo, Y He, Materials and Design **47**, 706-701 (2013)

Thermal effect on structural and magnetic properties of $Fe_{78}B_{13}Si_9$ annealed amorphous ribbons

Mohamed Larbi Soltani[1,*], *Abdelhay* Touares[1], *Tiburce A.M.* Aboki[2], and *Jean-Georges* Gasser[3]

[1]LPS, Dpt. de Physique-Annaba University, Route d'El Hadjar BP 12 Annaba, Algérie
[2]GMS-LPCS UMR 7045, Chimie Paristech, 11 rue Pierre et Marie Curie, F-75231 Paris Cedex 05, France
[3]LCP –A2MC Institut de Chimie, Physique et Matériaux, Université de Lorraine, 1 bd Dominique François Arago 57078, Metz Cedex 3, France

Abstract. In the present work, we study the influence of thermal treatments on the magnetic properties of as-quenched and pre-crystallized $Fe_{78}Si_9B_{13}$ after stress relaxation. The crystallization behavior of amorphous and treated $Fe_{78}Si_9B_{13}$ ribbons was revisited. The measurements were carried out by means of Differential Scanning Calorimetry, by X-ray diffraction and by Vibrating Sample Magnetometer, Susceptometer and fluxmeter. Relaxed samples were heated in the resistivity device up to 700°C and annealed near the onset temperature about 420°C for respectively 1, 3, 5, 8 hours. In as-quenched samples, two transition points occur at about 505°C and 564°C but in relaxed sample, the transition points have been found about 552°C and 568°C. Kinetics of crystallization was deduced for all studied samples. Annealing of the as-purchased ribbon shows the occurrence of α-Fe and tetragonal Fe_3B resulting from the crystallization of the remaining amorphous phase. The effects on magnetic properties were pointed out by relating the structural evolution of the samples. The magnetic measurements show that annealing change the saturation magnetization and the coercive magnetic field values, hence destroying the good magnetic properties of the material. The heat treatment shows that the crystallization has greatly altered the shape of the cycles and moved the magnetic saturation point of the samples. The effect of treatment on the magneto-crystalline anisotropy is also demonstrated.

1 Introduction

To obtain nanocrystalline structure, different techniques are available; the crystallization method is an effective one. The basic idea is to control the crystal nucleation and growth during the crystallization process by empirical adjusting of the heat treatment conditions. As an example, the isothermal heat treatment at 450°C leads to complete crystallization of

[*] Corresponding author: ml_soltani@yahoo.com

the amorphous Fe-B-Si in about 72 hours. By extrapolation, one can estimate that about 277 hours are required for the formation of nanograins of 400 nm size, as reported by [1]

The amorphous alloys based on Fe-B-Si are successfully applied as the core materials in distribution transformers due to their low core losses and relatively high saturation magnetization and low exciting current [2]. Therefore, it is necessary to determine both the conditions in which the properties of the as-received state of the metallic glass are conserved and the parameters of the kinetics of change. Knowledge of the crystallization kinetics of amorphous alloys offers the possibility for specific crystalline structures and hence for optimizing their properties. The elimination of both internal and external stresses improves the mobility of the Bloch walls of magnetic domains and facilitates reorientation of spins and magnetic saturation under an applied field H. Stress relaxation is observed by annealing the amorphous alloy at temperatures allowing sufficient atomic diffusion for structural relaxation but not enough to initiate crystallization [3]. The optimization of soft-magnetic properties of metallic glasses can be made by dynamical current annealing, One can obtain the structure relaxation, where the atoms rearrange to become more ordered with increasing annealing temperature, without bulk crystallization occurring. Authors reported that whereas there is no trace of crystalline phase for the samples annealed at 420°C, the as-quenched sample exhibits a broad peak from amorphous phase which is slightly larger than the corresponding peak of the relaxed sample, implying the occurrence of atomic rearrangement during annealing [4] Moreover, in the relaxed sample, the crystalline phase is still dispersed in the amorphous matrix and the small peak centered in the broad amorphous peak corresponds to α-Fe phase. The annealing effects on the thermomagnetic structural evolution of the $Fe_{78}B_{13}Si_9$ amorphous alloy have an effective disadvantage: the amorphous alloys such as Fe-B-Si are ductile in the as-cast state, but become brittle during conventional annealing [5]. After crystallization, the most attractive properties of an amorphous alloy, such as soft magnetic behavior, high strength, and ductility are usually lost. Thus, the limiting factor of many technological applications of nanocrystalline materials is their brittleness [6].

The magnetic saturation would depend on the internal stress states resulting from the as-cast conditions and further treatments of the alloy. Escobar *et* al. [3] have reported, also like Kronmüller quoted by Luborsky [7], the coercive field dependence on temperature [8]. H. Kronmüller, in [9], again pointed out that the defect concentration and fluctuations affect the magnetic properties of amorphous alloys. This behavior may be compared with the results of Escobar *et* al. [3] who optimized the magnetic properties of Fe-Si-B alloys by a very short annealing at 430°C. But it seems that the absolute magnetic saturation of this alloy occurs at higher field in accordance with the work of Pankhurst et al [10], who using Mossbauer studies did not reach the saturation state for the same composition with applied field of up to 5T.

The aim of this study is to analyze the thermal, microstructural and magnetic properties of Fe-B-Si metallic glass. As-quenched alloys exhibit the typical macroscopic behavior of fully amorphous structures, while annealing induces, besides strain relaxation, moderate growth of the nanocrystals.

Quenched state (as-cast) and relaxed ribbons which exhibit in a dispersed α-Fe crystallites, have been investigated, particularly, in point of view of kinetics of crystallization of Fe-B-Si. The non-isothermal Differential Scanning Calorimetry (DSC) was used with the effect of annealing procedure on the structural and magnetic properties.

DOI: 10.1051/epjconf/201715107002

2 Experimental

Specimen used in this work are cut from ribbons approximately 25-µm thick and 10-mm wide wired, purchased from Allied Chemical Ltd (2605-S2). The amorphous ribbon, which has undergone structural relaxation by heat treatment at about 250° C, for a period of 2 hours, is referred to as a relaxed amorphous alloy.

This treatment is carried out immediately after rapid quenching by Allied Corp either without the application of a magnetic field or under magnetic field according to the requirement. In the case of this study, the magnetic field which allows uniaxial magnetic anisotropy is not applied. This structural relaxation makes it possible to reorganize the amorphous structure by relaxing the constraints due to rapid solidification.

Differential Scanning Calorimetry (DSC) measurements are used to study the thermal behavior. The METLER TOLEDO DSC 822^{e} apparatus is used. Amorphous and relaxed samples were continuously heated in the device to temperatures at which different crystallization stages could be obtained.

The measurements were carried out at continuous heating rates between 5 and 80 K/minute up to 700°C with accuracy of ±1°C, for as-quenched, relaxed, fully crystallized and isothermally annealed samples at different times. Heat treatment conditions were selected according to the beginning of the pre-crystallization process resulting from the first heating and cooling. At 420 °C, measurements are performed as function of time. This temperature is known to bring the optimum of magnetic quality (near Curie Temperature Tc ≈ 415°C). Crystalline phase's characterization was obtained by means of a Siemens D5000 diffractometer using Cu-Kα radiation. Susceptometer-3Tesla and VSM-9Tesla Quantum Design-PPMS (Physical Property Measurement System) were used for magnetic saturation measurements with high applied magnetic field up to 2 Tesla in order to saturate the as-received and 420°C-annealed samples.

To observe, in experiments, the dynamics of the walls in ferromagnetic one choose a Fluxmeter [11] for hysteresis loop measurements at variable amplitude. One connected a magnetic circuit like a transformer, by taking the values of magnetic applied field H (T) from primary circuit and magnetic induction B (T) from secondary circuit simultaneously.

3 Results and discussion

3.1 DSC results

The Differential Scanning Calorimetry DSC was used to study the crystallization kinetics of $Fe_{78}Si_9B_{13}$ glasses at the heating rates of 5, 20, 40 and 80K/m. Differential scanning calorimetry (DSC) measurements of the as-cast $Fe_{78}Si_9B_{13}$ metallic glass are presented in Figure 1. In each DSC curve, two clearly separated exothermic peaks are observed; this indicates that during the transformation two principal crystalline phases appear. One can see that the peaks temperatures are higher at higher heating rates. 'The cusp-like features in the DCS scans at temperatures between 415-418°C indicate the Curie temperatures Tc. These values are close to the result by Harmlin et al. of Tc= 413.3 °C [12].

 DOI: 10.1051/epjconf/201715107002

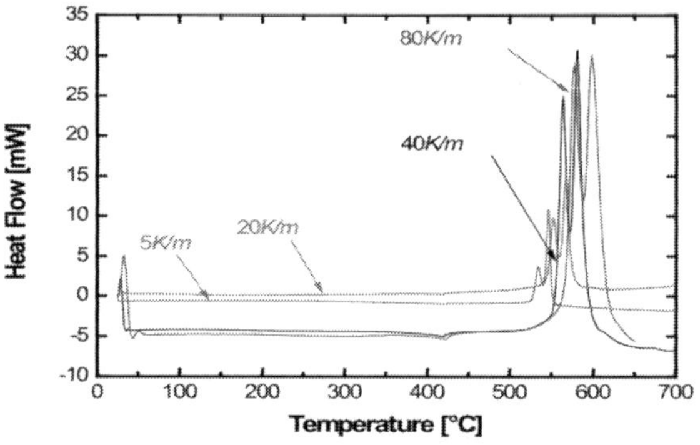

Fig. 1. DSC thermograms of the as-cast $Fe_{78}Si_9B_{13}$ metallic glass at heating rates of 5, 20, 40 and 80 K/min.

Differential scanning calorimetry (DSC) measurements of thermally treated and annealed samples at 420°C are presented in figure 2. One can observe from the DSC curve of the sample annealed at 420°C for 8h that only the second crystallization appears. Thus it explains that at this annealing time, the first stage of crystallization have been achieved completely. Figure 2 presents the continuous crystallization process and shows the fully crystallized structure without any transition observed in DSC measurement. However, the DSC curve in fully sample presents a solid-state phase transformation that is not reversible (not reported here).The kinetics indicates that during transformations two crystallization steps appear in table 1. The first crystallization temperatures observed in the thermograms are moved towards higher values with the increasing annealing time.

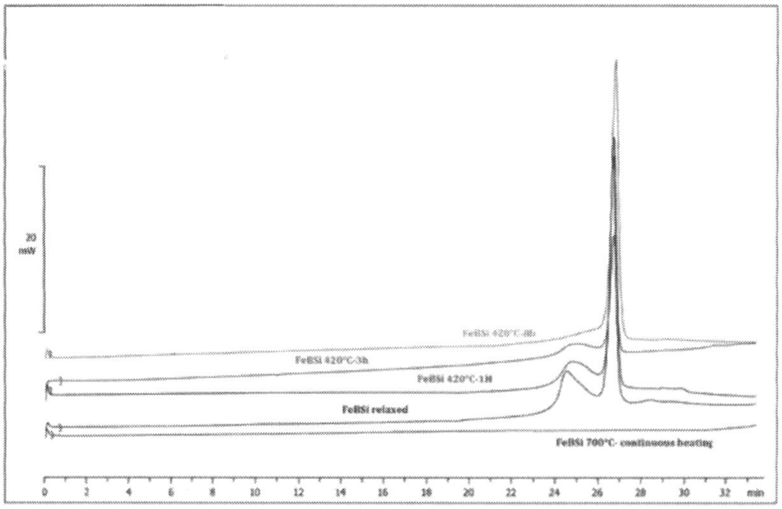

Fig. 2. DSC thermograms of the annealed at 420°C and continuous heated $Fe_{78}Si_9B_{13}$ samples.

DOI: 10.1051/epjconf/201715107002

3.2 Analysis of kinetics

These phase transformations may be assumed to be thermally activated processes. For thermally activated processes the heating rate β and the peak temperatures T_p are linked by the following equation:

$$\beta=\beta_0 \exp(-E_a/ R.T_p). \quad (1)$$

β_0 is the pre-exponential factor,
R: the gas constant and
E_a: the apparent activation energy.

The results for activation energies and pre-exponential factors of the two crystallization events observed in the DSC plots are summarized in table 1.

Table 1. Parameters describing the crystallization kinetics for all studied samples

Sample	E_a (T_1) kJ/mole	β_1 K/s	E_A(T_2)	β_2 K/s
As quenched	384	$5{,}9867\ 10^{23}$	308	$4{,}2779\ 10^{18}$
Relaxed	311	$1{,}2096\ 10^{20}$	304	$2{,}7550\ 10^{18}$
420°C/1H	295	$7{,}9441\ 10^{18}$	315	$1{,}8718\ 10^{19}$
420°C/3H	203	$5.91\ 10^{12}$	314	$1{,}5588 10^{19}$
420°C/8H	---	---	296	$1{,}3627\ 10^{18}$

Based on Kissinger's model, the activation energy of crystallization can be derived from:

$$\ln(\beta/ T_P^2) = E_c/RT_p + \text{Constant} \quad (2)$$

Figure 3 shows the plots of $\ln(\beta/ T_P^2)$ against $1000/T_p$. We can use the slope of the fitted data to calculate the activation energy E_c. Then the Kissinger analysis of the DSC scans gives the parameters compiled in table 2.

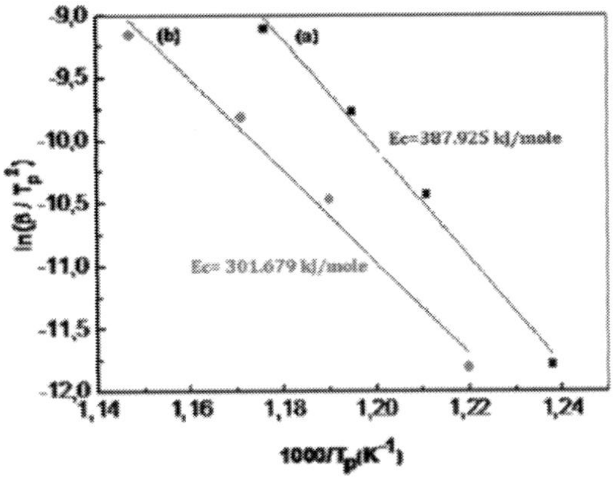

Fig. 3. $\ln(\beta/T_p^2)$ versus $1000/T_p$ plot for $Fe_{78}Si_9B_{13}$ metallic glass: (a) first peak and (b) second peak. Solid lines correspond to linear fit of the experimental data

Table 2. Values of activation energies E_{c1} and E_{c2} calculated by Kissinger's model for the $Fe_{78}Si_9B_{13}$ metallic glass.

Activation energy E_{c1} for the first crystallization stage ($kJmol^{-1}$)	**357.925**
Activation energy E_{c1} for the second crystallization stage ($kJmol^{-1}$)	**301.679**

By comparison, the Auguis-Bennet's model describes the kinetics by using Equation 3 to calculate the activation energy E_c of the crystallization as well as the frequency factor K_0.

$$\ln(\beta/T_p) = -E_c/RT_p + \ln(K_0) \quad (3)$$

The figure 4 and the table 3 describe the deduction of E_c and K_0 with the same procedure previously used. Since the Fe-Si grains determine the magnetic properties of the material, the study of the first peak is interesting. So, the apparent activation energy according to thermally activated process was deducted (E_{c1} = 384 KJ mol^{-1}) and agrees with a kinetic evaluation of the peak temperature according to Kissinger's and Auguis-Bennet's models [13-14]. More kinetic analysis can be described using different models like The Advance Model Free Kinetics (AMFK) model developed by Vyazovkin [15-16]. Results of kinetics for relaxed and annealed samples give further information concerning the crystallization process. It should be noted that the activation energy changes fundamentally during crystallization.

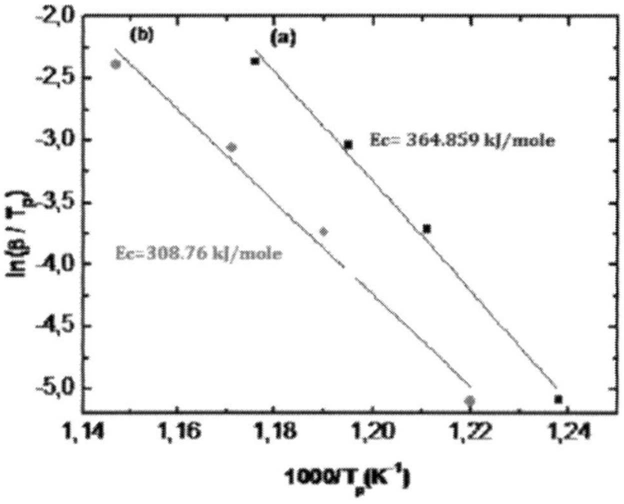

Fig. 4. $\ln(\beta/T_p)$ versus $1000/T_p$ plot for $Fe_{78}Si_9B_{13}$ metallic glass (a) first peak and (b) second peak. Solid lines correspond to linear function fit of experimental data

DOI: 10.1051/epjconf/201715107002

Table 3. Values of activation energies, E_{c1} and E_{c2}, calculated by Auguis-Bennet's model for the $Fe_{78}Si_9B_{13}$ metallic glass.

Activation energy E_{c1} for the first crystallization stage ($kJmol^{-1}$)	**364.859**
Activation energy E_{c1} for the second crystallization stage ($kJmol^{-1}$)	**308.76**
Frequency factor K_0	**lnK_0**
	49.348

3.3 X-Ray Diffraction results

Indeed, we are speaking of amorphous ribbons. We then stated that the samples are "partially crystalline". This can be contradictory. In reality, the ribbons exhibit, as we shall see with the X-ray diffraction, a small outgrowth indicating an appearance of α- Fe crystallites. The as-purchased material is a relaxed amorphous ribbon that contains some amounts of α-Fe of bcc structure whose (110) line emerges over the stretched amorphous halo. This demonstrates that the heat treatment slightly modified the structure which is not completely amorphous (see fig.5a).

The following annealing treatments on samples from the relaxed ribbon induce an increase of the amount of the α-Fe of bcc structure from the residual amorphous phase with the formation of borides.

The X-rays diffraction patterns (XRD) of the annealed ribbons show, in fig.5b, the main lines belong to bcc solid solution α-(Fe, Si) and tetragonal Fe_3B. The proportions of the crystalline phases related to the X-rays intensity indicate a different amount of phases within the different samples. At 420°C-1H, a (110) α–Fe peak corresponding to d_{hkl} at about 2.02 nm on the XRD spectrum can be observed along with a small peak of (301) tetragonal (t) Fe_3B at d_{hkl} = 2.33 nm. This latter increases following the annealing time. Moreover, the 2.02 nm peak shows a side peak at about 2.01 nm which is likely that of (121) of t-Fe_2B. However in the XRD spectrum of the sample heated to 700°C, the intensity of the peak at 2.02 nm is greater than that of the peak at 2.33 nm without any additionnal peak of t-Fe_2B. These observations indicate that the proportion of the α–Fe is greater in the sample heated than in the annealing samples. Thus long annealing time favours the borides formation.

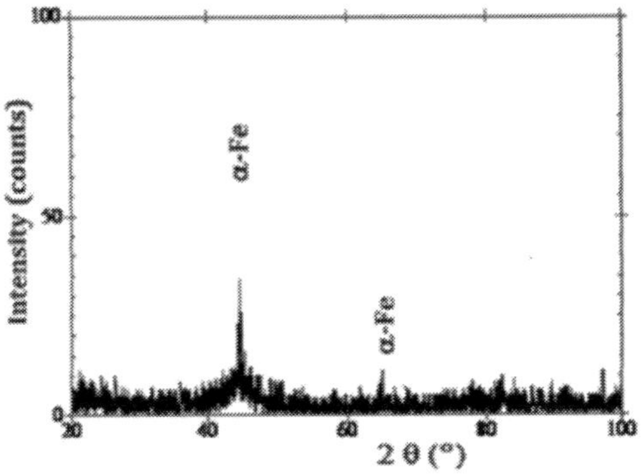

Fig. 5a. XRD pattern of relaxed ribbon

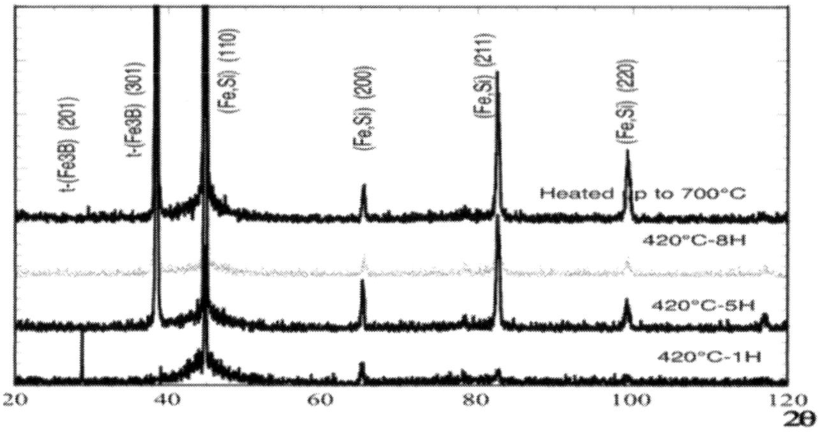

Fig. 5b. XRD patterns of crystallized $Fe_{78}Si_9B_{13}$ ribbons after annealing at 420°C for 1, 5, 8, hours, and for aribbon continuously heated up to 700°C.

3.4 Magnetic properties

We performed two separate experiments over time with the same samples. The first, to evaluate the saturation magnetization at high magnetic field and to confirm the role of the increase of the α–(Fe-Si) phase in the annealed material. Figure 6 shows the magnetization of annealed samples as a function of the applied magnetic field up 2 tesla (20000 Oersted). Progression to complete saturation seems faster for samples annealed during 5h and 8h with a tendency to unidirectional anisotropy. For the relaxed (amorphous) sample, the saturation is more difficult to achieve and is reached beyond the maximum applied field , this suggests that saturation is facilitated with the increase of volume fraction of the α-(Fe,Si) phase. For the relaxed sample, the structural relaxation may cause the value of the saturation relatively low compared to the treated samples from 5h to 8h when the soft magnetic phase (Fe, Si) is present in more large quantities.

In order to better define the remarkable properties of this soft magnetic material, we use a lower magnetic field experiment using a fluxmeter. Coercivity and anisotropy could be deduced from the hysteresis cycles for all samples. For a good comparison, as-cast amorphous samples show a low coercivity H_c and a fairly high magnetization with a low magnetocrystalline anisotropy indicating a rounded cycle characteristic of the amorphous structure.

An annealing procedure without applied magnetic field makes a somewhat rounded appearance. In this case, two processes are present: domain wall movement and rotation of moments.

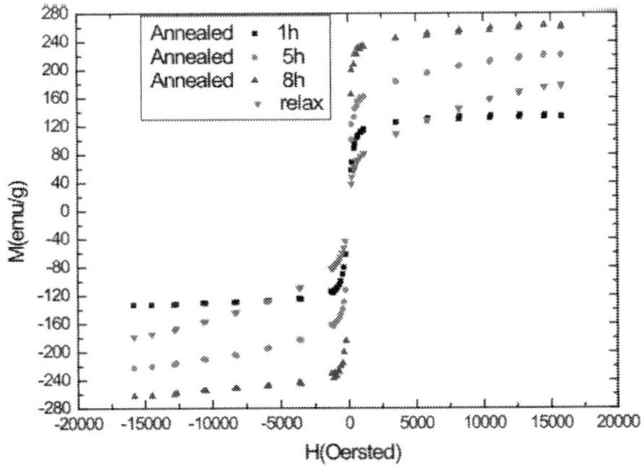

Fig. 6. Magnetization of annealed samples as a function of the applied magnetic field up 2 tesla

Table 4 presents some characteristic values inferred from hysteresis loops for maximum applied magnetic field H_{max}=1 kOe (VSM) giving magnetization M_s and hysteresis loops up to 27000 A/m (about 300 Oe) give the coercive field and relative anisotropy.

The amorphous Fe-Si-B alloys which are treated at 420°C for 1 hour remain amorphous. Structural relaxation takes place due to the formation of Fe_3Si, which is a soft magnetic phase. For the annealed alloys, we observe a strong increase in coercivity for the Fe-Si-B alloy with the annealing time. The growth of dendrites increases the coercivity by providing additional obstacles for the movement of the domain walls. In addition, the formation of hard magnetic phase Fe_3B results in a deterioration of soft magnetic properties. Indeed, the formation of the Fe_3B phase leads to an increase in the magneto-crystalline anisotropy, following which the magnetic curing takes place. For longer annealing times up to 8 hours, the saturation magnetization continues to increase due to the formation of the Fe_3Si phase. The Fe_3B phase continues to increase in parallel. When the sample is brought to 700 ° C, the magnetization continues to increase for the Fe-Si-B alloy due to the stopping of Fe_3B evolution as well as the (Fe,Si) phase progression (see section 3.3). Thus, the anisotropy decreases due to the reduction of the influence of the Fe_3B phase which no longer progresses.

Conclusion

We have studied the thermal behavior of amorphous and thermally treated $Fe_{78}B_{13}Si_9$ ribbons that contain a certain amount of bcc-αFe. DSC and XRD were used. Pre-annealed samples were annealed at 420°C during 1, 3, 5 and 8 hours.

The kinetics indicates that during transformations two crystallization steps appear.

XRD diffraction confirmed the presence of the latter phase and the stability of the bcc-α-Fe crystals due to the crystallization of the residual amorphous phase during the increase of the heating time.

Magnetic measurements indicate an evolution by annealing procedure which suggests improvement of saturation magnetization and occurring of directional anisotropy characterized by a growing remnant magnetization, but also destroying of mechanical properties as ductility and leads to magnetic hardening by increasing of coercive field H_c.

Table 4. Some properties inferred from hysteresis loops measured for the different samples by VSM and fluxmeter

Sample	Ms (emu/g) at 1 kOe VSM	Mr (emu/g) at 1kOe VSM	Hc (A/m) at 300 Oe Fluxmeter	Mr/Ms at 300 Oe Fluxmeter	Hc/Mr at 300 Oe Fluxmeter
As cast 1	116.89	4.13	22	0.0350	5.327
As cast 2	102.83	4.72	25	0.0459	5.297
Relaxed	51.415	17.884	460.834	0.348	25.773
1h	46.12	15.28	541.381	0.331	35.461
3h	124.575	52.519	602.44	0.422	11.479
5h	undefined	undefined	730.105	0.230	
8h	89.226	31.712	882.369	0.355	27.778
700°C	105.341	50.45	989.896	0.475	19.607

Acknowledgments

Project supported by the University of Annaba, in collaboration with the teams of Dr Aboki and Prof. Gasser. Special thanks for magnetic measurements, presented to Pr Tadeusz Kulik and Dr Jaroslaw Ferenc from Faculty of Materials Science and Engineering in Warsaw University of technology.

References

[1] Y. Tong, B.Z Ding. H.G. Jiang, K. Lu, J.T.Wang and Z.Q. .Hu, J. Appl. Phys., **75**, 1 (1994)

[2] R. Hasegawa, Applications of amorphous magnetic alloys, Material Science and Engineering A, **A375-377**, 90-7 (2004)

[3] M.-A. Escobar, A. Reza Yavari, R. Barrue and J.C. Perron, IEEE Trans. Magn. 2, **4**, 1911-1916 (1992).

[4] Hang Nam Ok and A.H. Morrish, Phys. Rev. B, **22**, 3471-3480 (1980).

[5] H. Sakamoto, T. Yamada, N. Okumura, T. Sato, Mater. Sci. Eng. **A 206**,150-153 (1996).

[6] H. Chiriac and C. Hison, , J. Mag. Mater. **254-255,** 475-476 (2003)

[7] F. Luborsky, in *Amorphous Metallic Alloys*, (*ed.* F.E. Luborsky, Butherworth, London, 1983) p.360

[8] E.P. Wolfarth, *in Amorphous Metallic Alloys*, (ed. F.E. Luborsky Butherworth, London, 1983) p.283.

[9] H. Kronmüller, Phil. Mag. **B48**, 127 (1983)

[10] P.A.Pankhurst, J.Z. Jiang, S. Betteridge, M.R.J. Gibbs, G.A.Gehring, J. Phys-Condens. Mat., **7**, 957 (1995)

[11] T. Kulik, H.T. Savage, and A. Hernando, J. Appl. Phys. 73, 10 (1993)

[12] M.Harmelin, E. Etchessahar, J. Debuigne and J. Bigot, Thermochimica Acta, **130**, 177-192 (1988)

[13] H. E. Kissinger, Anal. Chem. **29**, 1702 (1994).

[14] J.A. Augis, J. E. Bennet, J. Therm. Anal. **13**, 283 (1978)

[15] S.Vyazovkin , Thermochem. Acta, **194,** 375-383 (1992).

[16] S.Vyazovkin and C. A.Wight, Annu. Rev. Phys. Chem, **48,**125-149 (1997).

AUTHOR INDEX